Spin-Wave Theory and its Applications to Neutron Scattering and THz Spectroscopy

Spin-Wave Theory and its Applications to Neutron Scattering and THz Spectroscopy

Randy S Fishman and Jaime A Fernandez-Baca
Oak Ridge National Laboratory, Oak Ridge, Tennessee, USA

Toomas Rõõm
National Institute of Chemical Physics and Biophysics, Tallinn, Estonia

Morgan & Claypool Publishers

ISBN 978-1-64327-114-9 (ebook)
ISBN 978-1-64327-111-8 (print)
ISBN 978-1-64327-112-5 (mobi)

DOI 10.1088/978-1-64327-114-9

Version: 20181101

IOP Concise Physics
ISSN 2053-2571 (online)
ISSN 2054-7307 (print)

A Morgan & Claypool publication as part of IOP Concise Physics
Published by Morgan & Claypool Publishers, 1210 Fifth Avenue, Suite 250, San Rafael, CA, 94901, USA

IOP Publishing, Temple Circus, Temple Way, Bristol BS1 6HG, UK

To our mothers, who made this book possible.

Contents

Appendices

Acknowledgments

RF would like to acknowledge support from the US Department of Energy, Office of Basic Energy Sciences, Materials Sciences and Engineering Division. He would also like to acknowledge the hospitality of the Ulsan National Institute of Science and Technology (UNIST) in South Korea, where some of this book was written. JF would like to acknowledge support by the Scientific User Facilities Division, Office of Basic Energy Sciences, US Department of Energy. TR would like to acknowledge support from the Estonian Ministry of Education and Research with institutional research funding IUT23-3, and the European Regional Development Fund Project No. TK134.

A few words about the epigraphs, those witty (or not so much) quotations at the start of each chapter. We found a few quotations that were ready made for this book but, sadly, humor is often hard to come by in physics, so we had to write a few epigraphs ourselves. We leave it to the reader to decide which are the 'fake' epigraphs.

If you run into quotations in any language that might be suitable, by all means pass them on. They might make the next edition! You can email possible epigraphs, along with typos or plain old errors in the text to fishmanrs@ornl.gov. For a steep price, we might even give you hints about the solutions to exercises not provided with this book.

This book has benefited from the advice and knowledge of many colleagues. We would like to thank Sasha Balatsky, Clarence Beeks, Andy Christianson, Javier Campo, Bill Hamilton, Jason Haraldsen, Minseong Lee, Jan Musfeldt, Sai Mu, Urmas Nagel, Satoshi Okamoto, Charlie Parker, Karlo Penc, Wayne Saslow, Rogério deSousa, John Tranquada, Helen Walker, and Feng Ye.

Author biographies

Randy S Fishman

Randy Fishman was born New York City and received his PhD from Princeton University in 1985. After serving as a faculty member at North Dakota State University, he joined Oak Ridge National Laboratory in 1995 and is currently a Distinguished Research Scientist. He models the complex magnetic states of oxides and molecule-based magnets, often with neutron scatterers and spectroscopists. He is the author of over 160 scientific publications and a Fellow of the American Physical Society.

Jaime A Fernandez-Baca

Jaime Fernandez-Baca was born in Lima, Peru. He received training in neutron scattering at the National Bureau of Standards (now the National Institute of Standards and Technology) and at the University of Maryland-College Park, where he received his PhD in 1986 under Jeff Lynn. Afterwards, he joined ORNL as a postdoctoral fellow and is currently a Distinguished Research Scientist at ORNL. Fernandez-Baca's expertise is in the study of the magnetic ordering and spin dynamics of complex oxides and related alloys utilizing neutron scattering techniques. He is the author of over 150 technical articles. Fernandez-Baca is a Fellow of the American Physical Society and the Neutron Scattering Society of America.

Toomas Rõõm

Toomas Rõõm was born in Tallinn, Estonia and received his PhD from Tartu University in 1993. After working with Tom Timusk at McMaster University, he established a terahertz spectroscopy research group at the National Institute of Chemical Physics and Biophysics in Tallinn, where he has been research professor since 1998. His research interest is combining infrared and terahertz spectroscopy with high magnetic fields and low temperatures to study magnetic and correlated-electron materials.

Conversion factors, abbreviations, and symbols

Conversion factors:

1 THz = 33.36 cm^{-1} = 4.136 meV = 48.00 K
1 meV = 11.60 K = 8.064 cm^{-1} = 0.2418 THz
1 cm^{-1} = 1.439 K = 0.1240 meV = 0.029 98 THz
1 K = 0.020 84 THz = 0.6950 cm^{-1} = 0.086 17 meV
1 emu K mol^{-1} = 2.666 μ_B^2
1 μ_B = 0.057 89 meV T^{-1} = 5585 emu Oe mol^{-1}
1 T = 10^4 Oe

Abbreviations

Abbreviation	Definition
ac	alternating current
AF	antiferromagnetic or antiferromagnet
AFMR	antiferromagnetic resonance
BWO	backward-wave oscillator
BZ	Brillouin zone
CE	competing exchange
CL	collinear
D	dimension(s)
dc	direct current
DM	Dzyaloshinskii–Moriya
ESR	electron spin resonance
FEL	free-electron laser
FM	ferromagnetic
GF	geometric frustration
hf-ESR	high frequency ESR
HP	Holstein–Primakoff
INS	inelastic neutron scattering
MOFS	metal–organic frameworks
NC	non-collinear
NIR	near-infrared
RIXS	resonant inelastic x-ray scattering
rlu	reciprocal-lattice units
SC	spin current
SW	spin wave
TLA	triangular-lattice antiferromagnet

Symbols

Symbol	Definition
$a_i^{(r)}, a_i^{(r)\dagger}$	creation and annihilation operators on site i of sublattice r
$a_{\mathbf{q}}^{(r)}, a_{\mathbf{q}}^{(r)\dagger}$	Fourier transforms of $a_i^{(r)}$ and $a_i^{(r)\dagger}$
$\alpha_{\mathbf{q}}^{(r)}, \alpha_{\mathbf{q}}^{(r)\dagger}$	creation and annihilation operators that diagonalize $\underline{\mathcal{L}}$
$\alpha(\omega)$	optical absorption
$\mathbf{a}_n$	structural-lattice vectors ($n = 1, 2, 3$)
$\mathbf{b}_n$	reciprocal-lattice vectors ($n = 1, 2, 3$)
$\mathbf{B}$	external magnetic field
c	speed of light in vacuum, $c = 1/\sqrt{\epsilon_0\mu_0}$
D	usually a DM interaction, but could also be the SW stiffness
E, K	possibly anisotropy coefficients
E	energy
$\mathbf{E}^\omega$	THz electric field
ϵ	dielectric constant
ϵ_n	eigenvalues of $\underline{\mathcal{L}}$
$f(\kappa)$	magnetic form factor
$\Gamma_{rs}^{(u)}(\mathbf{q})$	couples sublattice r and s through exchange $J_{rs}^{(u)}$
(H, K, L)	reduced wavevector in reciprocal-lattice units
$\mathbf{B}^\omega$, $\mathbf{H}^\omega$	THz magnetic field
J_n	exchange interaction
$J_{rs}^{(u)}$	the uth exchange interaction coupling sublattices r and s
κ	imaginary part of complex index of refraction $\mathcal{N}$
$\boldsymbol{\kappa}$	SW wavevector in any BZ
$\mathbf{k}_{i,f}$	initial (i) and final (f) wavevector of a neutron or photon
$\underline{L}$	$2M$-dimensional equation-of-motion matrix
$\underline{\mathcal{L}}$	$2M$-dimensional matrix $\underline{L} \cdot \underline{N}$
$\{m_1, m_2, m_3\}$	parameterizes lattice site $\mathbf{R} = m_1\mathbf{a}_1 + m_2\mathbf{a}_2 + m_3\mathbf{a}_3$
$\mathbf{m}$	orientation of external magnetic field, but could also be magnetic moment
M	number of sites in magnetic unit cell
$\mathbf{M}$	magnetization, $\mathbf{M} = 2\mu_B \sum_i \mathbf{S}_i$ if $g_{\alpha\beta} = 2\delta_{\alpha\beta}$
μ	magnetic permeability
μ_B	Bohr magneton
$\mathcal{N}(\omega)$	complex index of refraction
n	real part of complex index of refraction $\mathcal{N}$
$n_B(\omega)$	boson distribution function $1/[\exp(\hbar\omega/k_B T) - 1]$
N	number of sites $N = N_u M$
N_u	number of magnetic unit cells
$\underline{N}$	$2M$-dimensional matrix with $N_{i\neq j} = 0$, $N_{ii} = 1$ for $1 \leqslant i \leqslant M$, and $N_{ii} = -1$ for $M + 1 \leqslant i \leqslant 2M$
ν	frequency, unit Hz
ω	frequency, unit rad s^{-1}

$\bar{\omega}$	wavenumber, number of waves in one cm, unit cm^{-1}
$\omega_n(\mathbf{q})$	SW frequency of mode n at wavevector $\mathbf{q}$
p	cycloidal turn number
$\mathbf{P}$	electric polarization
$\mathbf{q}$	SW wavevector shifted into the first structural BZ
$\mathbf{q}'$	SW wavevector measured from $\mathbf{Q}$
$\mathbf{Q}$	ordering wavevector of spin state, also called the magnetic propagation vector
r, s	magnetic sublattice index with $1 \leqslant r, s \leqslant M$
$\mathbf{R}_i$	lattice site
S	spin value
$\mathbf{S}_i$	spin on site i
$S_{\alpha\beta}(\mathbf{q},\omega)$	spin–spin correlation function in spin channels α and β
$S_{\alpha\beta}^{(n)}(\mathbf{q})$	spectral weight of mode n in spin channels α and β
$S^{(n)}(\boldsymbol{\kappa})$	spectral weight of mode n summed over spin channels given by $\sum_{\alpha,\beta}\{\delta_{\alpha\beta} - \kappa_\alpha\kappa_\beta/\kappa^2\} S_{\alpha\beta}^{(n)}(\boldsymbol{\kappa})$
$\boldsymbol{\tau}$	a structural reciprocal-lattice vector
$\underline{U}_i$	unitary matrix that rotates spin i into local frame
$\mathbf{v_q}$	$2M$-dimensional vector $(a_{\mathbf{q}}^{(1)}, \ldots, a_{\mathbf{q}}^{(M)}, a_{-\mathbf{q}}^{(1)\dagger}, \ldots a_{-\mathbf{q}}^{(M)\dagger})$
$\mathbf{w_q}$	$2M$-dimensional vector $(\alpha_{\mathbf{q}}^{(1)}, \ldots, \alpha_{\mathbf{q}}^{(M)}, \alpha_{-\mathbf{q}}^{(1)\dagger}, \ldots \alpha_{-\mathbf{q}}^{(M)\dagger})$
$\underline{X}$	eigenvectors of $\underline{\mathcal{L}}$
$\bar{\mathbf{x}}_i, \bar{\mathbf{y}}_i, \bar{\mathbf{z}}_i$	orthogonal axes in local frame with spin i aligned along $\bar{\mathbf{z}}_i$
$\chi^{\text{ab}}(\omega)$	electric (a or b = e) or magnetic (a or b = m) susceptibilities
$\chi_{\alpha\beta}^{\text{ss}}(\mathbf{q},\omega)$	spin susceptibility
$z_{rs}^{(u)}$	number of sites on sublattices r and s coupled by $J_{rs}^{(u)}$

IOP Concise Physics

Spin-Wave Theory and its Applications to Neutron Scattering and THz Spectroscopy

Randy S Fishman, Jaime A Fernandez-Baca and Toomas Rõõm

Chapter 1

Introduction

If you saw a spin wave, would you wave back?

—apologies to Steven Wright[1].

Magnetism is one of the oldest scientific fields [1], dating back to the discovery of lodestone (now known as magnetite) by the ancient Greeks. For about two millennia, the magnetic properties of an isolated substance were assumed to be static. Following the discovery of quantum mechanics, Bloch [2] proposed that quantized magnetic excitations or spin waves (SWs) explained the observed $T^{3/2}$ reduction [3, 4] of the ferromagnetic (FM) magnetization $M(T)$ with temperature T. The concept of SWs as the elementary excitations of an ordered magnet was quickly accepted by the physics community. Holstein and Primakoff (HP) [5] formulated the second quantization of the spins in a model Hamiltonian. Dyson [6, 7] later formulated the theory of SW interactions at finite temperature. But until the development of neutron-scattering techniques following the Manhattan project, SWs could only be indirectly observed by their effects on the static properties of materials, such as the specific heat and susceptibility.

Unlike x-rays, neutrons couple directly to the magnetic ions in a material. Proposed by Ernest Wollan in 1944, elastic neutron scattering proved to be a powerful tool for the investigation of crystal and magnetic structures [8]. Interestingly, the first observations of SWs in iron and magnetite were performed by diffraction. In the 1950s, Elliott and Lowde [9] showed that SWs can be observed as coherent diffuse peaks near the elastic reflections and that their dispersion

[1] Comedian Steven Wright actually said 'If you saw a heat wave, would you wave back?'.

doi:10.1088/978-1-64327-114-9ch1

relations can be directly obtained by measuring the changes in width with the angular deviation of the crystal from the Bragg condition. Using this method, Elliott and Lowde [9, 10] observed the quadratic SW dispersion of iron and magnetite at small wavevectors q, as expected for a FM. Also for small q, Goedkoop and Riste [11] observed the linear SW dispersion of hematite expected for an antiferromagnet (AF).

With the invention of the triple-axis neutron spectrometer by Bertram Brockhouse [12] at Chalk River Laboratories, SW frequencies ω and their dependence on wavevector q could be directly measured. Inelastic measurements on FM metals such as Fe and Ni by Herb Mook [13–15] and collaborators extracted the SW stiffness D from the relation $\hbar\omega(q) = Dq^2$, as shown by the 1969 measurements on Ni in figure 1.1. Since D is proportional to the exchange constant J, those measurements were able to estimate the exchange coupling between neighboring magnetic ions.

In the following decades, neutron scattering was used to study increasingly complex magnetic materials with progressively more complex spins states from collinear (CL) states like FMs and AFs to non-collinear (NC) states like cycloids and helices[2]. The magnetic unit cells of these materials grew from the unit cell of FM Fe or Ni with a single spin to the unit cell of multiferroic materials such as $BiFeO_3$ with several thousand spins comprising a long-wavelength cycloid.

The triple-axis spectrometer was followed by time-of-flight instruments that utilize neutron pulses like the ones produced by the bombardment of a heavy metal target at a spallation source or by a chopper at a reactor source. Whereas a triple-axis spectrometer focuses on a narrow range of wavevectors and frequencies, a time-of-flight instrument probes a wide range of $\{\mathbf{q},\omega\}$ space in a single measurement.

Almost since the first spin excitations were observed on a triple-axis spectrometer, optical measurements have been used to supplement inelastic neutron-scattering (INS) measurements. Because the ac magnetic field of light couples to the spin and the ac electric field couples to the electric-dipole moment, a photon can interact with the spins in a material to excite SWs [16, 17]. In Raman spectroscopy, an incoming photon of frequency ω_1 creates a SW of frequency ω and an outgoing photon of frequency $\omega_2 = \omega_1 - \omega$ [18]. In THz spectroscopy, the incoming photon is completely absorbed by the material to create a SW with the same frequency $\omega = \omega_1$.

Although confined to the wavevector $\mathbf{q} \approx 0$, the excitation frequencies provided by optical spectroscopy are much more precise than those measured using neutron scattering. Optical spectroscopy can also be readily applied to powders and to samples in large magnetic fields. Generally, optical spectroscopy cannot be used to study SWs in metals because electromagnetic waves do not penetrate far enough into the conducting material and the magnetic-dipole coupling of SWs to THz radiation is weak. In some metallic AFs [19], however, the current driven by the THz electric field can enhance the absorption strength by three orders of magnitude.

[2] For future reference, a helix (also called a spiral or a proper screw) is a chiral state with wavevector $\mathbf{Q}$ normal to the plane of the spins; a cycloid has $\mathbf{Q}$ in the plane of the spins. See chapter 6 for more details.

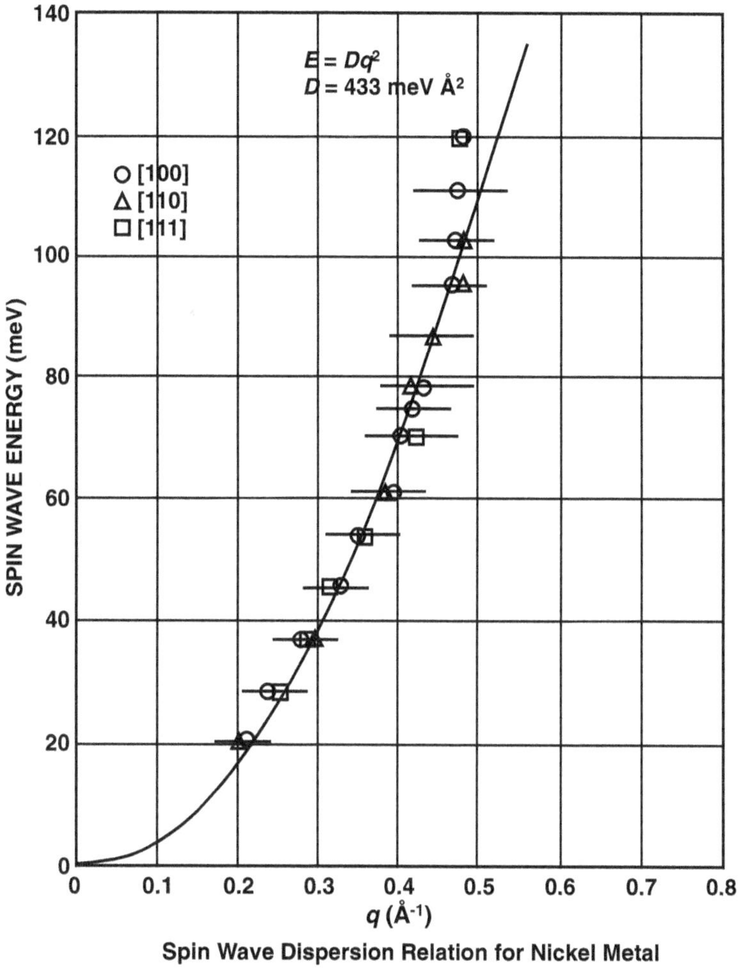

Figure 1.1. The SW frequency of Ni measured with the triple-axis spectrometer at the High Flux Isotope Reactor at Oak Ridge National Lab. Reproduced from [13] with permission of AIP publishing.

Raman spectroscopy excites ionic states with non-zero orbital angular momenta and strong electric-dipole moments [16]. The optical coupling to the electric-dipole moment dominates over the coupling to the magnetic-dipole moment when the incoming photon energy $\hbar\omega_1$ is of the same order as the energy $E_0 \gg \hbar\omega$ of the excited state. When $\hbar\omega_1 \ll E_0$, the indirect electric-dipole coupling can be neglected. Since this condition is satisfied by THz radiation, the incoming photon primarily couples to the magnetic-dipole moment of the spins through the ac magnetic field. A SW of energy $\hbar\omega = \hbar\omega_1$ is created by the magnetic-dipole coupling as the photon is absorbed.

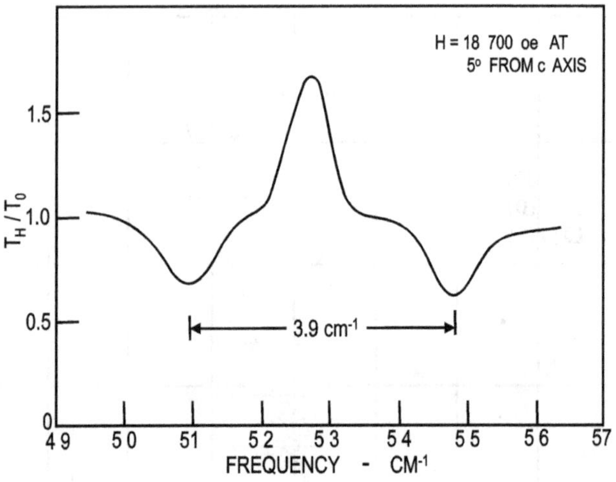

Figure 1.2. SW resonances in FeF_2 measured at 1.5 K with optical spectroscopy. Plotted is the ratio of the signal with a 1.87 T field compared to that with zero field as a function of wavenumber (1 cm^{-1} = 0.124 meV). Reprinted with permission from [20].

But electric-dipole coupling can also create a SW while absorbing a photon. Due to the spin–orbit interaction, magnetoelectric coupling mixes purely magnetic spin states with the charged states of electrons or optical phonons. A SW excited by the electric field of light is called an electromagnon [21]. Materials with electromagnons can exhibit non-reciprocal directional dichroism [22, 23], which means that the absorption of THz radiation changes when its propagation direction is reversed. To measure the magnetoelectric coupling, measurements of the SW absorption strength *and* its dependence on the radiation polarization must be performed.

In the late 1950s, the first THz studies of SWs in the AF MnF_2 were made using a klystron [24]. Shortly thereafter, Ohlmann and Tinkham [25] measured the SWs in the AF FeF_2 using a Hg arc lamp and grating spectrometer. While the central peak in figure 1.2 gives the inverse of the transmission in zero field, the sidepeak SW resonances are split by the magnetic field. Nowadays, a preferred technique is interferometric far-infrared spectroscopy, first used by P L Richards [26] to study SWs in NiF_2. In all of these studies, a small dc magnetic field is applied to shift the SW frequencies.

Assuming that all magnetic peaks can be measured, elastic neutron scattering can uniquely determine the ordered magnetic ground state of any material. In practice, only a subset of the elastic peaks can be detected so additional measurements are required to resolve the magnetic ground state. Even in principle, elastic scattering alone cannot ascertain all the interactions responsible for the magnetic state. For example, elastic scattering from a FM only determines the ordering wavevector $\mathbf{Q} = 0$ but not the different FM and AF exchange interactions between the aligned spins.

In contrast, the dynamical spectra $S(\mathbf{q},\omega)$ *overdetermine* the microscopic interactions of a material. Together with elastic scattering, analysis of $S(\mathbf{q},\omega)$ can determine both the spin structure and the microscopic interactions. Those

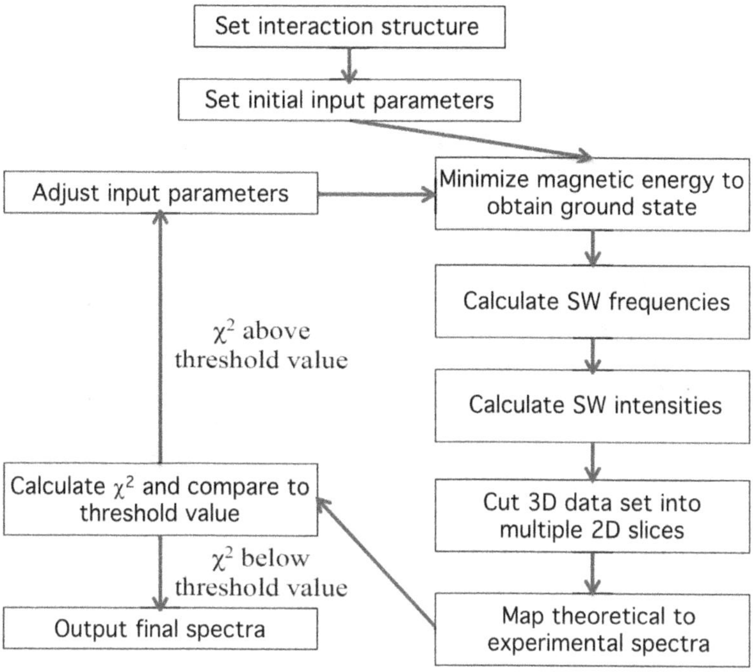

Figure 1.3. A flow diagram for the analysis of dynamical spectra.

interactions may include exchange, Dzyaloshinskii–Moriya (DM) interactions due to broken inversion symmetry, and crystal-field anisotropies.

The 'dynamical fingerprint' of a material has become increasingly significant as the complexity of important materials grows. To determine the spin-state and microscopic interactions in a material, we follow the flow chart in figure 1.3. First, assume a set of microscopic interactions in a microscopic Hamiltonian $\mathcal{H}$. Find the spin state by minimizing the energy $E = \langle \mathcal{H} \rangle$ over a set of spin angles or variational parameters. Based on this spin state, evaluate the dynamical spectra $S(\mathbf{q},\omega)$ and compare it to the observed spectra. Completing the loop, revise the microscopic interactions and repeat this process until the predicted and observed spectra agree.

Before going any further, let us clear up two sources of confusion. A SW is the same thing as a magnon: a well-defined magnetic excitation characterized by energy $\hbar\omega$ and wavevector $\mathbf{q}$. It is not the same thing as a spin-density wave, which refers to a static state where the spin amplitude is modulated from site to site. This book is concerned with magnetic excitations, not with static distortions of the spin amplitude. Indeed, SW theory assumes that the spin amplitude is fixed. One other clarification: this book discusses applications of conventional or 'linear' SW theory. Conventional because there is only one boson operator per site and linear because the equations-of-motion for each boson operator are linear in those operators. We shall briefly describe the more sophisticated multiboson SW theory in section 8.3.

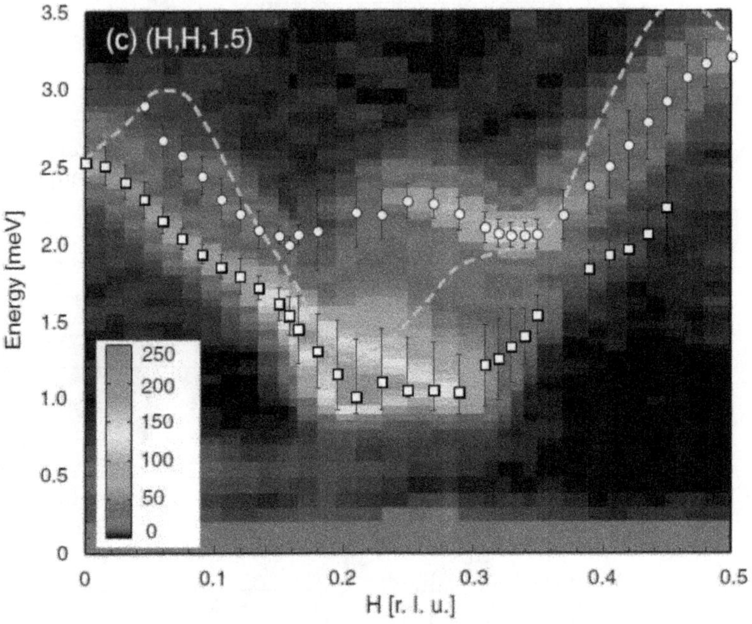

Figure 1.4. The INS spectrum of pure $CuFeO_2$ taken at the JRR-3 reactor at the Japanese Atomic Energy Agency. Circles and squares are the peak positions in constant-**q** scans. The dashed line was contributed by the twin in earlier studies [27]. Reproduced from [28]. Copyright 2011 by the Physical Society of Japan.

Several software packages are designed to automate one or more parts of the procedure described in figure 1.3. Those include McPhase[3], SpinW[4], Spinwave Genie[5], and Fullprof[6]. Convergence between the predicted and observed spectra is gauged by the agreement between the predicted and observed mode frequencies (see appendix A after chapter 9). In other words, the intensities of the mode frequencies are not directly considered in the convergence procedure. Rather, it is hoped with fingers crossed that the measured and predicted $S(\mathbf{q},\omega)$ look alike after the mode frequencies have converged!

As an example of a system where well-defined SWs were fit by the above procedure, figure 1.4 presents inelastic measurements [28] on pure $CuFeO_2$. To suppress the formation of twin domains, a small uniaxial strain was applied before the sample was cooled down. Whereas the twinned sample exhibits four SW modes [27], the untwinned sample exhibits only two. For this material, the relatively simple ↑↑↓↓ spin state can be obtained solely from the elastic spectrum. However, the exchange interactions and easy-axis anisotropy must be determined from the inelastic spectra.

[3] www.mcphase.de

[4] www.psi.ch/spinw/

[5] https://github.com/SpinWaveGenie/

[6] https://www.ill.eu/sites/fullprof/

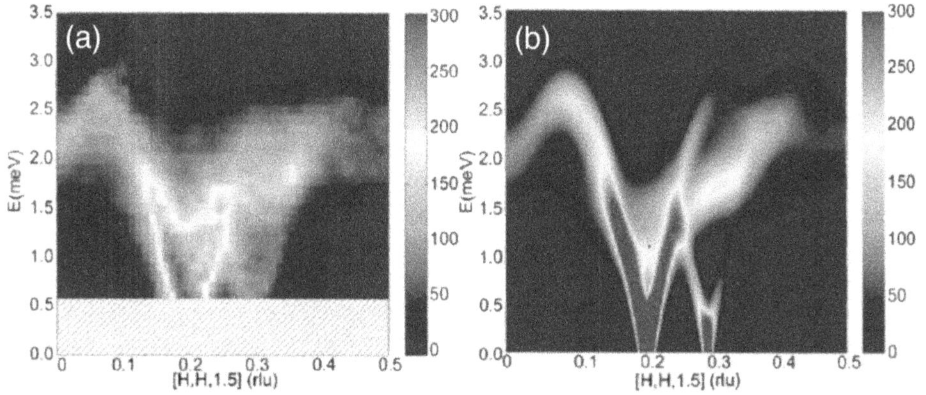

Figure 1.5. The inelastic neutron-scattering spectra of 3.5% Ga-doped $CuFeO_2$ taken at (a) the High Flux Isotope Reactor at Oak Ridge National Lab and (b) the predictions of SW theory. Reprinted figure with permission from [29]. Copyright 2010 by the American Physical Society.

With increasing computer power, this methodology has reached a turning point. We believe that future effort will be directed at converging the predicted and measured $S(\mathbf{q},\omega)$ rather than the predicted and measured mode frequencies $\omega(\mathbf{q})$. Future work will surely take advantage of machine-learning and pattern recognition techniques. Advanced analysis will be especially important for materials where the individual SWs cannot be separated. One such case is Ga-doped $CuFeO_2$ [29], where an incommensurate, distorted helix generates a large number of SWs that blend together to form a continuum, as seen in figures 1.5(a) and (b). Other such materials are powders, where all orientations of $\mathbf{q}$ with the same magnitude are averaged together. In both cases, convergence is often confirmed visually!

Standing at the brink of this threshold, what is the purpose of this book? That question has several answers. Before the process of analyzing dynamical spectra is relegated to a 'black box' (as is now the case for density-functional theory), we believe that it is useful to detail the ingredients of such a software package, at least insofar as SW theory is concerned. We are also convinced that many scientists will combine their own SW software with new machine-learning-based packages to gauge the convergence between the predicted and measured $S(\mathbf{q},\omega)$. In our own work, we have found that a one-size-fits-all approach is seldom adequate to predict the spin state and dynamics for a complex material. We imagine that this will remain true for many years to come and that some scientists will always prefer to 'roll their own' software. Finally, it may be convenient to have the SW machinery for NC spin systems readily available in one reference.

The linear SW theory developed in this book is based on the HP expansion of the Hamiltonian $\mathcal{H}$ about the classical limit. Whereas the spin along the local order axis is given by $S_{iz} = S - a_i^\dagger a_i$, the destruction and creation spin operators $S_{i+} = S_{ix} + iS_{iy}$ and $S_{i-} = S_{ix} - iS_{iy}$ are proportional to $\sqrt{S}a_i$ and $\sqrt{S}a_i^\dagger$, where a_i and $a_i^\dagger$ are boson operators. Since the Hamiltonian is bilinear in the spin operators, the HP expansion is really a $1/\sqrt{S}$ expansion in the boson operators. The zero-order term in this expansion, of order S^2, is the energy $E = \langle \mathcal{H} \rangle$. The first-order term, of order $S^{3/2}$, is linear in the boson operators and vanishes if the ground state is chosen correctly, just as the net force on an atom must vanish in equilibrium. The next-order term, of order S, is second order in the boson operators and determines the SW frequencies.

Within the equations-of-motion technique, the eigenvalues of the dynamical matrix $\underline{\mathcal{L}}(\mathbf{q})$ give the SW frequencies and the eigenvectors of $\underline{\mathcal{L}}(\mathbf{q})$ give the SW intensities. In principle, this is a straightforward technique; in practice, $\underline{\mathcal{L}}(\mathbf{q})$ can become quite complex for a spin state with a large unit cell or with a variety of spin interactions. For almost any material of interest, the eigenvalues and eigenvectors of $\underline{\mathcal{L}}(\mathbf{q})$ will be solved numerically rather than analytically.

Because it is based on a $1/S$ expansion, SW theory is most often applied to materials with $S > 1$. However, it has been productively employed to study spin-1 materials such as the nickelates [30–33] and even some spin-1/2 systems like the Cu divanadates [34]. We have elected to use this theory for several reasons. First, it is a tried-and-true technique that has been widely and successfully applied to a variety of materials. Second, most materials of current interest can be at least qualitatively modeled with a $1/S$ expansion. Finally, this technique can be systematically improved by including higher-order terms that incorporate the effects of quantum fluctuations.

SW theory is usually limited to zero temperature. While the spectral intensities can be evaluated as a function of temperature by incorporating Boson thermal factors, it is much more difficult to evaluate the temperature dependence of the SW frequencies themselves. There are 'work arounds' to incorporate the effect of temperature, such as replacing the spin S by the temperature-dependent quantity $m(T)/2\mu_B$, where $m(T)$ is the magnetic moment. However, these solutions are not very satisfactory because the basic form for the SW frequency is otherwise maintained.

By the same token, a SW expansion to order $1/S$ does not produce SW lifetimes. Rather, the SW frequencies $\omega_n(\mathbf{q})$ appear as delta functions $\delta(\omega - \omega_n(\mathbf{q}))$ in $S(\mathbf{q},\omega)$. To predict the observed spectra, we must smear the SW excitations by integrating over a resolution function in $\{\mathbf{q},\omega\}$ space. Ideally, the assumed widths $\Delta_\mathbf{q}$ and Δ_ω of the resolution function should be obtained from the instrumental resolution. In many cases, this procedure does not sacrifice any information because the observed SW peaks are 'resolution limited', meaning that the instrumental resolution exceeds the intrinsic widths of the SW modes.

It is also common to account for the reduction of the magnetic moment m from its maximal value of $2\mu_B S$ due to impurities or disorder by replacing the integer or half-integer spin S by the measured value $S' = m/2\mu_B$. This replacement is often required to study the field dependence of the SW frequencies. A trick to look out for is the replacement of the spin S by $\sqrt{S(S+1)}$, which is a 'poor man's' way of incorporating the effect of quantum spin fluctuations into the order S Hamiltonian. However, this replacement often introduces more problems than it solves, especially when studying AF resonance frequencies that are exactly linear in S.

When SW theory does fail, it usually does so catastrophically. In a one-dimensional (1D) AF with $S = 1/2$ or 1, the predictions of the SW expansion fail to account for the continuum of spin excitations and the spin gap [35]. Even for higher-spin systems with long-range magnetic order, SW theory can miss important effects. A fundamental assumption of the HP expansion is that the spin magnitude does not change. While spin fluctuations can suppress the expectation value of the spin along a given axis, the total spin on any site is assumed to be unchanged. This assumption fails when the spin stretches or contracts due to strong easy-plane anisotropy. Corrections to SW theory have been developed to provide a better description of these systems [36, 37]. For example, in $Ba_2CoGe_2O_7$ with $S = 3/2$ Co ions, strong easy-plane anisotropy suppresses the effective spin from 3/2 towards 1 [38]. For strong and moderate values of the anisotropy, oscillations of the spin amplitude can produce new SW modes in addition to those predicted by the $1/S$ expansion. As discussed in section 8.3, the extra SW modes in $Ba_2CoGe_2O_7$ can be described by a multiboson SW theory [39] that adds extra degrees of freedom to describe the oscillations of the spin amplitude.

A related assumption of SW theory is that the spins are localized. This assumption usually requires that the material is an insulator. Generally, the spin amplitudes in metals are suppressed from their classical values by the itinerancy of the electrons. For example, the magnitude of the spin in metallic Cr varies sinusoidally from site to site with a maximum of about 0.2 due to the formation of a spin-density wave [40]. This modulation invalidates the assumptions of the HP expansion.

Nevertheless, the excitations in metallic Fe and Ni have been modeled using SW theory since the early days of neutron scattering [13–15]. Remarkably, SW theory captures the essential features of the dispersion relations of those itinerant systems at low energies when the excitation wavelengths are long compared to the atomic spacing. So for small wavevectors and energies, the excitations of itinerant spins resemble the SWs of localized spins [41]. But it is well known that SW theory cannot explain the higher-energy features of the excitation spectra in metals [42, 43], which require a more realistic treatment that accounts for the itinerant nature of the electrons. As we shall see in the next chapter, applying SW theory to metallic systems like the manganites [44] must always be viewed with some degree of caution.

Because their spin states are disordered, 'spin liquids' of either the classical [45] or quantum [46] variety are not treated by this book. Nor shall we enter the fascinating world of high-entropy alloys [47]. However, we will use SW theory to study amorphous FMs in chapter 7.

The different wavevectors $\mathbf{k}_{i,f}$, $\mathbf{q}$, $\boldsymbol{\tau}$, $\mathbf{Q}$, and $\boldsymbol{\kappa}$ needed to discuss INS and THz spectroscopies are defined in table 1.1. Momentum conservation requires that $\boldsymbol{\kappa} = \mathbf{k}_i - \mathbf{k}_f$ for INS and that $\boldsymbol{\kappa} = \mathbf{k}_i$ for THz spectroscopy. We can shift $\boldsymbol{\kappa}$ into the first structural Brillouin zone (BZ) centered at the origin by subtracting a suitable reciprocal-lattice vector $\boldsymbol{\tau}$: $\mathbf{q} = \boldsymbol{\kappa} - \boldsymbol{\tau}$. For a chain of atoms separated by a in 1D, $\boldsymbol{\tau}$ can be any multiple n of $2\pi\mathbf{x}/a$. It is sometimes helpful to measure the wavevector from the ordering wavevector $\mathbf{Q}$. So we define $\mathbf{q}' = \mathbf{q} - \mathbf{Q} = \boldsymbol{\kappa} - \boldsymbol{\tau} - \mathbf{Q}$. For example, consider the helix with period $8a$ and ordering wavevector $Q = \pi/4a$ pictured in

Table 1.1. Wavevectors used in this book.

Wavevector	definition
$\mathbf{k}_{i,f}$	initial (i) and final (f) wavevector of a neutron or photon
$\boldsymbol{\tau}$	a structural reciprocal-lattice vector ($\boldsymbol{\tau} = 2\pi n\mathbf{x}/a$ in 1D)
$\mathbf{Q}$	ordering wavevector ($\mathbf{Q} = 2\pi\mathbf{x}/a$ for a FM in 1D), also called the propagation vector
$\boldsymbol{\kappa}$	SW wavevector in any BZ
$\mathbf{q}$	SW wavevector shifted into the first structural BZ
$\mathbf{q}'$	SW wavevector measured from $\mathbf{Q}$

figure 1.6(a). The magnetic BZ centered at the origin lies between $\pm Q/2 = \pm\pi/8a$. As shown in figure 1.6(b), the wavevector $\kappa = k_i - k_f$ can be translated into the first structural BZ by subtracting $\tau = 2\pi/a$ so that $q = \kappa - \tau$. Measured from Q, the wavevector is $q' = \kappa - \tau - Q$.

While it is our favorite part, there is much more to the theory of magnetism than SW theory. For general reviews of this topic, we recommend the books by Mattis [48] and Yosida [49], both titled *Theory of Magnetism*. Another great reference that sits on our bookshelves is *Quantum Theory of Magnetism* by White [50].

We hope to convince you that neutron scattering and optical spectroscopy are complementary techniques. Whereas INS is best suited to study spin states with short wavelengths so that the relevant wavevectors are spread throughout the BZ, optical spectroscopy works best for spin states with long wavelengths so that the important mode frequencies lie at very short wavevectors. Optical spectroscopy can also be used to probe small interaction energies that mostly affect the low-energy SW spectra.

Technical overviews of neutron scattering are contained in *Neutron Scattering with a Triple-Axis Spectrometer* by Shirane, Shapiro, and Tranquada [51] and *Theory of Magnetic Neutron and Photon Scattering* by Balcar and Lovesey [52]. A nice review of neutron scattering and its applications to the AF MnF_2 was given by Yamani, Tun, and Ryan [53].

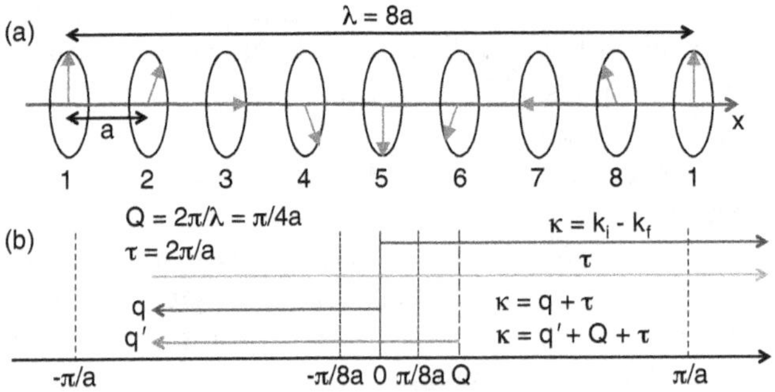

Figure 1.6. (a) A helix with period $8a$ and ordering wavevector $Q = \pi/4a$ in 1D. (b) Wavevector $\kappa = k_i - k_f$ is shifted by $\tau = 2\pi/a$ into the first structural BZ between $\pm\pi/a$. Measured from Q, $q' = q - Q$.

Our technical discussion of optical spectroscopy is limited to THz absorption. Although the $\mathbf{q} = 0$ SW frequencies obtained from Raman and THz spectroscopies are identical, their scattering and absorption intensities are quite different. As shown in chapter 4, the magnetic contribution to the THz absorption is simply related to the INS cross section at $\mathbf{q} = 0$. So THz spectroscopy can be treated using the same SW machinery developed for INS. Since no comprehensive review of THz spectroscopy in high magnetic fields has been previously written, our overview of this field will have to suffice.

Hopefully, this book will prove useful both to users at neutron-scattering and THz absorption facilities and to theorists trying to model dynamical spectra. Although a background in solid state physics would certainly help, the only strict requirements of this book are a familiarity with statistical mechanics and with the quantized harmonic oscillator! While this book can be profitably used by advanced undergraduates or graduate students, we hope that it also assists researchers of every level interested in neutron scattering or THz spectroscopy.

Chapters 2 and 3 provide reviews of neutron-scattering and THz absorption techniques, respectively. Chapter 4 develops the SW formalism and provides expressions for the neutron-scattering and THz absorption intensities. Several model examples are provided in chapters 5 and 6, which respectively treat CL and NC magnets. Exercises are given at the end of each of the experimental chapters 2 and 3 and at the end of the theory chapters 4–6. Especially demanding problems are marked by an asterisk. We then present several experimental case studies that use SW theory to analyze the dynamical spectra of materials. Chapter 7 describes the analysis of neutron-scattering measurements and chapter 8 describes the analysis of THz absorption measurements. A conclusion and prospectus for the future are provided in chapter 9. Abbreviations, conversion factors, and a partial list of the symbols used in this book are provided prior to this chapter.

References

[1] For a short history of magnetism, see Mattis D C 1965 *Theory of Magnetism* (New York: Harper and Row) ch 1

[2] Bloch F 1930 Zur Theorie des Ferromagnetismus *Z. Phys.* **61** 206–19

[3] Fallot M 1936 Ferromagntisme des alliages de fer *Ann. Phys.* **11** 305–87

[4] Argyle B E, Charap S H and Pugh E W 1963 Deviations from $T^{\frac{3}{2}}$ law for magnetization of ferrometals: Ni, Fe, and Fe^{+3} *Phys. Rev.* **132** 2051–62

[5] Holstein T and Primakoff H 1940 Field dependence of the intrinsic domain magnetization of a ferromagnet *Phys. Rev.* **58** 1098–13

[6] Dyson F J 1956 General theory of spin-wave interactions *Phys. Rev.* **102** 1217–30

[7] Dyson F J 1956 Thermodynamic behavior of an ideal ferromagnet *Phys. Rev.* **102** 1230–44

[8] Mason T E, Gawne T J, Nagler S E, Nestor M B and Carpenter J M 2013 The early development of neutron diffraction: science in the wings of the Manhattan Project *Acta Crystallogr. Sect.* A **69** 37–44

[9] Elliot R J and Lowde R D 1955 The inelastic scattering of neutrons by magnetic spin waves *Proc. R. Soc. Lond.* A **230** 46–73

[10] Lowde R D 1954 On the diffuse reflexion of neutrons by a single crystal *Proc. R. Soc. Lond.* A **221** 206–23
[11] Goedkoop J A and Riste T 1960 Neutron diffraction study of antiferromagnetic spin waves in α-ferric oxide *Nature* **185** 450
[12] Brockhouse B N 1957 Scattering of neutrons by spin waves in magnetite *Phys. Rev.* **106** 859–64
[13] Mook H A, Nicklow R M, Thompson E D and Wilkinson M K 1969 Spin-wave spectrum of nickel metal *J. Appl. Phys.* **40** 1450–1
[14] Mook H A and Nicklow R M 1973 Neutron scattering investigation of the magnetic excitations in iron *Phys. Rev.* B **7** 336–42
[15] Mook H A, Lynn J W and Nicklow R M 1973 Temperature dependence of the magnetic excitations in nickel *Phys. Rev. Lett.* **30** 556–9
[16] Fleury P A and Loudon R 1968 Scattering of light by one- and two-magnon excitations *Phys. Rev.* **166** 514–30
[17] Moriya T 1968 Theory of absorption and scattering of light by magnetic crystals *J. Appl. Phys.* **39** 1042–9
[18] Cottam M G and Lockwood D J 1986 *Magnetic Scattering in Solids* (New York: Wiley)
[19] Bhattacharjee N *et al* 2018 Néel spin–orbit torque driven antiferromagnetic resonance in Mn_2Au probed by time-domain THz spectroscopy *Phys. Rev. Lett.* **120** 237201
[20] Ohlmann R C and Tinkham M 1961 Antiferromagnetic resonance in FeF_2 at far-infrared frequencies *Phys. Rev.* **123** 425–34
[21] Pimenov A, Mukhin A A, Yu I V, Travkin V D, Balbashov A M and Loidl A 2006 Possible evidence for electromagnons in multiferroic manganites *Nat. Phys.* **2** 97–100
[22] Kézsmárki I, Kida N, Murakawa H, Bordács S, Onose Y and Tokura Y 2011 Enhanced directional dichroism of terahertz light in resonance with magnetic excitations of the multiferroic $Ba_2CoGe_2O_7$ oxide compound *Phys. Rev. Lett.* **106** 057403
[23] Kézsmárki I *et al* 2014 One-way transparency of four-coloured spin-wave excitations in multiferroic materials *Nat. Commun.* **5** 3203
[24] Johnson F M and Nethercot A H 1959 Antiferromagnetic resonance in MnF_2 *Phys. Rev.* **114** 705–16
[25] Ohlmann R C and Tinkham M 1961 Antiferromagnetic resonance in FeF_2 at far-infrared frequencies *Phys. Rev.* **123** 425–34
[26] Richards P L 1965 Far-infrared magnetic resonance in NiF_2 *Phys. Rev.* **138** A1769–75
[27] Ye F, Fernandez-Baca J A, Fishman R S, Ren Y, Kang H J, Qiu Y and Kimura T 2007 Magnetic interactions in the geometrically frustrated triangular lattice antiferromagnet $CuFeO_2$ *Phys. Rev. Lett.* **99** 157201
[28] Nakajima T, Mitsuda S, Haku T, Shibata K, Yoshitomi K, Noda Y, Aso N, Uwatoko Y and Terada N 2011 Spin-wave spectrum in single-domain magnetic ground state of triangular lattice antiferromagnet $CuFeO_2$ *J. Phys. Soc. Jpn.* **80** 014714
[29] Haraldsen J T, Ye F, Fishman R S, Fernandez-Baca J A, Yamaguchi Y, Kimura K and Kimura T 2010 Multiferroic phase of doped delafossite $CuFeO_2$ identified using inelastic neutron scattering *Phys. Rev.* B **82** 020404
[30] Mihály L, Dóra B, Ványolos A, Berger H and Forró L 2006 Spin-lattice interaction in the quasi-one-dimensional helimagnet $LiCu_2O_2$ *Phys. Rev. Lett.* **97** 067206
[31] Jensen T B S *et al* 2009 Anomalous spin waves and the commensurate-incommensurate magnetic phase transition in $LiNiPO_4$ *Phys. Rev.* B **79** 092413

[32] Ehlers G, Podlesnyak A A, Hahn S E, Fishman R S, Zaharko O, Frontzek M, Kenzelmann M, Pushkarev A V, Shiryaev S V and Barilo S 2013 Incommensurability and spin dynamics in the low-temperature phases of $Ni_3V_2O_8$ *Phys. Rev.* B **87** 214418

[33] Machado F L A, Ribeiro P R T, Holanda J, Rodrguez-Suárez R L, Azevedo A and Rezende S M 2017 Spin-flop transition in the easy-plane antiferromagnet nickel oxide *Phys. Rev.* B **95** 104418

[34] Gitgeatpong G, Zhao Y, Piyawongwatthana P, Qiu Y, Harriger L W, Butch N P, Sato T J and Matan K 2017 Nonreciprocal magnons and symmetry-breaking in the noncentrosymmetric antiferromagnet *Phys. Rev. Lett.* **119** 047201

[35] Haldane F D M 1983 Continuum dynamics of the 1-D Heisenberg antiferromagnet: Identification with the O(3) nonlinear sigma model *Phys. Lett.* A **93** 464–8

[36] Lindgård P-A and Kowalska A 1976 Bose operator expansions of tensor operators in the theory of magnetism. II *J. Phys. C: Solid State Phys.* **9** 2081

[37] Riseborough P S 1983 Quantum renormalizations in one dimensional easy plane ferromagnets *Solid State Commun.* **48** 901–5

[38] Penc K *et al* 2012 Spin-stretching modes in anisotropic magnets: Spin-wave excitations in the multiferroic $Ba_2CoGe_2O_7$ *Phys. Rev. Lett.* **108** 257203

[39] Romhányi J and Penc K 2012 Multiboson spin-wave theory for $Ba_2CoGe_2O_7$: A spin-3/2 easy-plane Néel antiferromagnet with strong single-ion anisotropy *Phys. Rev.* B **86** 174428

[40] Fawcett E 1988 Spin-density-wave antiferromagnetism in chromium *Rev. Mod. Phys.* **60** 209–83

[41] Nocera A, Patel N D, Fernandez-Baca J, Dagotto E and Alvarez G 2016 Magnetic excitation spectra of strongly correlated quasi-one-dimensional systems: Heisenberg versus Hubbard-like behavior *Phys. Rev.* B **94** 205145

[42] Furukawa N 1996 Spin excitation spectrum of $La_{1-x}A_xMnO_3$ *J. Phys. Soc. Jpn.* **65** 1174–7

[43] Fernandez-Baca J A, Dai P, Hwang H Y, Kloc C and Cheong S -W 1998 Evolution of the low-frequency spin dynamics in ferromagnetic manganites *Phys. Rev. Lett.* **80** 4012–5

[44] Perring T G, Aeppli G, Hayden S M, Carter S A, Remeika J P and Cheong S -W 1996 Spin waves throughout the brillouin zone of a double-exchange ferromagnet *Phys. Rev. Lett.* **77** 711–4

[45] Balents L 2010 Spin liquids in frustrated magnets *Nature* **464** 199–208

[46] Savary L and Balents L 2017 Quantum spin liquids: a review *Rep. Prog. Phys.* **80** 016502

[47] Miracle D B and Senkov O N 2017 A critical review of high entropy alloys and related concepts *Acta Mater.* **122** 448–511

[48] Mattis D C 1965 *Theory of Magnetism* (New York: Harper and Row)

[49] Yosida K 1996 *The Theory of Magnetism* (Berlin: Springer)

[50] White R M 1970 *Quantum Theory of Magnetism* (New York: Springer)

[51] Shirane G, Shapiro S M and Tranquada J M 2002 *Neutron Scattering with a Triple-Axis Spectrometer* (Cambridge: Cambridge University Press)

[52] Balcar E and Lovesey S W 1989 *Theory of Magnetic Neutron and Photon Scattering* (Oxford: Clarendon)

[53] Yamani Z, Tun Z and Ryan D H 2010 Neutron scattering study of the classical antiferromagnet MnF_2: a perfect hands-on neutron scattering teaching course *Can. J. Phys.* **88** 771–97

Chapter 2

Inelastic neutron scattering

A neutron walks into a bar and asks how much for a drink? The barman replies, for you no charge.—

Dr Sheldon Cooper

2.1 Introduction

Due to their unique properties, neutrons are a powerful probe of condensed-matter systems. Unlike electromagnetic radiation or electrons, neutrons interact with atoms mainly through weak short-range nuclear interactions. Consequently, neutrons penetrate into most samples far more effectively than charged particles or x-rays. Also unlike other common probes, neutrons can be produced with wavelengths and energies in the ideal range to study material properties. Neutrons with wavelengths comparable to the interatomic distances in solids and liquids are ideal for structural studies. With energies (≈25 meV) of the same order as thermal fluctuations in solids, neutrons are also well suited to study lattice dynamics. Because the neutron spin couples to the electronic spin, they are an ideal probe to study magnetic structures and SWs. While the neutron spin also couples to nuclear spins, this interaction is much weaker and, in most cases, negligible[1]. The scattering from nuclear spins is generally not important for measuring magnetic order and excitations in solids, but they can contribute to the background through incoherent scattering.

[1] While neutron spins couple to both nuclear and electronic spins, the nuclear magnetic moments are about 10^3 times weaker than the electronic magnetic moment. As in the case of their electronic counterparts, the nuclear spins can also order, but this happens at ultra-low temperatures (usually < 1 mK) and the observation of nuclear magnetic order using neutron scattering is very challenging. See for example, reference [1].

doi:10.1088/978-1-64327-114-9ch2

In scattering experiments, neutrons are produced either by fission in a nuclear reactor or by spallation at an accelerator. Spallation describes the process whereby an energetic beam of ions (generally protons) knocks neutrons from a heavy metal target. High-energy (~ MeV) neutrons produced in either fission or spallation are subsequently slowed down by multiple collisions in a moderator until they approach thermal equilibrium. The neutron velocities are then described by a Maxwellian distribution corresponding to the moderator temperature. Neutrons in equilibrium with a moderator at room temperature are referred to as 'thermal' neutrons; neutrons thermalized in moderators at low temperatures (such as liquid hydrogen at 20 K) are referred to as 'cold' neutrons. While reactors produce a continuous stream of neutrons, only those in a narrow band of wavelengths are selected, usually by a 'monochromator' crystal. Although a continuous supply of neutrons can be produced by spallation, most existing sources produce pulsed neutron beams. The energy of each neutron is then determined by its time of flight.

The foundations of neutron scattering were established in the 1940s and 50s, when significant neutron fluxes were available from the first nuclear reactors in the United States and Canada. Between 1948 and 1955, E O Wollan and Clifford Shull laid the basis for neutron diffraction [2] at Oak Ridge's Graphite Reactor. Inspired by the classic papers of Halpern and Johnston [3], they characterized the ferrimagnetic state in Fe_3O_4 [4]. First proposed by Néel [5], this state contains an AF arrangement of ferrous and ferric ions. Because their opposing moments do not completely cancel out, Fe_3O_4 exhibits a net magnetic moment. Wollan and Shull subsequently embarked on an extensive neutron-scattering study of paramagnetic and AF materials [6].

In the mid-1950s Bertram Brockhouse developed the technique of INS at the NRX reactor in Chalk River, Canada. Laying the groundwork for modern neutron spectroscopy, Brockhouse designed the triple-axis spectrometer shown in figure 2.1. This was the first instrument able to analyze the energy spectrum of neutrons after they scatter from a sample [7, 8]. Using Brockhouse's invention, INS became an essential tool to study elementary excitations like phonons [9] and SWs [10].

Interestingly, SWs in iron and magnetite were first observed in the mid-1950s without the energy-analysis capability developed by Brockhouse! At Harwell in the United Kingdom, Elliott and Lowde [11] showed that SWs appear as coherent diffuse peaks near elastic reflections in a neutron-diffraction measurement. The SW dispersion was determined by the change in peak widths with the angular offset of the crystal from the Bragg condition. Using this method to study iron and magnetite, Elliott and Lowde measured the quadratic dispersion of the SW energies for small wavevectors q. Employing the same method at the Kjeller reactor in Norway, Goedkoop and Riste [12] measured the linear SW dispersion of hematite for small q. While these were important results, the neutron-diffraction method for measuring SW frequencies was cumbersome and limited to small wavevectors q. On the other hand, the triple-axis neutron spectrometer allowed scientists to directly measure the SW frequencies ω and their full dependence on wavevector $\mathbf{q}$.

In 1994, Shull and Brockhouse were awarded the Nobel Prize in Physics for their 'pioneering contributions to the development of neutron scattering techniques for

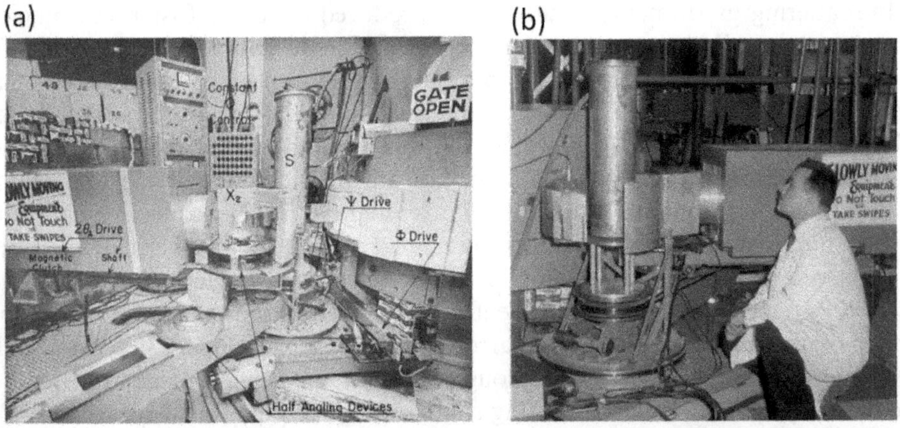

Figure 2.1. (a) The original neutron triple-axis spectrometer designed and built by Brockhouse. The monochromatic beam of neutrons coming from the right hits a sample (S) located in a cryostat. Scattered neutrons are then analyzed by a crystal (X_2) before they are counted at a neutron detector located inside the shield on the left [8]. (b) Brockhouse inspecting the first triple-axis spectrometer at the NRX reactor in the 1950s. Reprinted photographs with permission of the Atomic Energy Energy of Canada Limited (AECL). Canadian Crown Copyright, 1961.

studies of condensed matter'[2]. Unfortunately, Wollan had died in 1967 and could not share this honor.

2.2 Neutron scattering basics

Considering the special focus of this book, this section will concentrate on neutron-scattering measurements of the spin dynamics in a solid. Since we will only provide a general overview of INS, the interested reader may wish to consult more technical reviews [13–17] if needed.

Start with an incident beam of monochromatic neutrons of linear momentum $\hbar\mathbf{k}_i$ and energy $\hbar\omega_i = (\hbar k_i)^2/2m_n$ (the subscript i stands for 'initial' or prior to the scattering process, and m_n is the neutron mass). The energy and momentum of this beam are usually selected by Bragg diffraction from a monochromator crystal in a reactor or by time of flight using a neutron chopper in a spallation source. Neutrons are scattered by their interactions with the nuclei and electronic spins of the sample. Measuring the momentum and energy spectra of the scattered neutrons provides important information about the sample's atomic structure and dynamics.

A typical experiment measures both the momentum $\hbar\mathbf{k}_f$ and energy $(\hbar k_f)^2/2m_n$ of the scattered neutrons (the subscript f means final or after scattering). We define

$$\hbar\boldsymbol{\kappa} = \hbar(\mathbf{k}_i - \mathbf{k}_f), \tag{2.1}$$

[2] www.nobelprize.org/nobel_prizes/physics/laureates/1994/

$$\hbar\omega = \frac{\hbar^2}{2m_n}\left(k_i^2 - k_f^2\right), \tag{2.2}$$

as the momentum and energy transferred to the system by the incident neutrons. For *elastic* neutron scattering, no energy is imparted to the sample at the atomic level so that $\hbar\omega = 0$. For INS, the energy $\hbar\omega \neq 0$ transferred to the sample creates or destroys an elementary excitation like a SW or a phonon.

While the scattering wavevector $\boldsymbol{\kappa}$ is related to the momentum transferred to the sample under study, the relevant wavevector $\mathbf{q}$ associated with an excitation must lie within a given BZ, i.e. the excitation wavevector $\mathbf{q}$ can be measured around any allowed reciprocal-lattice vector $\boldsymbol{\tau}$. Bloch's theorem requires that the energy of any wave-like eigenstate in a periodic lattice must also be periodic in $\boldsymbol{\tau}$. Hence, the scattering wavevector $\boldsymbol{\kappa}$ defined in equation (2.1) can be expressed as $\boldsymbol{\kappa} = \boldsymbol{\tau} + \mathbf{q}$, where $\boldsymbol{\tau}$ is a reciprocal-lattice vector and $\mathbf{q}$ is the excitation wavevector (also referred to as the reduced wavevector), as illustrated in figure 2.2. For a special choice of $\boldsymbol{\tau}$, $\mathbf{q}$ will lie within the first BZ.

As discussed further in chapter 4, the periodicity of a magnetic system may be different from that of the lattice. Therefore, we introduce the magnetic ordering wavevector $\mathbf{Q}$ (often referred to as the magnetic propagation wavevector), which characterizes the periodicity of the magnetic lattice. The scattering wavevector $\boldsymbol{\kappa}$ for a magnetic system can then be generally written as

$$\boldsymbol{\kappa} = \boldsymbol{\tau} + \mathbf{q} = \boldsymbol{\tau} + \mathbf{Q} + \mathbf{q}', \tag{2.3}$$

where $\mathbf{q}' = \mathbf{q} - \mathbf{Q}$ is measured from the magnetic ordering wavevector, as shown by figure 1.6.

For a FM, $\mathbf{Q} = 0$ so the magnetic and lattice periodicities are the same. The excitation wavevector $\mathbf{q}' = \mathbf{q}$ is then measured from $\boldsymbol{\tau}$. For an AF or a ferrimagnet, $\mathbf{q}'$ is measured from $\mathbf{Q}$, which can be commensurate or incommensurate with the lattice. Unlike the SW intensities, the SW frequencies are periodic in both $\boldsymbol{\tau}$ and $\mathbf{Q}$, as discussed in section 4.3.

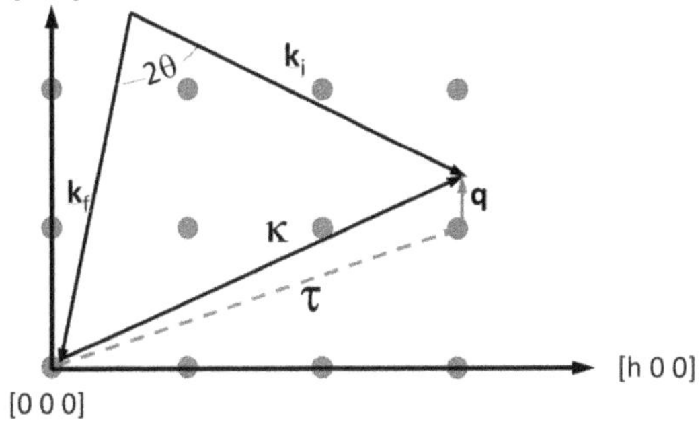

Figure 2.2. Schematic scattering diagram: $\boldsymbol{\kappa}$ is the scattering wavevector defined in equation (2.1), $\boldsymbol{\tau}$ is a reciprocal-lattice vector, and $\mathbf{q}$ is the reduced wavevector. Note that the angle between $\mathbf{k}_i$ and $\mathbf{k}_f$ is the scattering angle 2θ.

INS directly measures the scattering function $S_{\alpha\beta}(\mathbf{q},\omega)$, which contains all information about the spin correlations in the system. The partial differential cross section for the scattering of unpolarized neutrons from a system of electron spins is [13]:

$$\frac{d^2\sigma}{d\Omega\, d\omega} = \left(\frac{\gamma e^2}{m_e c^2}\right)^2 \left[\frac{g}{2} f(\boldsymbol{\kappa})\right]^2 \frac{k_f}{k_i} \sum_{\alpha,\beta} \left(\delta_{\alpha\beta} - \frac{\kappa_\alpha \kappa_\beta}{\kappa^2}\right) S_{\alpha\beta}(\boldsymbol{\kappa},\omega), \tag{2.4}$$

where $(\gamma e^2/m_e c^2)^2 = 0.291$ barns is the coupling constant of the neutron to the unpaired electron spins, g is the Landé factor, and $f(\boldsymbol{\kappa})$ is the magnetic form factor related to the Fourier transform of the radial part of the electronic wave function. The prefactor $\left(\delta_{\alpha\beta} - \kappa_\alpha\kappa_\beta/\kappa^2\right)$ in equation (2.4) is produced by the dipole–dipole interaction between neutrons and electrons. It implies that neutrons only probe magnetic fluctuations perpendicular to their momenta $\hbar\boldsymbol{\kappa}$. The scattering function $S_{\alpha\beta}(\boldsymbol{\kappa},\omega) = S_{\alpha\beta}(\mathbf{q},\omega)$ is the space and time Fourier transform of the time-dependent spin–spin correlation function:

$$S_{\alpha\beta}(\mathbf{q},\omega) = \frac{1}{2\pi N} \sum_{i,j} \int dt\; e^{-i\omega t} e^{-i\mathbf{q}\cdot(\mathbf{R}_i - \mathbf{R}_j)} \left\langle S_{i\alpha}(0)\, S_{j\beta}(t)\right\rangle, \tag{2.5}$$

where N is the number of sites. Hence, neutrons scattered with momentum $\hbar\boldsymbol{\kappa}$ and energy transfer $\hbar\omega$ directly measure a single Fourier component of the spin–spin correlation function.

Since we will be concerned with dynamical processes, $S_{\alpha\beta}(\mathbf{q},\omega)$ will denote the inelastic ($\omega \neq 0$) part of the scattering function although the elastic part ($\omega = 0$) will be needed to satisfy sum rules. The fluctuation–dissipation theorem states that

$$S_{\alpha\alpha}(\mathbf{q},\omega) = [n_{\mathrm{B}}(\omega) + 1]\, \mathrm{Im}\, \chi^{\mathrm{ss}}_{\alpha\alpha}(\mathbf{q},\omega), \tag{2.6}$$

where $n_{\mathrm{B}}(\omega) = 1/[\exp(\hbar\omega/k_{\mathrm{B}}T) - 1]$ is the Boson distribution function. The generalized spin susceptibility $\chi^{\mathrm{ss}}_{\alpha\beta}(\mathbf{q},\omega)$ is the equilibrium response of the α component of the spin to the β component of a time- and space-dependent external field with wavevector $\mathbf{q}$ and frequency ω. It is derived in section 4.11 (appendix E of chapter 4).

When the total z component of the spin is a constant of the motion, such as for a Heisenberg FM or AF with spins aligned along $\pm\mathbf{z}$, the off-diagonal $\alpha \neq \beta$ terms in the cross section vanish and equation (2.4) simplifies to

$$\begin{aligned} \frac{d^2\sigma}{d\Omega\, d\omega} &= \left(\frac{\gamma e^2}{m_e c^2}\right)^2 \left[\frac{g}{2} f(\boldsymbol{\kappa})\right]^2 \frac{k_f}{k_i} \\ &\times \left\{\left(1 + \frac{\kappa_z^2}{\kappa^2}\right) S_{xx}(\mathbf{q},\omega) + \left(1 - \frac{\kappa_z^2}{\kappa^2}\right) S_{zz}(\mathbf{q},\omega)\right\}, \end{aligned} \tag{2.7}$$

which uses the equivalence of spin fluctuations along x and y to set $S_{xx}(\mathbf{q},\omega) = S_{yy}(\mathbf{q},\omega)$. This cross section contains two terms, denoted as transverse (perpendicular to **z**) and longitudinal (parallel to **z**). Because the longitudinal term involving $S_{zz}(\mathbf{q},\omega)$ usually produces only elastic or quasi-elastic scattering, it will be neglected for the purposes of this book. Corresponding to the creation and annihilation of SWs, the transverse term $S_{xx}(\mathbf{q},\omega)$ is the one we are interested in. Keep in mind that when the spins are not all parallel to **z**, the z component of the spin is no longer a constant of motion and all three components of $S_{\alpha\alpha}(\mathbf{q},\omega)$ become important.

In the absence of disorder and intrinsic SW damping, the imaginary susceptibility can be evaluated based on SW theory:

$$\mathrm{Im}\,\chi^{\mathrm{ss}}_{\alpha\alpha}(\mathbf{q},\omega) = S_{\alpha\alpha}(\mathbf{q})\,\delta(\omega - \omega(\mathbf{q})) - S_{\alpha\alpha}(-\mathbf{q})\,\delta(\omega + \omega(-\mathbf{q})), \tag{2.8}$$

which corresponds to the creation and annihilation of SWs with energy $\hbar\omega(\pm\mathbf{q})$. We will derive an explicit form for $S_{\alpha\alpha}(\mathbf{q})$ in section 4.3. Since $n_{\mathrm{B}}(-\omega) + 1 = -n_{\mathrm{B}}(\omega)$ vanishes as $T \to 0$ for $\omega > 0$, only the $\delta(\omega - \omega(\mathbf{q}))$ term contributes to $S_{\alpha\alpha}(\mathbf{q},\omega)$ at zero temperature. The annihilation term does not survive because there are no SWs to annihilate at zero temperature.

Generally, disorder or interactions between SWs shorten the SW lifetime and introduce damping. In those cases, the delta functions are often replaced by Lorentzian functions. Even in the presence of damping or disorder, various sum rules constrain the integrals of $\omega^n\,\mathrm{Im}\,\chi_{\alpha\alpha}(\mathbf{q},\omega)$ over ω and $\mathbf{q}$. We shall briefly discuss one of these sum rules (with $n = 0$) in section 4.3.

2.3 Some practical considerations

INS measurements on single crystals typically require a few grams of material. Since the excitation wavevector $\mathbf{q}$ can be measured around any allowed reciprocal-lattice vector $\boldsymbol{\tau}$, experiments on single crystals have great flexibility. The most advantageous BZs are usually those closest to the origin $\boldsymbol{\tau} = 0$ because they yield the highest values for the magnetic form factor $f(\boldsymbol{\kappa})$ in the cross section of equation (2.4)[3].

With some limitations, INS measurements can also be performed on polycrystalline [18, 19] or even amorphous specimens [20, 21]. Since amorphous materials do not have a reciprocal lattice, the origin is the only available reciprocal-lattice vector. So the accessible wavevectors in these experiments are $\boldsymbol{\kappa} = \mathbf{q}$ averaged over all possible directions. These measurements must be performed at small scattering angles 2Θ, which severely restricts the range of accessible wavevectors $\mathbf{q}$ and imposes other experimental challenges [20] discussed in Chapter 7. Such experiments are best suited to study isotropic (or nearly isotropic) FMs [20, 21]. For polycrystalline samples, neutron scattering measures the average cross section over many crystallites oriented in random directions and the accessible wavevectors $\boldsymbol{\kappa}$ are averaged over all possible directions. Usually, the most practical reciprocal-lattice vector $\boldsymbol{\tau}$ in these experiments is also the origin [18], although recent INS experiments [19]

[3] This is exactly the opposite from the case of phonon measurements. The cross section for the measurement of phonons has a form factor proportional to q^2. This means that phonon measurements are best made far away from the origin where the values of q are high.

have obtained useful information at other reciprocal-lattice vectors as shown in section 7.4.

For practical purposes, it is also important to consider the instrumental resolution. Until now, we have assumed perfect resolution so that all energies and wavevectors are determined exactly. In a real experiment, a finite instrumental resolution is created by the uncertainties in the physical measurements and by the integration over a region of $\{\mathbf{q},\omega\}$ space needed to accumulate a significant signal. All measurements are then smeared by the instrumental resolution function $R(\mathbf{q},\omega)$, which is normally approximated by a 4D ellipsoid [22]. Hence, the scattering intensity at a particular point $\{\mathbf{q},\omega\}$ really measures

$$I_{\alpha\beta}(\mathbf{q},\omega) = \int R(\mathbf{q} - \mathbf{k}, \omega - \omega')\, S_{\alpha\beta}(\mathbf{k}, \omega')\, d\omega'\, d^3\mathbf{k}, \tag{2.9}$$

which is the convolution of $S_{\alpha\beta}(\mathbf{q},\omega)$ over the resolution function.

Figure 2.3 illustrates the effect of the instrumental resolution on the measurement of low-q SWs in $La_{0.70}Ca_{0.30}MnO_3$ at $T = 80$ K [23]. The peaks on the positive and negative side of the energy axis, corresponding to the creation and annihilation of SWs, satisfy the detailed balance condition introduced in section 4.3. At this low temperature, $S_{\alpha\beta}(\mathbf{q},\omega)$ is expected to have the double delta-function structure of equation (2.8). The observed broadening of the peaks is produced by the instrumental resolution, confirming that the intrinsic broadening of the SWs is negligible. A typical value of the energy resolution is 1 meV for thermal neutrons and 0.1 meV for cold neutrons.

While neutrons penetrate into most samples far more effectively than charged particles or x-rays, a few materials containing isotopes such as ^{113}Cd, ^{10}B, ^{157}Gd,

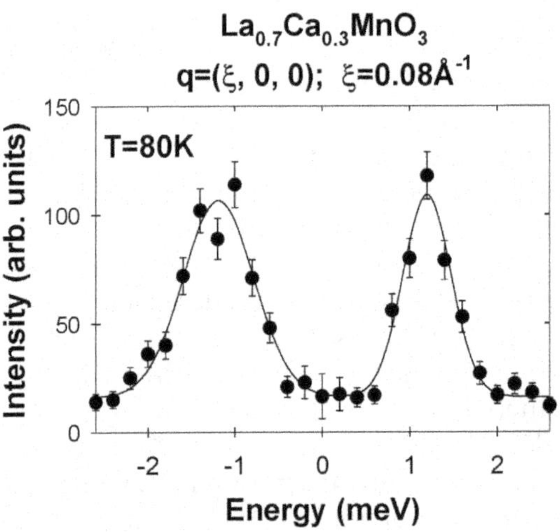

Figure 2.3. SWs in $La_{0.70}Ca_{0.30}MnO_3$ at $T = 80$ K with spectral weight modeled by a double Dirac delta function broadened by the instrumental energy resolution. The intrinsic broadening of the SWs is negligible. A temperature-independent, non-magnetic component at $\omega = 0$ has been subtracted. Reprinted from [23] with permission from the authors. Copyright 2001 by the American Physical Society.

and ^{149}Sm strongly absorb neutrons and cannot be used as samples for neutron-scattering experiments. Those isotopes strongly absorb neutrons mainly due to neutron radiative capture by their nuclei. The neutron-absorption cross sections do not depend on the number of electrons of the chemical elements but rather vary somewhat capriciously among them and even among isotopes of the same element. When preparing a sample for neutron scattering, it is often possible to choose an isotope that has a low absorption cross section, a practice known as isotope substitution. Isotope substitution is also a common practice for materials containing hydrogen because the incoherent scattering from ^{1}H is so large. In those cases, ^{2}H (also referred to as ^{2}D) can be substituted for ^{1}H in the sample preparation. Called deuteration, this procedure will be discussed further in the INS case studies of chapter 7.

2.4 Instruments for INS

This section describes the two main instruments employed in INS measurements: the triple-axis spectrometer and the chopper spectrometer.

2.4.1 The triple-axis spectrometer

The triple-axis spectrometer was originally conceived by Brockhouse to analyze the energy spectrum of neutrons scattered by a sample. A technical discussion of this instrument and its capabilities can be found in the book 'Neutron scattering with a Triple-axis Spectrometer' by Shirane, Shapiro and Tranquada [17].

Sketched in figure 2.4, the triple-axis spectrometer is so named because the neutrons are scattered three times before they are detected and counted: first at the monochromator (M), then at the sample (S), and finally at the analyzer (A). The first bounce at the monochromator (M) uses Bragg diffraction to select neutrons with linear momentum $\hbar\mathbf{k}_i$. These neutrons then exchange momentum and energy with the nuclei and electronic spins of the sample (S). Also by Bragg diffraction, an analyzer crystal (A) selects neutrons that are scattered from the sample at an angle 2Θ with linear momentum $\hbar\mathbf{k}_f$ before they are counted at the detector. By selecting incident and scattered neutrons with wavevectors $\mathbf{k}_i$ and $\mathbf{k}_f$ (depicted as thick red arrows in figure 2.4), the triple-axis spectrometer effectively selects and counts neutrons with scattering wavevector $\mathbf{q}$ and energy ω (see equations (2.1) and (2.2)). Hence, a triple-axis spectrometer directly measures the scattering function $S_{\alpha\alpha}(\mathbf{q},\omega)$, which contains all information about the spin correlations of the system under study.

Although it measures one point in $\{\mathbf{q},\omega\}$ space at a time, a triple-axis spectrometer can be used to map the full SW dispersion by fixing $|\mathbf{k}_f|$ (constant E_f) and scanning in $\mathbf{k}_i$ (i.e. scanning in E_i) for a particular wavevector transfer $\mathbf{q}$. This is commonly known as a constant-$\mathbf{q}$ scan. Figure 2.3 provides an example of a constant-$\mathbf{q}$ scan for $La_{0.7}Ca_{0.3}MnO_3$ with $\mathbf{q} = (0, 0, 0.08)$ Å^{-1}. The scan in E_i was performed by fixing $E_f = 14.7$ meV and using a pyrolytic graphite filter before the analyzer to eliminate higher-order Bragg reflections. The x-axis provides the energy $E = E_i - E_f$ transferred to the electronic spins to create ($E > 0$) or annihilate ($E < 0$)

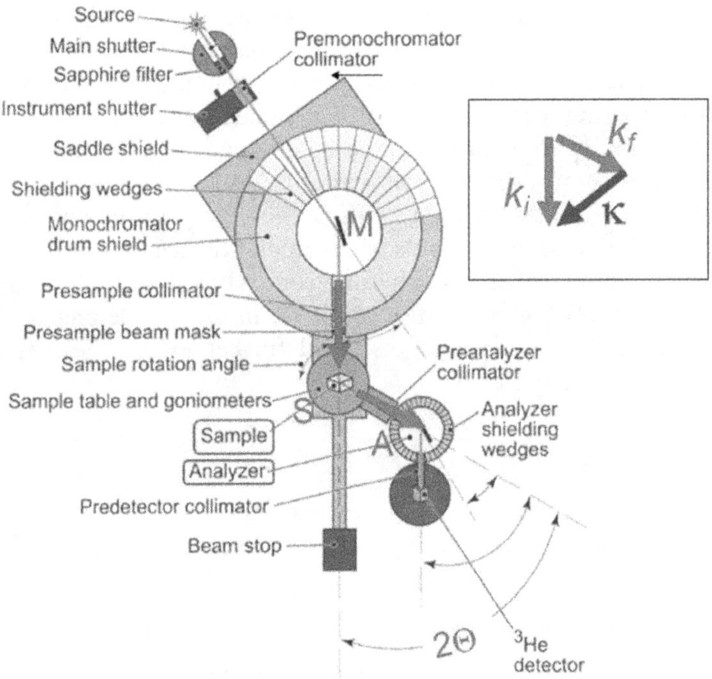

Figure 2.4. Schematic diagram of a triple-axis spectrometer. The monochromator (M) and analyzer (A) select the wavevectors $\mathbf{k}_i$ and $\mathbf{k}_f$ represented by the bold red arrows. For clarity the insert shows the 'scattering triangle' formed by the wavevectors $\mathbf{k}_i$, $\mathbf{k}_f$ and $\boldsymbol{\kappa}$, the scattering wavevector defined in equation (2.1).

a SW. The full SW dispersion of a material can be mapped by repeating this measurement for many wavevectors along a particular high-symmetry direction.

Another example, figure 2.5 plots the SW dispersion of three different manganites along the high-symmetry directions $\mathbf{q} = (0, 0, \xi)$ and $(\xi, \xi, 0)$ [24]. SW energies are plotted as open symbols and selected phonon energies are plotted as full symbols. Assuming a nearest-neighbor Heisenberg model, the solid lines are the expected SW dispersions. There are clear departures from the predicted dispersions as the wavevectors approach the BZ boundary. Hence, the simple Heisenberg model is only adequate for small wavevectors or large wavelengths.

Because it measures only one value of $S(\mathbf{q},\omega)$ at a time, the triple-axis spectrometer is best suited to precisely measure $S(\mathbf{q},\omega)$ over a limited region in $\{\mathbf{q},\omega\}$ space. However, several recently designed instruments with multiplexed back ends, mostly in the cold-neutron range, allow the simultaneous measurement of many $\{\mathbf{q},\omega\}$ points [25–27]. A more effective way to probe an extended region of $\{\mathbf{q},\omega\}$ space is with the neutron-chopper spectrometer described in the next section.

2.4.2 The chopper spectrometer

The previous section explained why the reactor-based triple-axis spectrometer is best suited to make high-precision measurements of $S_{\alpha\beta}(\mathbf{q},\omega)$ in limited regions of $\{\mathbf{q},\omega\}$ space. This section describes the chopper spectrometer, a neutron time-of-flight

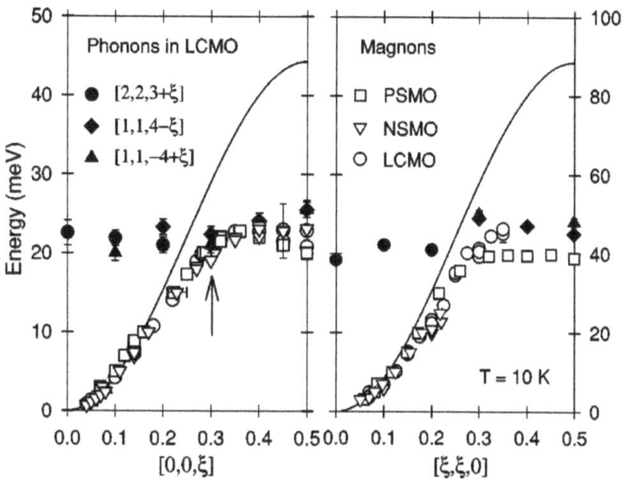

Figure 2.5. SW dispersions for manganites $La_{1-x}Ca_xMnO_3$, $Nd_{1-x}Sr_xMnO_3$, and $Pr_{1-x}Sr_xMnO_3$, near $x = 0.3$. SW energies are plotted as open symbols and selected phonon energies are plotted as full symbols. Predictions of the nearest-neighbor Heisenberg model for FM SWs, plotted as the solid lines, do not account for the softening of the SW energies at the BZ boundaries. Reprinted from [24] with permission from the authors. Copyright 2000 by the American Physical Society.

instrument that can be based at either a reactor or a spallation source and can measure $S_{\alpha\beta}(\mathbf{q},\omega)$ over extended regions of $\{\mathbf{q},\omega\}$ space. As its name implies, the time-of-flight technique determines the energy of a neutron by measuring the time it takes to travel a known distance. Developed in 1935 [28], this technique was first applied to solids in the 1960s. For a more detailed description of this technique, the reader is referred to the excellent article by Copley and Udovic [29].

In a reactor source, the initial energy of the neutrons is selected by Bragg reflection from a monochromator crystal. One or more choppers is used to determine the shape and phase of the neutron pulses. In a pulsed, spallation neutron source, neutrons are slowed down by a suitable moderator and the neutron energy is determined by one or more choppers. The most common chopper arrays are Fermi and disk choppers. A Fermi chopper array spins about a vertical axis perpendicular to the direction of the beam. Fitted with a set of slits, the curvature of the slits and the speed of the chopper determine the transmitted energy. A disk chopper array contains one or more disk pairs that counter rotate about a horizontal axis parallel to the direction of the beam. The size of the opening and the speed of rotation then determine the transmitted energy.

The sample chamber can accommodate a wide variety of sample environments such as cryostats, cryomagnets, and furnaces. An oscillating radial collimator between the sample and the detectors blocks the neutrons from scattering off components like the heat shields and vacuum cans surrounding the sample. A large array of neutron detectors is located at a fixed distance, typically a few meters, from the sample. The space between the sample and detectors is usually evacuated to eliminate scattering from air.

Modern chopper spectrometers have pixilated arrays of detectors that cover a large solid angle up to one or two *sr* [30]. The final energy E_f for each neutron that reaches a pixel in a detector is determined by its time of flight. Simultaneously, $\mathbf{k}_i$ is determined by the incident energy selected by the chopper assembly and by the direction of the incident beam. Then, $\mathbf{k}_f = -\boldsymbol{\kappa} + \mathbf{k}_i$ is fixed by E_f and the direction of the scattered neutron. Unlike for a conventional triple-axis spectrometer, the range of accessible wavevectors $\boldsymbol{\kappa}$ is not limited to the equatorial plane. Thus, the combination of time-of-flight measurements and a large, fixed pixilated detector array allows the measurement of $S_{\alpha\beta}(\mathbf{q},\omega)$ over an extended region of $\{\mathbf{q},\omega\}$ space.

At the Spallation Neutron Source (SNS) in Oak Ridge, four chopper spectrometers measure excitations in a broad range of energies from meV to eV [30–34]. Figure 2.6 shows the Cold Neutron Chopper Spectrometer (CNCS) [31], a multi-chopper instrument designed for high-resolution neutron spectroscopy with incident energies between 1 and 50 meV. Neutrons from the spallation source are slowed down by a cold-coupled moderator. One high-speed chopper shapes the neutron pulse from the moderator and a second compresses the pulse length at the sample position. These two choppers provide an adjustable energy resolution from 1.2% to

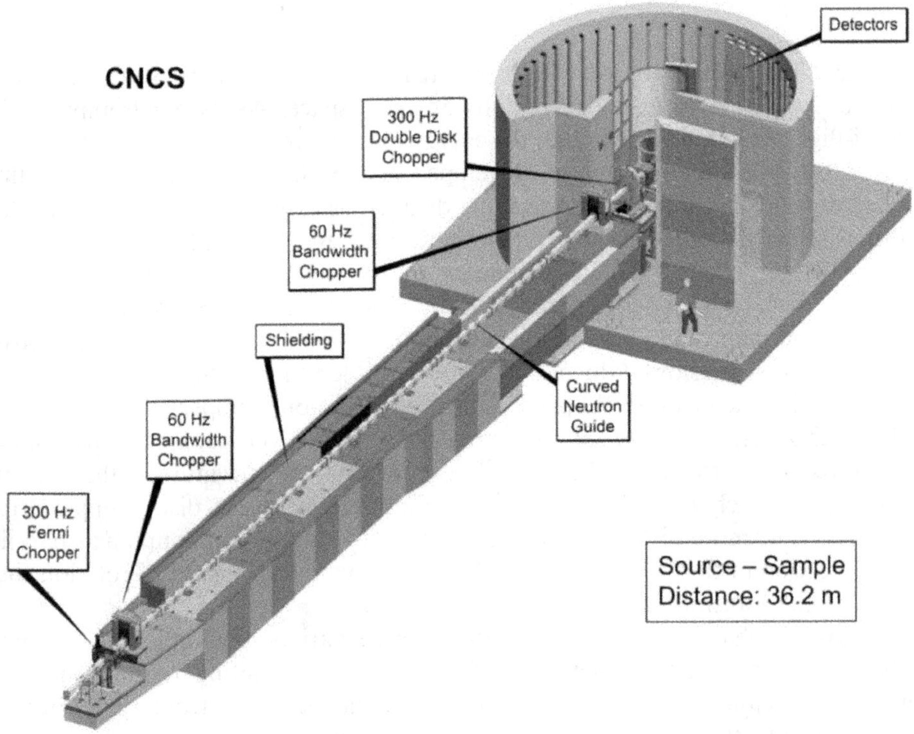

Figure 2.6. Layout of the Cold Neutron Chopper Spectrometer (CNCS) at ORNL's SNS. Reprinted from [31], with permission of AIP Publishing.

about 10% of the incident energy. The distance between the source and the sample is 36.2 m and a curved neutron guide is used to reduce the neutron background. In the secondary flight path, with a length of 3.5 m, neutrons encounter a highly pixilated detector covering scattering angles between 50° and 135° in the equatorial plane and ±16° out of the equatorial plane. With a total solid angle of 1.7 *sr*, the detector array consists of 400 individual 2 m long position-sensitive tubes filled with ^{3}He gas.

2.5 Neutron scattering at large user facilities

Since the early days at Oak Ridge and Chalk River, significant advances have been made in the development of intense neutron sources and neutron-scattering instruments. However, the construction of reactors and spallation neutron sources requires enormous investment. Consequently, most major facilities serve diverse scientific and engineering communities around the world. Among the major reactor-based user facilities are the High Flux Isotope Reactor (HFIR)[4] (Oak Ridge, Tennessee) and the NIST Center for Neutron Research (NCNR)[5] (Gaithersburg, Maryland) in the US, the Institut Laue-Langevin (ILL)[6] in France, the Research Neutron Source Heinz Maier-Leibnitz[7] at the FRM-II reactor in Germany, and the Australian Center for Neutron Scattering (ACNS)[8] at ANSTO. The major spallation-based user facilities include the Spallation Neutron Source (SNS)[9] in the US, the ISIS Neutron and Muon Source[10] in the United Kingdom, the Materials and Life Science Experimental Facility (MLF)[11] at the J-PARC complex in Japan, and the continuous spallation neutron source SINQ in Switzerland[12]. All these facilities have diverse instruments to probe the spin dynamics in solids. At each facility, a neutron-scattering user program awards neutron beam time, free of charge, to qualified experimental proposals on a competitive basis. Recently the China Spallation Neutron Source (CSNS)[13] in Dongguan City has started operations. A new facility, the European Spallation Source (ESS)[14] in Sweden is expected to be completed in 2023. In Oak Ridge, planning is under way for the design and construction of a second target station (STS)[15] that will complement the capabilities of the first target station at the SNS.

[4] https://neutrons.ornl.gov
[5] www.ncnr.nist.gov
[6] www.ill.eu
[7] www.frm2.tum.de/en/home
[8] www.ansto.gov.au/acns
[9] https://neutrons.ornl.gov/sts
[10] www.isis.stfc.ac.uk
[11] http://j-parc.jp/MatLife/en
[12] www.psi.ch/sinq/
[13] http://english.ihep.cas.cn/csns
[14] https://europeanspallationsource.se
[15] https://neutrons.ornl.gov

2.6 Exercises

1. Use equations (2.6) and (2.8) to show that for CL spins aligned along $\pm\mathbf{z}$,

$$S_{xx}(\mathbf{q},\omega) = S_{yy}(\mathbf{q},\omega) \propto [n_B(\omega(\mathbf{q})) + 1]\delta(\omega - \omega(\mathbf{q})) + n_B(\omega(\mathbf{q}))\delta(\omega + \omega(\mathbf{q})). \tag{2.10}$$

   Use $n_B(-\omega) = -1 - n_B(\omega)$. Assume that there is no easy-plane anisotropy and that the system is inversion symmetric.
2. At temperatures $k_B T \gg \hbar\omega(\mathbf{q})$, use the above result to show that $S_{xx}(\mathbf{q},\omega)$ and $S_{yy}(\mathbf{q},\omega)$ have two positive and nearly equal peaks at $\pm\omega(\mathbf{q})$, as in figure (2.3).

References

[1] Steiner M 2004 Neutron scattering and ordering of nuclear spins *J. Low Temp. Phys.* **135** 545–78

[2] Wollan E O and Shull C G 1948 The diffraction of neutrons by crystalline powders *Phys. Rev.* **73** 830–41

[3] Halpern O and Johnson M H 1939 On the magnetic scattering of neutrons *Phys. Rev.* **55** 898–923

[4] Shull C G, Wollan E O and Strauser W A 1951 Magnetic structure of magnetite and its use in studying the neutron magnetic interaction *Phys. Rev.* **81** 483–4

[5] Néel L 1952 Antiferromagnetism and ferrimagnetism *Proc. Phys. Soc.* A **65** 869

[6] Shull C G, Strauser W A and Wollan E O 1951 Neutron diffraction by paramagnetic and antiferromagnetic substances *Phys. Rev.* **83** 333–45

[7] Brockhouse B N 1955 Energy distribution of neutrons scattered by paramagnetic substances *Phys. Rev.* **99** 601–3

[8] Brockhouse B N 1961 Methods for neutron spectrometry *Proc. of the Symp. on Inelastic Scattering in Solids and Liquids (Vienna, October 11–14, 1960)* (Vienna: International Atomic Energy Agency) p 113; also available as Brockhouse B N 1961 Methods for neutron spectrometry *Technical Report* Atomic Energy of Canada, AECL-1183 www.osti.gov/servlets/purl/4025288

[9] Brockhouse B N and Stewart A T 1955 Scattering of neutrons by phonons in an aluminum single crystal *Phys. Rev.* **100** 756–7

[10] Brockhouse B N 1957 Scattering of neutrons by spin waves in magnetite *Phys. Rev.* **106** 859–64

[11] Elliott R J and Lowde R D 1955 The inelastic scattering of neutrons by magnetic spin waves *Proc. R. Soc. Lond.* A **230** 46–73

[12] Goedkoop J A and Riste T 1960 Neutron diffraction study of antiferromagnetic spin waves in α-ferric oxide *Nature* **185** 450–2

[13] Lovesey S W 1984 *Theory of Neutron Scattering from Condensed Matter* vol II (Oxford: Clarendon)

[14] Balcar E and Lovesey S W 1989 *Theory of Magnetic Neutron and Photon Scattering* (Oxford: Clarendon)

[15] Squires G L 2012 *Introduction to the Theory of Thermal Neutron Scattering* 3rd edn (Cambridge: Cambridge University Press)

[16] Price D L and Sköld K 1986 Introduction to neutron scattering *Methods of Experimental Physics, Neutron Scattering* vol 23A ed K Sköld and D L Price (Orlando, FL: Academic), pp 1–97

[17] Shirane G, Shapiro S M and Tranquada J M 2002 *Neutron Scattering with a Triple-Axis Spectrometer* (Cambridge: Cambridge University Press)

[18] Lynn J W, Erwin R W, Borchers J A, Huang Q, Santoro A, Peng J-L and Li Z Y 1996 Unconventional ferromagnetic transition in $La_{1-x}Ca_xMnO_3$ *Phys. Rev. Lett.* **76** 4046–9

[19] Walker H C, Duncan H D, Le M D, Keen D A, Voneshen D J and Phillips A E 2017 Magnetic structure and spin-wave excitations in the multiferroic magnetic metal–organic framework $(CD_3)_2ND_2[Mn(DCO_2)_3]$ *Phys. Rev.* B **96** 094423

[20] Lynn J W and Fernandez-Baca J A 1995 Neutron scattering studies of the spin dynamics of amorphous alloys *The Magnetism of Amorphous Metals and Alloys* ed J A Fernandez-Baca and W-Y Ching (Singapore: World Scientific), pp 221–60

[21] Fernandez-Baca J A, Lynn J W, Rhyne J J and Fish G E 1987 Long-wavelength spin-wave energies and linewidths of the amorphous invar alloy $Fe_{100-x}B_x$ *Phys. Rev.* B **36** 8497–511

[22] Cooper M J and Nathans R 1967 The resolution function in neutron diffractometry I. The resolution function of a neutron diffractometer and its application to phonon measurements *Acta Crystallogr.* **23** 357–67

[23] Dai P, Fernandez-Baca J A, Plummer E W, Tomioka Y and Tokura Y 2001 Magnetic coupling in the insulating and metallic ferromagnetic $La_{1-x}Ca_xMnO_3$ *Phys. Rev.* B **64** 224429

[24] Dai P, Hwang H Y, Zhang J, Fernandez-Baca J A, Cheong S-W, Kloc C, Tomioka Y and Tokura Y 2000 Magnon damping by magnon-phonon coupling in manganese perovskites *Phys. Rev.* B **61** 9553–7

[25] Rodriguez J A *et al* 2008 MACS—a new high intensity cold neutron spectrometer at NIST *Meas. Sci. Technol.* **19** 034023

[26] Groitl F, Graf D, Birk J O, Marko M, Bartkowiak M, Filges U, Niedermayer C, Ruegg C and Rønnow C 2016 CAMEA a novel multiplexing analyzer for neutron spectroscopy *Rev. Sci. Instrum.* **87** 035109

[27] Groitl F *et al* 2017 MultiFLEXX—the new multi-analyzer at the cold triple-axis spectrometer FLEXX *Sci. Rep.* **7** 13637

[28] Dunning J R, Pegram G B, Fink G A, Mitchell D P and Segrè E 1935 Velocity of slow neutrons by mechanical velocity selector *Phys. Rev.* **48** 704

[29] Copley J R D and Udovic T J 1993 Neutron time-of-flight spectroscopy *Res. Natl Inst. Stand. Technol.* **98** 71–87

[30] Stone M B *et al* 2014 A comparison of four direct geometry time-of-flight spectrometers at the spallation neutron source *Rev. Sci. Instrum.* **85** 045113

[31] Ehlers G, Podlesnyak A A, Niedziela J L, Iverson E B and Sokol P E 2011 The new cold neutron chopper spectrometer at the spallation neutron source: design and performance *Rev. Sci. Instrum.* **82** 085108

[32] Abernathy D L, Stone M B, Loguillo M J, Lucas M S, Delaire O, Tang X, Lin J Y Y and Fultz B 2012 Design and operation of the wide angular-range chopper spectrometer arcs at the spallation neutron source *Rev. Sci. Instrum.* **83** 015114

[33] Granroth G E, Kolesnikov A I, Sherline T E, Clancy J P, Ross K A, Ruff J P C, Gaulin B D and Nagler S E 2010 Sequoia: a newly operating chopper spectrometer at the SNS *J. Phys.: Conf. Ser.* **251** 012058

[34] Winn B *et al* 2015 Recent progress on HYSPEC, and its polarization analysis capabilities *EPJ Web Conf.* **83** 03017

Chapter 3

THz spectroscopy

Launch photon torpedos!

—Captain James T Kirk

3.1 Introduction

The main optical techniques used to probe SWs are THz absorption, Raman scattering, and resonant inelastic x-ray scattering (RIXS). In THz absorption, one photon is absorbed and one SW is created or vice versa. Since the momentum k of a photon at THz frequencies is small compared to the size of the BZ, the SW has nearly zero momentum, $q = k \approx 0$. Although the spin cycloid in $BiFeO_3$ is among the largest known spin structures with a period of 62 nm [1], it is still an order of magnitude shorter than the wavelength of a visible photon.

Raman scattering and RIXS are similar to INS except that photons rather than neutrons are scattered by SWs. The smallest SW wavevector detected is $q = k_i - k_f$ (forward scattering) and the largest is $q = k_i + k_f$ (back scattering), where k_i and k_f are the wavevectors of the incident and scattered photons. With photon wavelengths from 0.4 to 1 μm (from the ultraviolet to the infrared), Raman scattering detects SWs with q close to zero. RIXS uses very short wavelength photons, e.g. 1.3 nm at the energy 931 eV of the Cu L_3 peak [2]. The largest SW wavevector $q = k_i + k_f$ is then comparable to the size of the structural BZ. Due to its element sensitivity and ability to measure the dispersion of magnetic excitations, RIXS is a very powerful spectroscopic technique [3].

While INS and RIXS can measure the dispersion of SW energies, THz and Raman spectroscopies can only measure SW energies close to the center of the BZ or at $q \approx 0$. The routine energy resolution of THz and Raman spectroscopies is 0.01 meV compared to 0.1 meV for INS measurements with cold neutrons and >1 meV for RIXS. On the other hand, the wavevector resolution of all three optical techniques is far better than the typical resolution 10^{-2} Å^{-1} of INS. THz spectroscopy has a wavevector resolution of $\nu/c \approx 3 \times 10^{-10}$ Å^{-1} for $\nu = 1$ THz or $\hbar\omega = 4.1$ meV. With a 500 nm wavelength (600 THz) visible photon, Raman scattering has a wavevector resolution of 4×10^{-4} Å^{-1}. While the wavevector resolutions of THz and Raman spectroscopy are simply given by the wavevector of the incoming photon, the wavevector resolution of RIXS is only limited by instrumentation. Some RIXS measurements [4] have reported a resolution of 4×10^{-3} Å^{-1}.

In contrast to the sample requirements of INS, the dimensions rather than the mass of the sample are important for THz spectroscopy. The lateral dimension of a sample must be larger than the longest wavelength (corresponding to the lowest frequency) of interest. To reach a frequency of 0.1 THz, for example, the sample should be wider than 3 mm. However, the required thickness for transmission measurements is between 0.1 and 1 mm (see section 3.3).

Unlike THz spectroscopy, Raman spectroscopy does not require so wide a sample because a focused laser beam has a spot size of about 500 nm in the visible spectrum. For RIXS, the lateral resolution is only limited by instrumentation and typically lies between 10 and 100 μm. The greatest advantage of both RIXS and Raman over THz spectroscopy is that they can be used to study much thinner films. For example, RIXS was used to measure the SW dispersion in 30 nm thick NiO films [4] and Raman spectroscopy was used to study SWs in 70 nm thick $BiFeO_3$ films [5].

Photons are oscillations of the electric $\mathbf{E}^{\omega}$ and magnetic $\mathbf{B}^{\omega}$ fields and, unlike neutrons, do not carry a magnetic moment. The magnetic field $\mathbf{B}^{\omega}$ of a photon directly interacts with the magnetic moment $\mathbf{m} = \mathbf{M}/N$ of a spin through the magnetic-dipole interaction $-\mathbf{B}^{\omega} \cdot \mathbf{M}$. To first order, the photon's electric field does not interact with the spin's magnetic moment. Hence, SWs detected by THz spectroscopy are usually magnetic-dipole active. However, spin–orbit or spin–phonon coupling may 'dress' the spins with charge, producing an effective electric-dipole moment $\mathbf{P}/N$. In THz spectroscopy, SWs that are electric-dipole active are called electromagnons [6] or magnetoelectric excitations [7].

In Raman scattering and RIXS, the electric-dipole scattering is four orders of magnitude stronger than the magnetic-dipole scattering [8]. But the theoretical treatment of electric-dipole scattering depends on how spin–orbit coupling dresses the spins with charge. Therefore, the following discussion will concentrate on THz spectroscopy, where the magnetic contribution to the optical absorption is simply related to the INS cross section, as shown in section 4.4. The Raman scattering of magnetic excitations is discussed in [9–11]. For a review of RIXS, see [12, 13].

Since 1 THz = 33.36 cm^{-1}, the wavelength of 1 THz radiation is 0.2998 mm. Experimental frequencies typically range from 0.1 to 10 THz. Below 0.1 THz, the propagation of electromagnetic waves in spectrometers is confined to waveguides, transmission lines, and coaxial cables. Diffraction effects are less important above 0.1 THz, so pipes, lenses, and mirrors can be used. The term 'THz spectroscopy' arose with the invention of photoconductive antennas that generate and detect THz radiation after excitation by a sub-picosecond laser pulse [14]. Before the invention of photoconductive antennas, the THz range was covered by sub-millimeter and far-infrared spectroscopies. In this book, 'THz spectroscopy' refers to all spectroscopic techniques that use electromagnetic radiation in the THz spectral range. Overviews of THz techniques in solid-state spectroscopy are given by [15–17].

3.2 THz spectroscopy in high magnetic fields

Magnetic fields are used to shift the SW frequencies and produce new magnetic ground states. For an isolated electron spin **S**, the Zeeman interaction $2\mu_B\mathbf{B}\cdot\mathbf{S}$ shifts the Larmor frequency by 0.93 cm^{-1} T^{-1}. So the electron spin-resonance frequency is about 1 THz in a magnetic field of 33 T.

Magnets can be either direct-current (dc) or pulsed field. A dc field is produced by superconducting magnets, Bitter magnets, and hybrids of Bitter and superconducting magnets (see table 3.1). For Bitter and hybrid magnets, the dc field can be maintained from minutes to hours; for superconducting magnets, it can be maintained for years.

For a pulsed field above 45 T, the maximum field may last only a few milliseconds or less. Pulsed fields are useful not only to extend the field beyond 45 T but also as a more affordable alternative to superconducting and Bitter magnets [18, 19]. THz measurements in fields above 20 T must be performed in one of several high magnetic-field laboratories located in Europe, Japan, or the United States, see table 3.1.

A THz absorption experiment in a magnetic field has four main components: the THz source, a THz detector, the sample cryostat, and the magnet. In the Faraday configuration, light propagates parallel to the magnetic field; in the Voigt configuration, it propagates perpendicular to the magnetic field. The magnetic field may be either horizontal or vertical and the magnet can be either a solenoid or a split coil. Two configurations with a solenoid and two with a split coil are shown in figure 3.1.

3.2.1 THz spectroscopy in pulsed and swept magnetic fields

At some field B, the SW energy may satisfy the condition $E(B) = h\nu_s$, where ν_s is the monochromatic radiation frequency in Hz. In a simple transmission configuration, a

Table 3.1. High-field THz user facilities. $\Delta\nu$ is the frequency range and B_{max} is the maximum field. SC = superconducting, B = Bitter, H = hybrid, and p = pulsed magnet, FT-IR = Fourier-transform infrared spectrometer, ss = solid-state. The Tallahasse 32 T magnet uses coils made of low-T_c and high-T_c superconductors; magnets made exclusively from low-T_c superconductors typically reach 20 T.

Location	$\Delta\nu$ (THz)	B_{max} (T)	THz sources
KOFUC Osaka	0.02–6	70(p)	Monochromatic
NHMFL Tallahassee	0.07–1.2	32(SC), 41.5(B), 45(H)	BWO
EMFL Nijmegen	0.25–120	33(B)	Hg lamp+FT-IR, FEL
EMFL Dresden	0.1–9	70(p)	ss oscillators, FEL

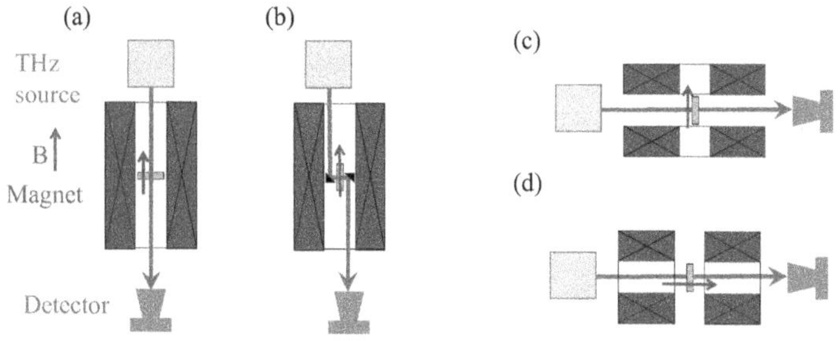

Figure 3.1. Four THz absorption configurations in a magnetic field. (a) Faraday and (b) Voigt configurations with a vertical-field solenoid. (c) Voigt and (d) Faraday configurations with a split-coil magnet. The Faraday configuration in (d) can be turned into the Voigt configuration by rotating the source and the detector in the horizontal plane by 90°, i.e. with the detector facing towards or away from the viewer. A sample cryostat (not shown) is inserted through the vertical bore of the magnet in (a) and (b). Both vertical and horizontal cryostats are possible in (c) and (d).

SW is detected by the drop of the detected THz radiation intensity when this resonance condition is satisfied during a field sweep.

Because a field pulse is short, pulsed fields require fast detectors. A field pulse usually repeats in tens of minutes or longer so the magnet can cool between pulses. Since it is not practical to accumulate data over such a long time, the THz radiation source must be intense enough to produce a meaningful spectrum during a single pulse. To acquire the SW spectra versus magnetic field, the field pulse is repeated at different source frequencies.

The field sweep of a dc magnet is usually performed with a fixed monochromatic source. Because the field sweep is slow, the signal-to-noise ratio can be improved [22] by using a lock-in technique, such as modulating the source intensity or the dc magnetic field. A pulsed or dc field sweep combined with a monochromatic THz source is often referred to as high-field electron-spin resonance (hf-ESR). See [23] and references therein for examples.

The main difference between AF resonance (AFMR) and ESR is that the AFMR resonance frequency is nonzero when $B = 0$. As an example, the AFMR spectrum of the $S = 2$ spin-chain compound $MnCl_3$(bpy) is shown in figure 3.2 [20]. The reference sample, DPPH (2,2-diphenyl-1-picrylhydrazyl) with $g = 2.0036$, exhibits a typical ESR signal, as shown in figure 3.2(a).

Monochromatic sources of THz radiation include vacuum tubes such as a backward-wave oscillator (BWO), far-infrared lasers, solid-state devices (e.g. products made by Virginia Diodes, Inc), and free-electron lasers (FEL). Liquid He cooled bolometers and room-temperature solid-state devices are commonly used as detectors. A THz-range FEL is available for experiments with pulsed fields up to 63 T in Dresden [24] and with dc fields up to 33 T in Nijmegen [25].

A magnetic-field sweep with a monochromatic THz source cannot detect SWs if their frequencies are independent of the magnetic field. Moreover, a quantitative analysis of the absorption line intensity is often not possible due to standing waves in the optical path. An example of the distorted line shape measured with a monochromatic source in a pulsed magnetic field is given in figure 3.2(b).

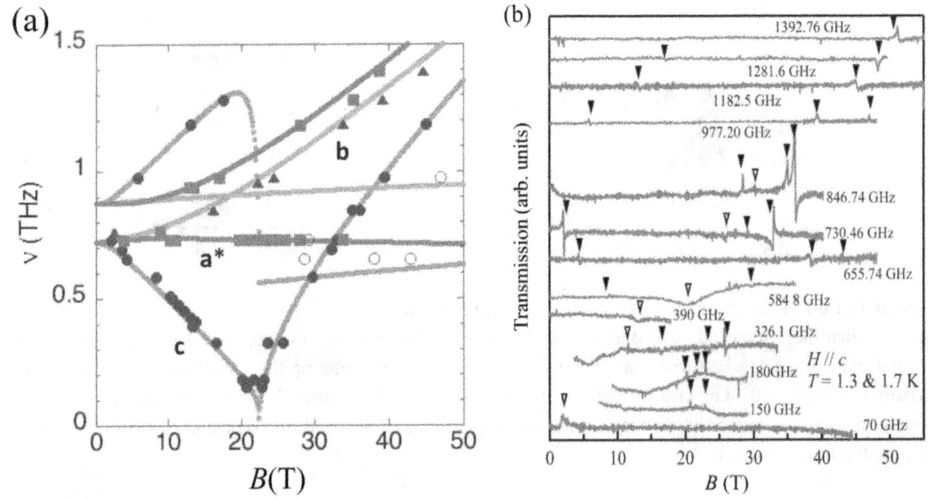

Figure 3.2. (a) The AFMR frequencies of $MnCl_3$(bpy) at $T \approx 1.3$–1.7 K as a function of magnetic field. Data points are the observed resonances [20] for fields parallel to the $\mathbf{a}^*$ (red squares), **b** (blue triangles), and **c** (green circles) axes. Open circles are 'outlier points' and solid lines are the theoretical fit described in [21]. (b) Transmission spectra in field **B** || **c**. Open and closed symbols mark the resonance fields of DPPH and $MnCl_3$(bpy), respectively. hf-ESR measurements in pulsed magnetic fields up to 50 T with a duration of about 7 ms utilized a far-IR laser or Gunn oscillators with frequency doublers to generate THz radiation. The detector was a magnetically tuned InSb hot-electron bolometer. All of these instruments and resources are located at the Center for Advanced High Magnetic Field Science of Osaka University. Reprinted with permission from [20, 21].

If $\mathbf{E}^{\omega}$ vanishes at the sample position in a resonant cavity, then hf-ESR detects only magnetic-dipole active transitions. But in the transmission mode without a resonant cavity, both $\mathbf{E}^{\omega}$ and $\mathbf{B}^{\omega}$ oscillate at the sample position and hf-ESR can also detect electric-dipole active resonances. THz absorption spectroscopy can also be used to detect both magnetic-dipole and electric-dipole active excitations without a resonant cavity.

Because it can be tuned over a limited frequency range, a spectrometer with a BWO can record the SW spectrum in a steady magnetic field. A Mach–Zehnder type interferometer can be used to detect both the amplitude and phase of radiation [26, 27]. However, several BWOs are then needed to cover the frequency range from 0.1 to 1 THz.

3.2.2 THz spectroscopy in dc magnetic fields with broad-band sources

A field sweep is not needed if a THz spectrum can be recorded over a broad continuous range. Broad-band THz sources can cover the range from 0.1 to several THz. We will discuss time- and frequency-domain THz systems that use photoconductive antennas and Fourier-transform interferometers with a synchrotron or mercury arc lamp as the radiation source.

While a time-domain THz system generates a broad-band spectrum in a single pulse [28], a frequency-domain THz system scans the frequency [29]. The time- and frequency-domain THz systems shown in figures 3.3 and 3.4 can detect both the amplitude and phase of THz radiation. Time-domain signal and spectra are shown in figure 3.5.

Due to their phase sensitivity, these setups require direct optical access to the sample, which is easiest with a split-coil magnet. However, a split coil cannot compete with a solenoid in field strength because it is essentially a solenoid with the central part of the coil removed. For a solenoid, direct optical access to the sample space is only possible along its narrow bore, which must also accommodate the cryostat. Some drawbacks of a solenoid can be overcome by placing a horizontal bore magnet with a dry cooling system in one arm of a time-domain THz spectrometer [30, 31].

The transmitter and receiver of a frequency-domain system are coupled to infrared lasers through single-mode fibers as shown in figure 3.4. Therefore, direct optical access from the laser to the transmitter/receiver is not required. In principle, it is possible to place the transmitter and receiver a few centimeters away from each other inside the cryogenic chamber of a solenoid [32].

A mercury arc lamp provides a broad-band source of THz radiation. Below 3 THz, the quartz envelope of the lamp becomes transparent to the radiation emitted by an arc plasma [34]. Although the plasma temperature is close to 5000 K, the intensity of black-body radiation falls off rapidly with frequency as ν^{-2}. Therefore,

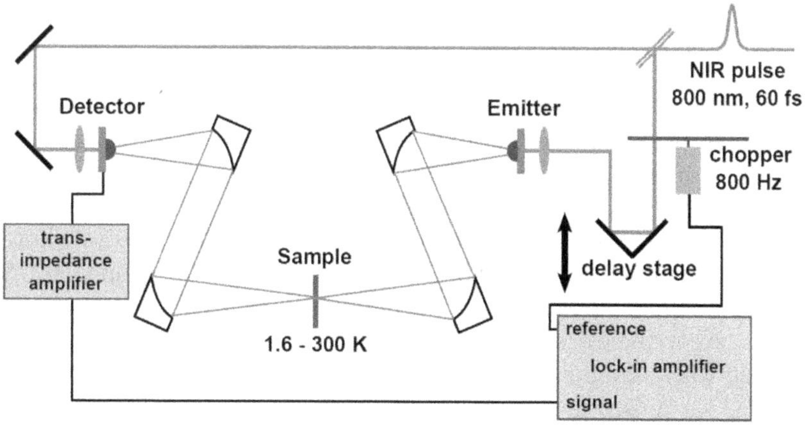

Figure 3.3. A time-domain THz spectroscopy setup. A Ti:Sapphire mode-locked laser generates 60 fs near-IR (NIR) pulses centered at 820 nm with an 80 MHz repetition rate. A beam splitter divides each pulse into two different paths. The first part of the pulse excites a photoconductive Auston-switch antenna, which accelerates electrons across the small antenna gap and emits a few ps THz pulse. This THz pulse is collimated and focused onto the sample with a silicon lens and a pair of off-axis parabolic mirrors. The transmitted pulse is then focused by a second pair of off-axis parabolic mirrors onto a detector antenna. The second part of the NIR pulse arrives at the receiver antenna at the same time as the THz pulse and photoexcites electrons in the antenna gap. The electric field of the THz pulse drives the photoexcited electrons across the antenna gap, producing a current that is converted to a voltage using a transimpedance amplifier and measured by a lock-in amplifier. A delay stage varies the arrival time of the THz pulse at the detector antenna to map out the full THz electric field as a function of time. A mechanical chopper modulates the NIR light incident on the emitter at 800 Hz. The sample cryostat and magnet are not shown. Reprinted with permission from supporting material of [33].

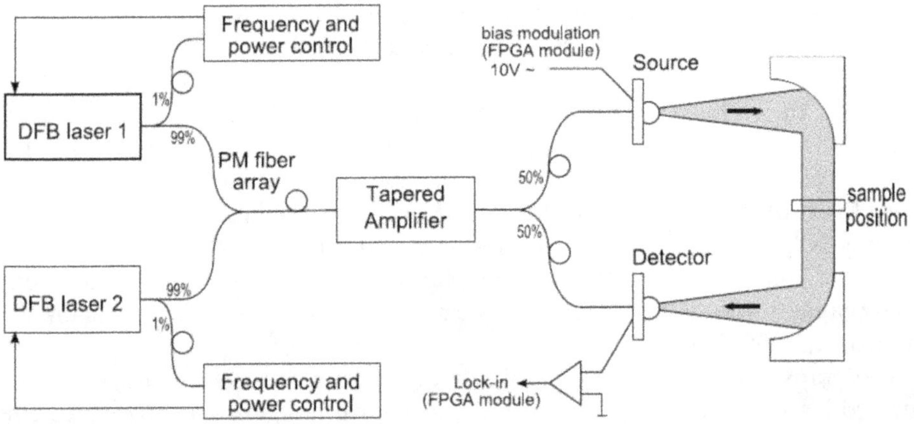

Figure 3.4. A frequency-domain THz spectroscopy setup. Two distributed-feedback diode lasers are tuned to around 853 and 855 nm. Using an interferometric frequency control, this setup achieves both spectral resolution and long-term frequency stability at the MHz level, allowing precise determination of phase. The upper limit of the difference frequency is obtained by cooling the first laser diode close to 0 °C while heating the second one to 50 °C. The laser frequency can then be varied continuously from 60 GHz to 1.8 THz. Superimposed laser beams are fed into a tapered amplifier. Two-color laser light is amplified before coupling to a fiber-optical 50:50 splitter with outputs connected to photoconducting antennas, i.e. to the THz source and detector. Modulation of the bias voltage of the source and lock-in detection of the preamplified signal of the detector are realized by a digital lock-in module based on a field-programmable gate array. Figure from [29]. Copyright Deutsche Physikalische Gesellschaft. Reproduced by permission of IOP Publishing. All rights reserved.

sensitive bolometers operating at or below liquid He temperature are required in the THz range [35]. InSb hot-electron and transition-edge superconducting bolometers do not work in high magnetic fields [36] and should be kept several meters away from the magnet. Less sensitive composite Si or Ge bolometers can be inserted directly into the sample cryostat. This has several advantages. First, it shortens the optical path from the sample to the bolometer and reduces radiation losses. Second, it eliminates thermal load on the bolometer from room-temperature optical components. Nevertheless, the magnetoresistance of composite bolometers must be considered when designing a high-field spectrometer [37]. Composite bolometers are not suited for pulsed-field experiments because their time constants are usually longer than a millisecond.

Bolometers operating at or below 0.3 K have been used mainly in astronomy [38]. Drew and Sievers [39] were the first to use a ^{3}He-cooled bolometer in solid-state THz spectroscopy. To our knowledge, bolometers operating below 0.3 K have not been used for solid-state THz spectroscopy.

Although orders of magnitude brighter than a mercury arc lamp [40], synchrotrons are not widely used in SW spectroscopy because high magnetic fields are not typically available at synchrotron sites. Nonetheless, Mihály *et al* [41] studied the SWs in $LaMnO_3$ at BNL's National Synchrotron Light Source. Measurements of mono- and multinuclear high spin ($S > 1/2$) systems using coherent synchrotron radiation in ESR up to 11 T were recently reviewed by J Nehrkorn *et al* [42].

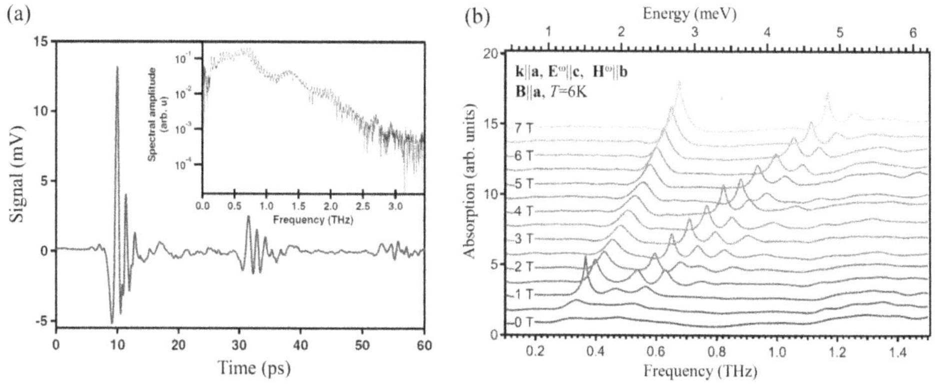

Figure 3.5. Time-domain THz spectroscopy of the 1D FM Ising chain compound $CoNb_2O_6$. (a) The time-domain signal and its Fourier transform (inset) in zero magnetic field at 6 K. The transmitted pulse at 10 ps is followed by two pulses reflected from the sample surfaces and delayed by $\Delta t = 22$ ps, which is the time for the THz pulse to pass the sample twice. Delayed pulses produce Fabry–Perot resonances seen in the Fourier-transformed signal as oscillations below 2.5 THz. (b) TD-THz absorption spectra of SW resonances as a function of applied magnetic field $\mathbf{B} \parallel \mathbf{a}$ for $\mathbf{E}^{\omega} \parallel \mathbf{c}$ and light propagating along **a**. Removal of the non-magnetic background absorption and interference fringes from the absorption spectra are described in the supplementary material of [33]. Reprinted with permission from [33].

A Michelson-type Fourier-transform interferometer with a dielectric beam splitter is not well suited for the THz range because its efficiency vanishes at zero frequency. To avoid this deficiency, a Fourier-transform interferometer [43] with a polarizing wire-grid beam splitter was developed by Martin and Puplett [44, 45]. The THz Fourier-transform interferometer is usually a vacuum instrument for two reasons. First, water and carbon monoxide have permanent electric-dipole moments and produce rotation lines in the THz range. Second, gas in the interferometer causes 'microphonic noise' due to the acoustically driven vibration of a thin film or wire-grid beam splitter.

THz measurements commonly use mirrors with metallic surfaces or plastic lenses. If direct optical access to the sample is not possible, then light pipes must be used. However, the incident intensity at the sample is then reduced [46] by multiple reflections from the pipe walls and bends [46]. Radiation from a light pipe can be concentrated on the sample or on the bolometer with either a Winston cone [47] or a straight cone [48].

Unfortunately, polarization is not preserved in a light pipe. Hence, a polarizer consisting of a free standing wire grid or a wire grid on a dielectric film [49] must be placed before or after the sample. A wire grid is transparent (reflective) when $\mathbf{E}^{\omega}$ is perpendicular (parallel) to the conducting wires of the grid. Even for an ideal polarizer, rays not parallel to the light pipe axis introduce the wrong polarization along that axis. This can be avoided by collimating the beam with a short section of non-reflecting tube with inner diameter approximately equal to the sample diameter. However, the intensity is then reduced even further than the factor of two loss in the polarizer.

3.2.3 THz spectroscopy under pressure in a magnetic field

Applied pressure is useful to study magnetic systems, especially when combined with high magnetic fields. But the smaller sample volume required for the pressure cell reduces the optical sensitivity. Smaller windows increase diffraction losses that become important when the wavelength of radiation is larger than the window size (recall that the wavelength of 0.1 THz radiation is 3 mm). Consequently, THz spectroscopy under pressure is a challenge. An overview of high pressure/hf-ESR was given by Sakurai *et al* [50] with a list of such facilities.

3.2.4 THz spectroscopy using the TeslaFIR system with a 17 T superconducting solenoid

This section describes the spectrometer system TeslaFIR for THz spectroscopy in high magnetic fields. The setup in figure 3.6 was designed for absorption and reflection spectroscopy between 3 and 200 cm^{-1} (0.1–6 THz) in high magnetic fields at liquid He temperatures. Low frequencies are reached with sensitive ^{3}He-cooled bolometers operating at 0.3 K and a Martin–Puplett interferometer. The system measures the intensity of transmitted or reflected radiation and is not phase-sensitive.

The light source is a mercury arc lamp in the vacuum housing of a Martin–Puplett type interferometer. Using roof mirrors and a polarizing beam splitter, this interferometer guides light to the sample and detectors with 16 mm diameter light pipes.

A separate vacuum chamber inside a liquid ^{4}He cryostat hosts two composite Si bolometers, each mounted on a 5 mm diameter sapphire disc. The sensitivity of these bolometers operating at 0.3 K is 4.5×10^7 V W^{-1}, two orders of magnitude greater than that of a 4.2 K bolometer. However, the high-frequency cutoff of 30 Hz is more than 30 times lower than the typical cutoff frequency of a 4.2 K bolometer. Consequently, the system is susceptible to low-frequency mechanical vibrations. A magnetic field has only a moderate effect on the bolometer, which is placed about 29 cm below the field center (14.5 cm below the coil end) of a 17 T solenoid. We have observed that the maximum field reduces the bolometer cutoff frequency by about 20%.

A filter wheel with eight positions is immersed in liquid helium between the sample chamber and the bolometer. The filter wheel contains low-pass multi-mesh filters with 0.3, 0.6, 2, 3, and 6 THz cutoffs and a scatter-absorbing fluorogold [51] with a 1.2 THz cutoff. An additional fixed-position 6 THz cutoff multi-mesh filter lies on the sample chamber window. The overall spectral range of this setup extends from 0.1 to 6 THz.

In the Faraday configuration with $\mathbf{k} \parallel \mathbf{B}$, up to five samples and an open-hole reference (see below) can be mounted. In the Voigt configuration with $\mathbf{k} \perp \mathbf{B}$, one sample position is available with the option to rotate the sample about the wavevector of light to study the dependence on the orientation of the magnetic field [52]. Reflectance is measured near normal incidence $\mathbf{k} \parallel \mathbf{B}$ with five samples and a reference mirror mounted on a piezo-driven rotator [53].

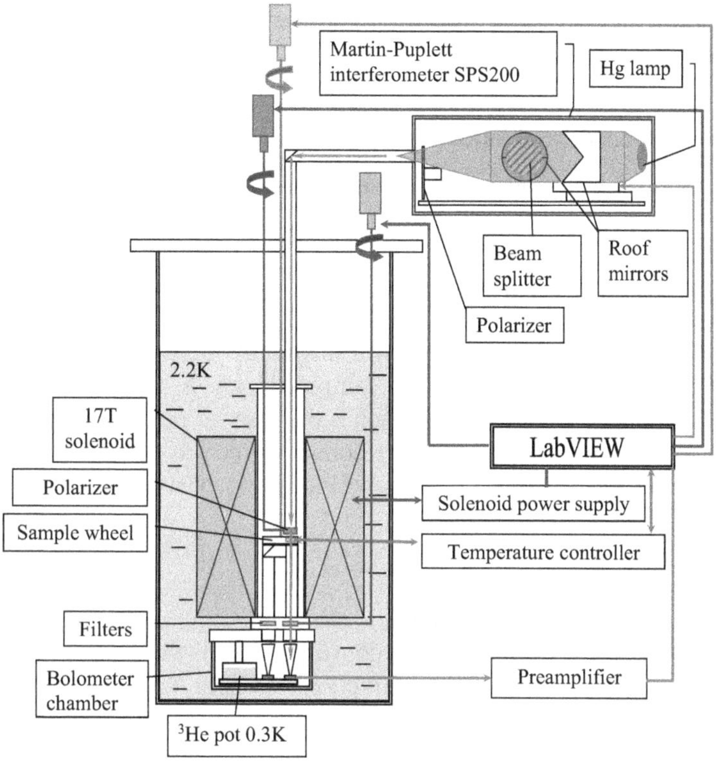

Figure 3.6. The magneto-optical cryostat TeslaFIR consisting of a Martin–Puplett type interferometer and a liquid He bath cryostat. The cryostat contains a sample chamber, low-pass THz filters, the bolometer chamber, and a 17 T solenoid. Light is guided from the interferometer to the sample and detectors with light pipes. Stepper motors rotate the polarizer, sample wheel, and filter wheel. The Faraday configuration is shown.

3.3 Acquisition and analysis of single-crystal SW spectra

Designing a THz experiment requires several considerations. What is the spectral range of interest? Will the experiment measure the polarization dependence of the absorption intensities? Are both the real and imaginary parts of optical constants required?

As for INS, single crystals are preferred for THz spectroscopy. While the $\mathbf{q} = 0$ SW frequencies of a powder sample are the same as those of a single crystal, the polarization dependence of the absorption intensity is averaged over all crystal orientations.

The lateral dimension of a single crystal in the plane perpendicular to the direction of light propagation is set by the diffraction limit: the size of the sample should be approximately equal to the wavelength corresponding to the lowest desired frequency. The sample thickness d is determined by the optical transmission of the sample and is analyzed in the next subsection. It is assumed that the rotation of the polarization of THz radiation within the sample is negligible and that the attenuation of linearly polarized radiation is solely due to absorption. A more rigorous treatment including rotation can be found in [54, 55] and in the supplementary information of [56].

3.3.1 The complex index of refraction and the absorption

The attenuation of radiation over a distance z is called the transmittance:

$$\mathcal{T}(z) = \exp(-\alpha z), \tag{3.1}$$

where α is the absorption coefficient. As a rough guide, $\mathcal{T}(z)$ is small when $d > \alpha^{-1}$. For a typical SW absorption of $\alpha = 10\ \mathrm{cm}^{-1}$, the optimal sample thickness is $d \approx 1$ mm. If a 1 mm thick sample is not transparent due to background absorption or high reflectance, then reducing the sample thickness below α^{-1} is not helpful because SWs would be barely visible. Such is the case for metallic samples and in the spectral range of strong electric-dipole resonances such as optical phonons.

THz spectroscopy measures the real and imaginary parts of the complex index of refraction

$$\mathcal{N} = n + \imath\kappa, \tag{3.2}$$

where n is the index of refraction and κ is the extinction coefficient. In the simplest case, $\mathcal{N} = \sqrt{\epsilon\mu}$, where ϵ is the dielectric constant and μ is the magnetic permeability. Averaged over the $\pm\mathbf{k}$ propagation directions of light (see equation (4.60)), the absorption coefficient is given by

$$\alpha = \frac{4\pi\nu}{c}\kappa = \frac{4\pi\nu}{c}\,\mathrm{Im}\,\mathcal{N}(\nu), \tag{3.3}$$

where ν is the radiation frequency in units of Hz and c is the speed of light in vacuum. Expressed in wavenumbers, $\bar{\omega} \equiv \lambda^{-1} = \nu/c$ and $\alpha = 4\pi\bar{\omega}\,\mathrm{Im}\,\mathcal{N}$. Note that $[\bar{\omega}] = \mathrm{cm}^{-1}$ and $[\alpha] = \mathrm{cm}^{-1}$.

Under normal incidence, the reflectivity from the sample is

$$\mathcal{R} = \left|\frac{\mathcal{N} - 1}{\mathcal{N} + 1}\right|^2. \tag{3.4}$$

Suppose that a SW with frequency ω_0 contributes $\Delta\epsilon$ and $\Delta\mu$ to the dielectric constant ϵ and magnetic permeability μ:

$$\epsilon(\omega) = \epsilon^b(\omega) + \Delta\epsilon(\omega_0 - \omega), \tag{3.5}$$

$$\mu(\omega) = \mu^b(\omega) + \Delta\mu(\omega_0 - \omega). \tag{3.6}$$

Dropping the explicit frequency dependence, we obtain the complex index of refraction

$$\mathcal{N} = \sqrt{\epsilon\mu} = \sqrt{\epsilon^b\mu^b}\left\{1 + \frac{\Delta\epsilon}{\epsilon^b} + \frac{\Delta\mu}{\mu^b} + \frac{\Delta\epsilon\Delta\mu}{\epsilon^b\mu^b}\right\}^{1/2}. \tag{3.7}$$

For $\Delta\epsilon \ll \epsilon^b$ and $\Delta\mu \ll \mu^b$, the square root can be expanded:

$$\mathcal{N} \approx \sqrt{\epsilon^b \mu^b} + \sqrt{\frac{\mu^b}{\epsilon^b}} \frac{\Delta\epsilon}{2} + \sqrt{\frac{\epsilon^b}{\mu^b}} \frac{\Delta\mu}{2}. \tag{3.8}$$

To lowest order, the refractive index $\mathcal{N} \approx \sqrt{\epsilon^b \mu^b}$ does not depend on the SW optical constants. Neglecting cross terms between the electrical and magnetic response, information about the SWs is contained in $\Delta\epsilon$ and $\Delta\mu$, which will be studied further in section 4.4.

3.3.2 Measuring SW resonances

The intensity of radiation (but not its amplitude or phase) can be measured using a bolometer and a Fourier-transform interferometer. Since every sample has two reflecting surfaces, the ratio of intensity $I(\omega)$ at the detector to the incident intensity $I_0(\omega)$ on the sample depends on both the transmittance $\mathcal{T}$ and the reflectivity $\mathcal{R}$.

A sample with two parallel surfaces acts like a Fabry–Perot resonator where radiation travels back and forth between two parallel mirrors. The interference of light beams within the sample appears as a periodic modulation of $I(\omega)$ called ‘interference fringes’. Expressions for $\mathcal{T}$ and $\mathcal{R}$ for a parallel-plane sample are found in [17, 59]. Since the phase accumulated at each passage of light through the sample depends on the sample’s index of refraction n and thickness d, the measured transmittance can be used to determine both $\kappa(\omega)$ and $n(\omega)$. However, this method requires a highly parallel light beam, which is not always possible.

An alternative method is to construct a wedge-shaped sample that suppresses interference fringes because the two surfaces are not parallel. Ideally, only reflections from the front and back surfaces attenuate the light reaching the detector and

$$\mathcal{T}_m \equiv \frac{I}{I_0} = \mathcal{T}(1 - \mathcal{R})^2, \tag{3.9}$$

where $\mathcal{T}_m$ is the measured transmission. For a sample of average thickness d, the absorption can be written

$$\alpha = -\frac{1}{d} \ln \frac{\mathcal{T}_m}{(1 - \mathcal{R})^2}, \tag{3.10}$$

where $\mathcal{R}$ from equations (3.4) and (3.7) depends only on ϵ^b and μ^b when $\Delta\epsilon \ll \epsilon^b$ and $\Delta\mu \ll \mu^b$.

The transmission $\mathcal{T}_m = I/I_0$ can be obtained by first measuring the intensity I through an open hole below the sample. After removing the sample, the intensity I_0 is again measured through the open hole. Another less accurate method is to keep the sample on one open hole and measure I_0 on another (the required apparatus is somtimes called a two-holer). This procedure measures the absolute absorption containing both the SW and background absorption. Figure 3.7 shows the absolute absorption spectrum of $BiFeO_3$.

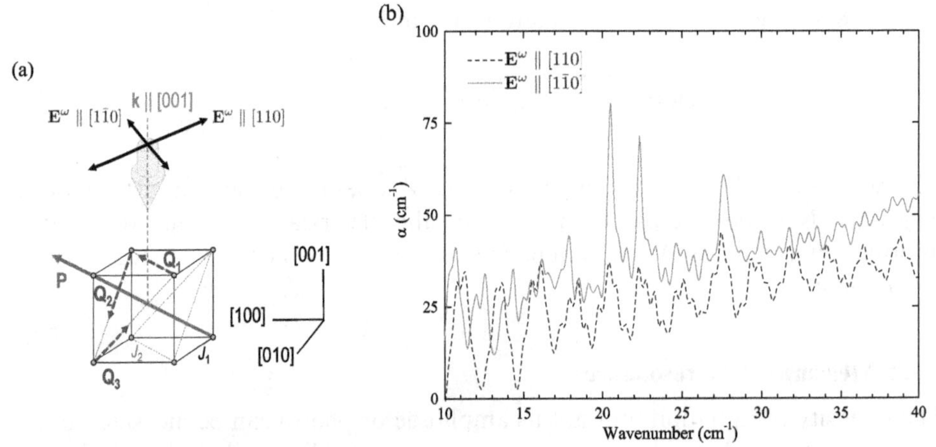

Figure 3.7. (a) Pseudo-cubic unit cell of $BiFeO_3$ showing the Fe ions, the ferroelectric polarization **P**, three equivalent directions of the cycloidal ordering wavevector $\mathbf{Q}_k$, and the wavevector of incident light **k** together with the electric field ($\mathbf{E}^\omega$) component of light in two orthogonal polarizations. J_1 and J_2 are the nearest- and next-nearest-neighbor exchange interactions between $S = 5/2$ Fe spins. (b) Absolute spectrum of $BiFeO_3$ crystal relative to an open hole at 3 K in zero magnetic field measured with TeslaFIR for $\mathbf{E}^\omega \parallel [1\bar{1}0]$ and $\mathbf{E}^\omega \parallel [110]$. Interference fringes with a periodicity of 2.5 cm^{-1} are observed for $\mathbf{E}^\omega \parallel [110]$ while they are less pronounced for $\mathbf{E}^\omega \parallel [1\bar{1}0]$ due to stronger background absorption. The same 0.37 mm thick crystal was used to study the magnetic-field dependence [57] and non-reciprocal directional dichroism [58] of the SW modes.

However, the SW absorption measured using the above procedure can be partially masked by the background absorption and by interference fringes, which may be incompletely suppressed even for a wedge-shaped sample. A more accurate method is to measure the intensity $I(B, T)$ through the sample in different magnetic fields and at different temperatures. The differential absorption spectrum is then (exercise 3.1)

$$\alpha(B, T) - \alpha(B_r, T_r) = -\frac{1}{d} \ln \frac{I(B, T)}{I(B_r, T_r)}, \tag{3.11}$$

where B_r and T_r are the magnetic field and temperature of the reference spectrum. This is a more accurate method than the open-hole technique because the sample is not moved in and out of the THz beam. It works best when $T_r > T_N$ or $B > B_c$ so that the reference spectrum does not contain SW resonances. In figure 3.8, a high-T reference was used to extract SW resonances in zero magnetic field for the Shastry–Sutherland model [60] compound $SrCu_2(BO_3)_2$. Since the singlet-to-triplet excitations broaden and lose intensity as T increases, the 15 K spectrum is suitable as a high-T reference [62].

What if neither T_N nor B_c is conveniently accessible? For example, both the Néel temperature of 640 K and the spin-flop field of 18 T are rather high in $BiFeO_3$. In that case, a series of spectra can be measured at different fields but at constant temperature $T = T_r$. Plots of $I(B, T)$ for $BiFeO_3$ in 0 T and 10 T at 2 K are given in figure 3.9. In most cases, $B_r = 0$ is chosen as the reference.

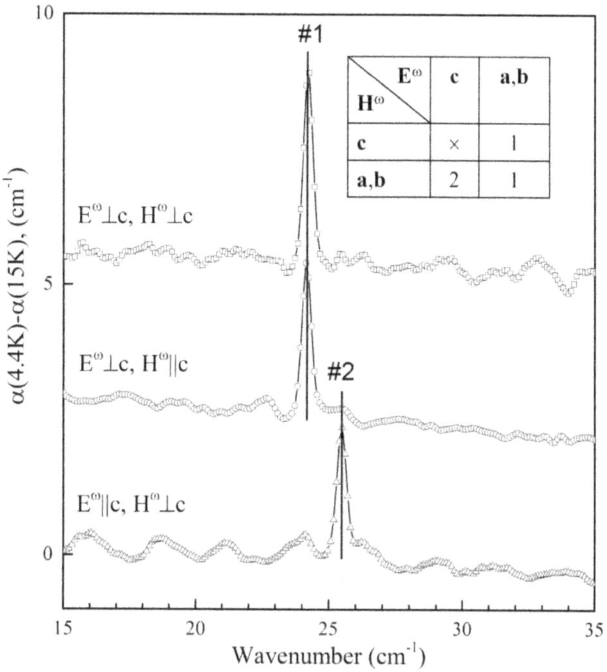

Figure 3.8. THz absorption spectra of $SrCu_2(BO_3)_2$ [60], showing excitations from the ground-state singlet $S = 0$ state to the excited-state triplet $S = 1$ in zero magnetic field [61]; spectra are offset in the vertical direction. The $S = 1$ triplet state is split into modes #1 and #2 by the DM interaction. $SrCu_2(BO_3)_2$ is tetragonal with isotropic properties in the *ab* plane. The two bottom spectra (circles and triangles) were measured on the same crystal cut in polarizations $\mathbf{E}^{\omega} \perp \mathbf{c}$ and $\mathbf{E}^{\omega} \parallel \mathbf{c}$ while the top spectrum (squares) was measured on a different cut with $\mathbf{k} \parallel \mathbf{c}$. The inset table shows modes that are electric-dipole active.

Of course, this method only works when the SW frequencies change with field. Since the spectral lines of the reference spectrum appear as negative features in $\alpha(B)-\alpha(B_r)$, it is easy to extract the reference spectrum by finding a minimum of the difference spectra at each frequency. Adding the extracted reference to the difference spectra in equation (3.11) gives the SW spectrum in each field, including the reference field. Provided that ϵ_b and μ_b are the same at B and B_r, the $(1 - \mathcal{R})^2$ term in equation (3.10) cancels out in the difference spectra.

Figure 3.10(a) plots the field dependence of the difference spectra $\alpha(B) - \alpha(0)$ for the SW modes in $BiFeO_3$ with polarizations $\mathbf{E}^{\omega} \parallel [1\bar{1}0]$ and $\mathbf{E}^{\omega} \parallel [110]$. The spectra $\alpha(B) > 0$ was plotted in figure 3.10(b) after extracting $\alpha(0)$. Comparing the zero-field spectra calculated from the difference spectra with the absolute spectra measured using an open-hole reference in figure 3.7(b), we see that the former method gives an excellent baseline. In particular, the SW modes with polarization $\mathbf{E}^{\omega} \parallel [110]$ are clearly visible in the difference spectra while they are not visible in the spectra that use an open-hole reference.

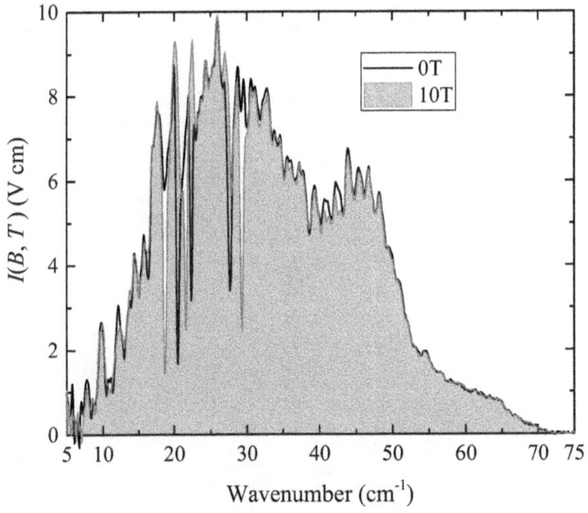

Figure 3.9. Two spectra of $BiFeO_3$ after Fourier-transforming interferograms measured with a 0.3 K bolometer on spectrometer TeslaFIR. Spectra were measured in 0 and 10 T fields at 2 K with $\mathbf{E}^{\omega} \parallel [1\bar{1}0]$. Part of the dataset calculated from $I(B, T)$ with equation (3.11) is shown in figure 3.10(a). Sample thickness is 0.037 cm; spectrometer resolution is 0.3 cm^{-1}. The acquisition time for each field is 800 s.

3.4 Selection rules

The selection rules for THz absorption determine the dependence of the absorption strength on the polarizations of $\mathbf{E}^{\omega}$ and $\mathbf{B}^{\omega}$ relative to the crystal axis and to the applied magnetic field **B**. Of course, $\mathbf{E}^{\omega}$ and $\mathbf{B}^{\omega}$ are perpendicular to each other and to the light propagation wavevector $\mathbf{k} \parallel \mathbf{E}^{\omega} \times \mathbf{B}^{\omega}$. Analysis of the selection rules is rather straightforward if the SWs are only magnetic- or electric-dipole active but not both.

To illustrate the selection rules, assume that a crystal has three orthogonal axes [100], [010], and [001]. Consider three thin crystal slabs with faces (100), (010), and (001). For the sample with $\mathbf{k} \parallel [001]$ propagating into the (001) face, the polarizations can be either $\{\mathbf{E}^{\omega} \parallel [100], \mathbf{B}^{\omega} \parallel [010]\}$ or $\{\mathbf{E}^{\omega} \parallel [010], \mathbf{B}^{\omega} \parallel [100]\}$. Altogether there are $3 \times 2 = 6$ combinations to measure on three crystal cuts. For two orthogonal propagation directions, there is always a combination where $\mathbf{E}^{\omega}$ for one propagation direction is parallel to $\mathbf{E}^{\omega}$ for the orthogonal propagation direction, and the same for $\mathbf{B}^{\omega}$. If the SW mode has the same intensity in two orthogonal propagation directions for the same direction of $\mathbf{B}^{\omega}$ or $\mathbf{E}^{\omega}$ with respect to the crystal axes, then it is magnetic- or electric-dipole active, respectively. In a magnetic field, the number of combinations of field directions and polarizations is still six in the Faraday configuration but another six configurations are added in the Voigt configuration where $\mathbf{E}^{\omega}$ or $\mathbf{B}^{\omega}$ can be parallel to **B**.

A zero-field example with two electric-dipole active modes is shown in figure 3.8 for $SrCu_2(BO_3)_2$ [61, 63]. The singlet-to-triplet transition is not electric-dipole active in the absence of mixing. Because the optical phonon modulates the DM interaction, the singlet and triplet states mix and the singlet-to-triplet transition becomes electric-dipole active.

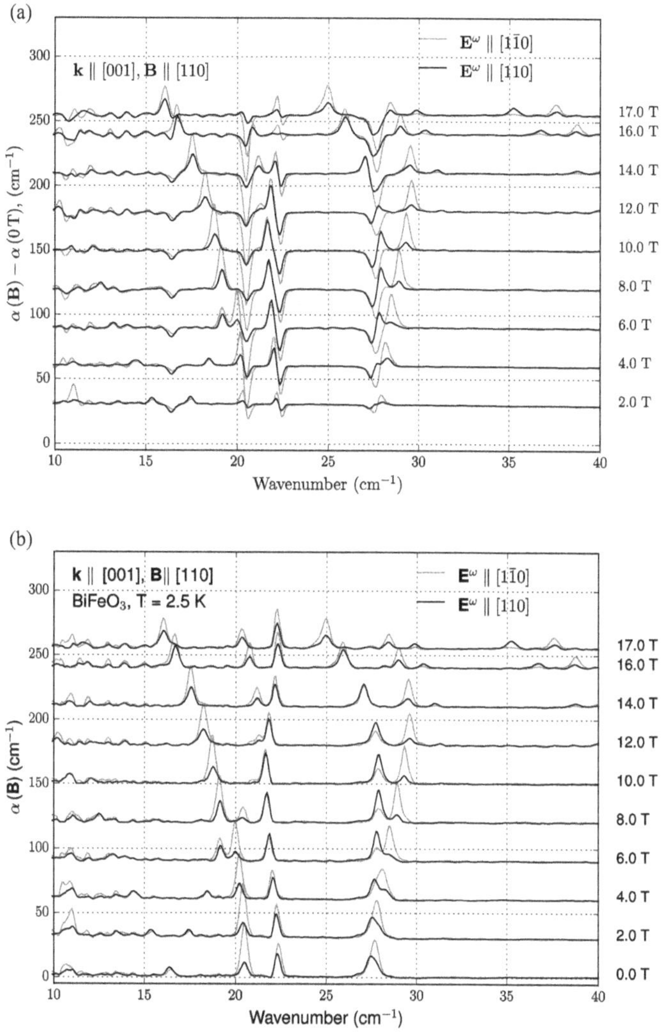

Figure 3.10. Magnetic-field dependence of THz absorption spectra of $BiFeO_3$ at $T = 2.5$ K in polarizations $\mathbf{E}^\omega \parallel [1\bar{1}0]$ and $\mathbf{E}^\omega \parallel [110]$. (a) The difference spectra calculated from equation (3.11) using the 0 T spectrum as a reference. (b) The spectra $\alpha(B) \geqslant 0$ calculated from the difference $\alpha(B) - \alpha(0)$ after extracting $\alpha(0)$. Crystal and axis notation are the same as in figure 3.7(a).

3.5 Summary

To summarize, THz spectroscopy measures both the resonance SW frequencies and their absorption strengths. Because a magnetic field shifts the SW frequencies, analysis of the SW spectra measured in different magnetic fields provides more accurate mode frequencies and intensities. High-field user facilities that support THz spectroscopy are listed in table 3.1.

3.6 Exercises

1. Derive equation (3.11) for $\Delta\alpha = \alpha(B, T) - \alpha(B_r, T_r)$. Does this expression have the right dimensions?
2. On a single-crystal sample, devise an experiment to show that a particular mode is an electromagnon with polarization $\mathbf{E}^{\omega}$ along [010].
3. The bolometer sensitivity may depend on magnetic field. If the magnetic-field dependence of the SW absorption is given by $I(B, T)$, then the measured intensity is $I_m(B, T) \approx a_0(B)\, I(B, T)$, where $a_0(B)$ is independent of frequency. Using equation (3.11), show that this correction does not change the SW absorption amplitude $\alpha_{\rm SW}(\omega - \omega_0)$ if the total absorption $\alpha(\omega) = \alpha_b(\omega) + \alpha_{\rm SW}(\omega - \omega_0)$ includes the magnetic-field independent background absorption $\alpha_b(\omega)$.
4. The mirror of a FT interferometer moves at a constant speed v in the rapid scan mode. The frequency of radiation $\bar{\omega}$ measured as a periodic modulation in the interferogram is transformed into a periodic signal in time with frequency ν_t at the intensity detector. Derive the formula relating the mirror speed v, the THz frequency $\bar{\omega}$ (in cm^{-1}), and the frequency ν_t (in Hz) of the corresponding time-dependent signal at the detector. Show that the 50 Hz main cycle produces a spectral frequency of 50 cm^{-1} if $v = 0.5$ cm s^{-1}. What is the optimal speed if the high-frequency cutoff of the detector is 30 Hz and the desired high-frequency cutoff of the spectrum is 2 THz?
5. Assume that the absorption of THz radiation below 100 cm^{-1} is dominated by a magnetic-dipole active SW resonance at $\bar{\omega}_0 = 50$ cm^{-1} and that the background $\epsilon^b = 20$ and $\mu^b = 1$ are both real. If the absorption due to this SW mode is $\alpha_{SW}(\bar{\omega}_0) = 50$ cm^{-1}, show that Im $\Delta\mu \ll \mu^b$ so that the expansion of equation (3.8) is justified.

References

[1] Sosnowska I, Peterlin-Neumaier T and Steichele E 1982 Spiral magnetic ordering in bismuth ferrite *J. Phys. C: Solid State Phys.* **15** 4835

[2] Kotani A and Shin S 2001 Resonant inelastic x-ray scattering spectra for electrons in solids *Rev. Mod. Phys.* **73** 203–46

[3] Braicovich L *et al* 2010 Magnetic excitations and phase separation in the underdoped $La_{2-x}Sr_xCuO_4$ superconductor measured by resonant inelastic x-ray scattering *Phys. Rev. Lett.* **104** 077002

[4] Betto D, Peng Y Y, Porter S B, Berti G, Calloni A, Ghiringhelli G and Brookes N B 2017 Three-dimensional dispersion of spin waves measured in NIO by resonant inelastic x-ray scattering *Phys. Rev.* B **96** 020409

[5] Sando D *et al* 2013 Crafting the magnonic and spintronic response of $BiFeO_3$ films by epitaxial strain *Nat. Mater.* **12** 641–6

[6] Pimenov A, Mukhin A A, Yu I V, Travkin V D, Balbashov A M and Loidl A 2006 Possible evidence for electromagnons in multiferroic manganites *Nat. Phys.* **2** 97–100

[7] Kézsmárki I, Kida N, Murakawa H, Bordács S, Onose Y and Tokura Y 2011 Enhanced directional dichroism of terahertz light in resonance with magnetic excitations of the multiferroic $Ba_2CoGe_2O_7$ oxide compound *Phys. Rev. Lett.* **106** 057403

[8] Fleury P A and Loudon R 1968 Scattering of light by one- and two-magnon excitations *Phys. Rev.* **166** 514–30
[9] Cottam M G and Lockwood D J 1985 *Light Scattering in Magnetic Solids* (New York: Wiley)
[10] Benfatto L and Silva Neto M B 2006 Field dependence of the magnetic spectrum in anisotropic and Dzyaloshinskii–Moriya antiferromagnets. I. Theory *Phys. Rev.* B **74** 024415
[11] Benfatto L, Silva Neto M B, Gozar A, Dennis B S, Blumberg G, Miller L L, Komiya S and Ando Y 2006 Field dependence of the magnetic spectrum in anisotropic and Dzyaloshinskii–Moriya antiferromagnets. II. Raman spectroscopy *Phys. Rev.* B **74** 024416
[12] van den Brink J 2016 Resonant inelastic X-ray scattering on elementary excitations *Quantum Materials: Experiment and Theory, Modeling and Simulation*vol 6 ed J van den Brink, E Pavarini, E Koch and G Sawatzky (Germany: Forschungszentrum Jülich) ch 12
[13] Baron A Q R 2015 Introduction to high-resolution inelastic x-ray scattering, arXiv: 1504.01098
[14] Smith P R, Auston D H and Nuss M C 1988 Subpicosecond photoconducting dipole antennas *IEEE J. Quantum Electron.* **24** 255–60
[15] Kozlov G and Alexander V 1998 Coherent source submillimeter spectroscopy *Millimeter and Submillimeter Wave Spectroscopy of Solids, Topics in Applied Physics* vol 74 (Berlin: Springer) pp 50–109
[16] Grüner G (ed) 1998 *Millimeter and Submillimeter Wave Spectroscopy of Solids, Topics in Applied Physics* vol 74 (Berlin: Springer)
[17] Dressel M and Grüner G 2002 *Electrodynamics of Solids* (Cambridge: Cambridge University Press)
[18] Motokawa M, Ohta H, Nojiri H and Kimura S 2003 The role of ESR in research of low-dimensional antiferromagnets *J. Phys. Soc. Jpn.* **72** 1–11
[19] Nojiri H, Motokawa M, Okuda K, Kageyama H, Ueda Y and Tanaka H 2003 THz-ESR system by using single shot and repeating pulsed magnetic fields *J. Phys. Soc. Jpn.* **72** 109–16
[20] Shinozaki S-I, Okutani A, Yoshizawa D, Kida T, Takeuchi T, Yamamoto S, Risset O N, Talham D R, Meisel M W and Hagiwara M 2016 Antiferromagnetic order in single crystals of the $S = 2$ quasi-one-dimensional chain $MnCl_3$(bpy) *Phys. Rev.* B **93** 014407
[21] Fishman R S, Shinozaki S-I, Okutani A, Yoshizawa D, Kida T, Hagiwara M and Meisel M W 2016 Long-range magnetic order and interchain interactions in the $S = 2$ chain system $MnCl_3$(bpy) *Phys. Rev.* B **94** 104435
[22] Krzystek J, Zvyagin S A, Ozarowski A, Trofimenko S and Telser J 2006 Tunable-frequency high-field electron paramagnetic resonance *J. Magn. Reson.* **178** 174–83
[23] Nagy K L, Quintavalle D, Fehér T and Jánossy A 2011 Multipurpose high-frequency ESR spectrometer for condensed matter research *Appl. Magn. Reson.* **40** 47–63
[24] Zvyagin S A, Ozerov M, Čižmár E, Kamenskyi D, Zherlitsyn S, Herrmannsdörfer T, Wosnitza J, Wünsch R and Seidel W 2009 Terahertz-range free-electron laser electron spin resonance spectroscopy: techniques and applications in high magnetic fields *Rev. Sci. Instrum.* **80** 073102
[25] Ozerov M, Bernth B, Kamenskyi D, Redlich B, van der Meer A F G, Christianen P C M, Engelkamp H and Maan J C 2017 A THz spectrometer combining the free electron laser FLARE with 33 T magnetic fields *Appl. Phys. Lett.* **110** 094106
[26] Volkov A A, Goncharov Y G, Kozlov G V, Lebedev S P and Prokhorov A M 1985 Dielectric measurements in the submillimeter wavelength region *Infrared Phys.* **25** 369–73

[27] Kuzmenko A M, Dziom V, Shuvaev A, Pimenov A, Schiebl M, Mukhin A A, Yu I V, Gudim I A, Bezmaternykh L N and Pimenov A 2015 Large directional optical anisotropy in multiferroic ferroborate *Phys. Rev.* B **92** 184409

[28] Naftaly M, Clarke R G, Humphreys D A and Ridler N M 2017 Metrology state-of-the-art and challenges in broadband phase-sensitive terahertz measurements *Proc. IEEE* **105** 1151–65

[29] Roggenbuck A, Schmitz H, Deninger A, Cámara Mayorga I, Hemberger J, Güsten R and Grüninger M 2010 Coherent broadband continuous-wave terahertz spectroscopy on solid-state samples *New J. Phys.* **12** 043017

[30] Yu S *et al* 2017 Terahertz frequency magnetoelectric effect in Ni doped $CaBaCo_4O_7$, arXiv:1708.04345

[31] Yu S, Gao B, Kim J W, Cheong S W, Man M K L, Madéo J, Dani K M and Talbayev D 2018 High-temperature terahertz optical diode effect without magnetic order in polar $FeZnMo_3O_8$ *Phys. Rev. Lett.* **120** 037601

[32] Langenbach M, Thirunavukkuarasu K, Camara Mayorga I, Roggenbuck A, Deninger A, Hemberger J and Grueninger M 2013 Broadband continuous-wave THz spectroscopy at low temperature and high magnetic field *38th Int. Conf. on Infrared, Millimeter, and Terahertz Waves (IRMMW-THZ)* (New York: IEEE)

[33] Morris C M, Valdés Aguilar R, Ghosh A, Koohpayeh S M, Krizan J, Cava R J, Tchernyshyov O, McQueen T M and Armitage N P 2014 Hierarchy of bound states in the one-dimensional ferromagnetic Ising chain $CoNb_2O_6$ investigated by high-resolution time-domain terahertz spectroscopy *Phys. Rev. Lett.* **112** 137403

[34] Kimmitt M F, Walsh J E, Platt C L, Miller K and Jensen M R F 1996 Infrared output from a compact high pressure arc source *Infrared Phys. Technol.* **37** 471–7

[35] Richards P L 1994 Bolometers for infrared and millimeter waves *J. Appl. Phys.* **76** 1

[36] Hijmering R A, Khosropanah P, Ridder M, Gao J R, Hoevers H, Jackson B, Goldie D, Withington S and Kozorezov A G 2014 Comparison of the effects of magnetic field on low noise MoAu and TiAu TES bolometers *J. Low Temp. Phys.* **176** 316–22

[37] De Moor P *et al* 1996 Calorimetric single particle detection in high magnetic fields at low temperatures *J. Appl. Phys.* **79** 3811–5

[38] Naylor D A, Gom B G, Ade P A R and Davis J E 1999 Design and performance of a dual polarizing detector system for broadband astronomical spectroscopy at submillimeter wavelengths *Rev. Sci. Instrum.* **70** 4097–109

[39] Drew H D and Sievers A J 1969 A ^{3}He-cooled bolometer for the far infrared *Appl. Opt.* **8** 2067–71

[40] Abo-Bakr M, Feikes J, Holldack K, Kuske P, Peatman W B, Schade U, Wüstefeld G and Hübers H-W 2003 Brilliant, coherent far-infrared (THz) synchrotron radiation *Phys. Rev. Lett.* **90** 094801

[41] Mihály L, Talbayev D, Kiss L F, Zhou J, Fehér T and Jánossy A 2004 Field-frequency mapping of the electron spin resonance in the paramagnetic and antiferromagnetic states of $LaMnO_3$ *Phys. Rev.* B **69** 024414

[42] Nehrkorn J, Holldack K, Bittl R and Schnegg A 2017 Recent progress in synchrotron-based frequency-domain Fourier-transform THz-EPR *J. Magn. Reson.* **280** 10–9

[43] Genzel L 1998 Far-infrared Fourier transform spectroscopy *Millimeter and Submillimeter Wave Spectroscopy of Solids, Topics in Applied Physics* vol 74 ed G Grüner (Berlin: Springer), ch 5, pp 169–220

[44] Martin D H and Puplett E 1969 Polarised interferometric spectrometry for the millimetre and submillimetre spectrum *Infrared Phys.* **10** 105–9
[45] Martin D H 1982 Polarizing (Martin–Puplett) interferometric spectrometers for the near- and submillimeter spectra *Systems and Components, Infrared and Millimeter Waves* vol 6 ed K J Button (London: Academic) ch 2, pp 66–149
[46] Hawthorn D G and Timusk T 1999 Transmittance of skew rays through metal light pipes *Appl. Opt.* **38** 2787–94
[47] Harper D A, Hildebrand R H, Stiening R and Winston R 1976 Heat trap: an optimized far infrared field optics system *Appl. Opt.* **15** 53–60
[48] Witte W 1965 Cone channel optics *Infrared Phys.* **5** 179–85
[49] Miica M, Bucko V, Postava K, Vanwolleghem M, Lampin J-F and Pitora J 2016 Analysis of wire-grid polarisers in terahertz spectral range *J. Nanosci. Nanotechnol.* **16** 7810–3
[50] Sakurai T, Okubo S and Ohta H 2017 High-field/high-pressure ESR *J. Magn. Reson.* **280** 3–9
[51] Halpern M, Gush H P, Wishnow E and De Cosmo V 1986 Far infrared transmission of dielectrics at cryogenic and room temperatures: glass, fluorogold, eccosorb, stycast, and various plastics *Appl. Opt.* **25** 565–70
[52] Fishman R S *et al* 2017 Competing exchange interactions in multiferroic and ferrimagnetic $CaBaCo_4O_7$ *Phys. Rev.* B **95** 024423
[53] Nagel U, Hüvonen D, Joon E, Kim J S, Kremer R K and Rõõm T 2008 Far-infrared signature of the superconducting gap in intercalated graphite CaC_6 *Phys. Rev.* B **78** 041404
[54] Miyahara S and Furukawa N 2014 Theory of magneto-optical effects in helical multiferroic materials via toroidal magnon excitation *Phys. Rev.* B **89** 195145
[55] Kuzmenko A M, Shuvaev A, Dziom V, Pimenov A, Schiebl M, Mukhin A A, Yu I V, Bezmaternykh L N and Pimenov A 2014 Giant gigahertz optical activity in multiferroic ferroborate *Phys. Rev.* B **89** 174407
[56] Bordács S *et al* 2012 Chirality of matter shows up via spin excitations *Nat. Phys.* **8** 734–8
[57] Nagel U, Fishman R S, Katuwal T, Engelkamp H, Talbayev D, Yi H T, Cheong S-W and Rõõm T 2013 Terahertz spectroscopy of spin waves in multiferroic $BiFeO_3$ in high magnetic fields *Phys. Rev. Lett.* **110** 257201
[58] Kézsmárki I, Nagel U, Bordács S, Fishman R S, Lee J H, Yi H T, Cheong S -W and Rõõm T 2015 Optical diode effect at spin-wave excitations of the room-temperature multiferroic $BiFeO_3$ *Phys. Rev. Lett.* **115** 127203
[59] Born M and Wolf E 1999 *Principles of Optics* (Cambridge: Cambridge University Press)
[60] Shastry B S and Sutherland B 1981 Exact ground state of a quantum antiferromagnet *Physica* **108B** 1069–70
[61] Rõõm T, Hüvonen D, Nagel U, Hwang J, Timusk T and Kageyama H 2004 Far-infrared spectroscopy of spin excitations and Dzyaloshinskii–Moriya interactions in the Shastry–Sutherland compound $SrCu_2(BO_3)_2$ *Phys. Rev.* B **70** 144417
[62] Rõõm T, Nagel U, Lippmaa E, Kageyama H, Onizuka K and Ueda Y 2000 Far-infrared study of the two-dimensional dimer spin system $SrCu_2(BO_3)_2$ *Phys. Rev.* B **61** 14342–45
[63] Cépas O and Ziman T 2004 Theory of phonon-assisted forbidden optical transitions in spin-gapped systems *Phys. Rev.* B **70** 024404

Randy S Fishman, Jaime A Fernandez-Baca and Toomas Rõõm

Chapter 4

Spin-wave theory

All solvable problems are like the harmonic oscillator, every unsolvable problem is unlike the harmonic oscillator in a different way.—

Dr Bill Hamilton

Here comes the pain.

—Al Pacino in 'Carlito's Way'

4.1 Introduction

This chapter describes how to evaluate the SW spectra and optical absorption. Although things are going to get a bit technical, the resulting methodology is quite versatile and can handle any NC spin state. Much of this formalism was developed in the late 1970s by Walker and Walstedt [1, 2] to study the excitations of spin glasses. Those are still good references to consult in the unlikely event that you remain confused after reading this chapter. Another very nice recent review of applications of SW theory to incommensurate spin states is by Tóth and Lake [3].

4.2 SW formalism

The local reference frame for the spin $\mathbf{S}_i$ on site i is defined in terms of the unitary matrix $\underline{U}_i$ by $\bar{\mathbf{S}}_i = \underline{U}_i \cdot \mathbf{S}_i$. In the 'laboratory frame', the spin on site i is parameterized as

$$\mathbf{S}_i = S\big(\sin\theta_i \cos\phi_i,\ \sin\theta_i \sin\phi_i,\ \cos\theta_i\big). \tag{4.1}$$

Then $\underline{U}_i$ and $\underline{U}_i^{-1}$ can be taken as

doi:10.1088/978-1-64327-114-9ch4

$$\underline{U}_i = \begin{pmatrix} \cos\theta_i \cos\phi_i & \cos\theta_i \sin\phi_i & -\sin\theta_i \\ -\sin\phi_i & \cos\phi_i & 0 \\ \sin\theta_i \cos\phi_i & \sin\theta_i \sin\phi_i & \cos\theta_i \end{pmatrix}, \tag{4.2}$$

$$\underline{U}_i^{-1} = \underline{U}_i^{t} = \begin{pmatrix} \cos\theta_i \cos\phi_i & -\sin\phi_i & \sin\theta_i \cos\phi_i \\ \cos\theta_i \sin\phi_i & \cos\phi_i & \sin\theta_i \sin\phi_i \\ -\sin\theta_i & 0 & \cos\theta_i \end{pmatrix}, \tag{4.3}$$

where $\underline{U}_i^{-1}$ and $\underline{U}_i^{t}$ are the inverse and transpose of $\underline{U}_i$, respectively.

From the expression $\bar{\mathbf{S}}_i = \underline{U}_i \cdot \mathbf{S}_i$ and equation (4.2) for $\underline{U}_i$, we find the expected result $\bar{\mathbf{S}}_i = S(0, 0, 1)$ in the local reference frame of the spin. The axes $\bar{\mathbf{y}}_i$ and $\bar{\mathbf{z}}_i$ that define the local reference frame of the spin $\mathbf{S}_i$ are sketched in figures 4.1(a) and (b). In the lab frame, $\bar{\mathbf{z}}_i$ is given by

$$\bar{\mathbf{z}}_i = \underline{U}_i^{-1} \cdot \mathbf{z} = (\sin\theta_i \cos\phi_i, \sin\theta_i \sin\phi_i, \cos\theta_i), \tag{4.4}$$

which is parallel to $\mathbf{S}_i = S\,\bar{\mathbf{z}}_i$. While

$$\bar{\mathbf{y}}_i = \underline{U}_i^{-1} \cdot \mathbf{y} = (-\sin\phi_i, \cos\phi_i, 0) \tag{4.5}$$

lies in the xy plane,

$$\bar{\mathbf{x}}_i = \underline{U}_i^{-1} \cdot \mathbf{x} = \bar{\mathbf{y}}_i \times \bar{\mathbf{z}}_i = (\cos\theta_i \cos\phi_i, \cos\theta_i \sin\phi_i, -\sin\theta_i) \tag{4.6}$$

points out of the xy plane. Other unitary rotation matrices corresponding to the same $\bar{\mathbf{z}}_i$ but different $\bar{\mathbf{x}}_i$ and $\bar{\mathbf{y}}_i$ are also allowed, as demonstrated in exercise 4.25. Of course, the SW spectra cannot depend on the choice of $\underline{U}_i$.

An HP transformation [4] is used to express the local spin operators $\bar{\mathbf{S}}_i$ in terms of the boson creation and annihilation operators $a_i^\dagger$ and a_i, which satisfy the boson commutation relation

$$[a_j, a_i^\dagger] = a_j a_i^\dagger - a_i^\dagger a_j = \delta_{ij}, \tag{4.7}$$

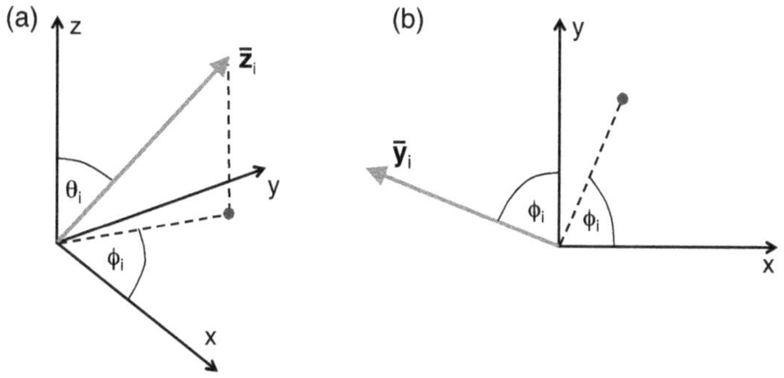

Figure 4.1. Local reference frame of the spin $\mathbf{S}_i = S\,\bar{\mathbf{z}}_i$ indicating the local (a) $\bar{\mathbf{z}}_i$ and (b) $\bar{\mathbf{y}}_i$ axes corresponding to our choice for $\underline{U}^i$.

$$[a_i, a_j] = [a_i^\dagger, a_j^\dagger] = 0. \tag{4.8}$$

The exact expansion of the spin operators in terms of the boson operators is

$$\bar{S}_{iz} = S - a_i^\dagger a_i, \tag{4.9}$$

so that the z component of the spin in its local reference frame is suppressed by the number $a_i^\dagger a_i$ of SWs at that site. The transverse components of $\bar{\mathbf{S}}_i$ are given by

$$\bar{S}_{i+} = \bar{S}_{ix} + i\bar{S}_{iy} = \sqrt{2S}\sqrt{1 - \frac{a_i^\dagger a_i}{2S}}\, a_i, \tag{4.10}$$

$$\bar{S}_{i-} = \bar{S}_{ix} - i\bar{S}_{iy} = \sqrt{2S}\, a_i^\dagger \sqrt{1 - \frac{a_i^\dagger a_i}{2S}}. \tag{4.11}$$

Using the Boson commutation relations, we then recover the spin commutation relations (exercise 4.1)

$$[\bar{S}_{i\alpha}, \bar{S}_{j\beta}] = i\delta_{ij}\epsilon_{\alpha\beta\gamma}\bar{S}_{i\gamma}. \tag{4.12}$$

Notice that spin operators on different sites commute. The idea of transforming into the local reference frame of each spin is sometimes called 'Kübler's trick' [5].

To transform from real space to momentum space and back, we use

$$a_{\mathbf{q}}^{(r)} = \frac{1}{\sqrt{N_u}} \sum_i^{(r)} e^{-i\mathbf{q}\cdot\mathbf{R}_i} a_i^{(r)}, \tag{4.13}$$

$$a_{\mathbf{q}}^{(r)\dagger} = \frac{1}{\sqrt{N_u}} \sum_i^{(r)} e^{i\mathbf{q}\cdot\mathbf{R}_i} a_i^{(r)\dagger}, \tag{4.14}$$

$$a_i^{(r)} = \frac{1}{\sqrt{N_u}} \sum_{\mathbf{q}}{}' e^{i\mathbf{q}\cdot\mathbf{R}_i} a_{\mathbf{q}}^{(r)} \tag{4.15}$$

$$a_i^{(r)\dagger} = \frac{1}{\sqrt{N_u}} \sum_{\mathbf{q}}{}' e^{-i\mathbf{q}\cdot\mathbf{R}_i} a_{\mathbf{q}}^{(r)\dagger}, \tag{4.16}$$

where N_u is the number of magnetic unit cells in the material and the sums over i are restricted to a particular site r in each unit cell. As indicated by the prime, the sum over $\mathbf{q}$ is restricted to the BZ associated with the magnetic unit cell. With M spins in each unit cell, the total number of magnetic sites is then $N = N_u M$. The Fourier transforms given above are sometimes defined by replacing N_u by N/M. The Fourier-transformed boson operators $a_{\mathbf{q}}^{(r)}$ and $a_{\mathbf{q}}^{(r)\dagger}$ obey the commutation relations

$$[a_{\mathbf{q}}^{(r)}, a_{\mathbf{q}'}^{(s)\dagger}] = \delta_{r,s}\delta_{\mathbf{q},\mathbf{q}'}, \tag{4.17}$$

$$[a_{\mathbf{q}}^{(r)}, a_{\mathbf{q}'}^{(s)}] = 0, \tag{4.18}$$

where $\mathbf{q}$ is confined to the first magnetic BZ.

The Hamiltonian is then expanded in powers of $1/\sqrt{S}$ as

$$\mathcal{H} = E_0 + \mathcal{H}_1 + \mathcal{H}_2 + \cdots, \tag{4.19}$$

where $\mathcal{H}_n$ is of order $S^{2-n/2}$. While $E_0 = \langle \mathcal{H} \rangle$ is the classical energy, the first-order term $\mathcal{H}_1$ vanishes if the spin state minimizes E_0. The spin dynamics is obtained from the second-order term

$$\mathcal{H}_2 = \sum_{\mathbf{q}}{}' \mathbf{v}_{\mathbf{q}}^{\dagger} \cdot \underline{L}(\mathbf{q}) \cdot \mathbf{v}_{\mathbf{q}}, \tag{4.20}$$

where

$$\mathbf{v}_{\mathbf{q}} = \left(a_{\mathbf{q}}^{(1)}, \ldots, a_{\mathbf{q}}^{(M)}, a_{-\mathbf{q}}^{(1)\dagger}, \ldots, a_{-\mathbf{q}}^{(M)\dagger}\right) \tag{4.21}$$

is a $2M$-dimensional vector and $\underline{L}(\mathbf{q})$ is a $2M$-dimensional Hermitian matrix. The second-order Hamiltonian $\mathcal{H}_2$ can be obtained rather simply by ignoring the $a_i^{\dagger} a_i / 2S$ terms in equations (4.10) and (4.11) for $\bar{S}_{i\pm}$.

Using the boson commutation relations for $a_{\mathbf{q}}^{(r)}$ and $a_{\mathbf{q}'}^{(s)\dagger}$, we find that $\mathbf{v}_{\mathbf{q}}$ and $\mathbf{v}_{\mathbf{q}}^{\dagger}$ satisfy the commutation relations

$$[\mathbf{v}_{\mathbf{q}}, \mathbf{v}_{\mathbf{q}'}^{\dagger}] = \underline{N} \delta_{\mathbf{q},\mathbf{q}'} \tag{4.22}$$

$$[\mathbf{v}_{\mathbf{q}}, \mathbf{v}_{\mathbf{q}'}] = [\mathbf{v}_{\mathbf{q}}^{\dagger}, \mathbf{v}_{\mathbf{q}'}^{\dagger}] = 0, \tag{4.23}$$

where

$$\underline{N} = \begin{pmatrix} \underline{I} & 0 \\ 0 & -\underline{I} \end{pmatrix} \tag{4.24}$$

and $\underline{I}$ is the M-dimensional unit matrix. Based on these commutation relations, we obtain the equations of motion for $\mathbf{v}_{\mathbf{q}}$:

$$i\hbar \frac{d\mathbf{v}_{\mathbf{q}}}{dt} = -[\mathcal{H}_2, \mathbf{v}_{\mathbf{q}}] = \underline{\mathcal{L}}(\mathbf{q}) \cdot \mathbf{v}_{\mathbf{q}}, \tag{4.25}$$

where $\underline{\mathcal{L}}(\mathbf{q}) = \underline{L}(\mathbf{q}) \cdot \underline{N}$. While the SW frequencies are obtained from the eigenvalues of $\underline{\mathcal{L}}$, the strengths or spectral weights of the SW frequencies depend on the eigenvectors of $\underline{\mathcal{L}}$. This is not quite as easy as it sounds because unlike $\underline{L}(\mathbf{q})$, $\underline{\mathcal{L}}(\mathbf{q})$ is *not* Hermitian!

The diagonal form for $\mathcal{H}_2$ is given by

$$\mathcal{H}_2 = \sum_{\mathbf{q}}{}' \mathbf{w}_{\mathbf{q}}^{\dagger} \cdot \underline{L}'(\mathbf{q}) \cdot \mathbf{w}_{\mathbf{q}}, \tag{4.26}$$

where $\underline{L}'(\mathbf{q}) = \underline{\mathcal{L}}'(\mathbf{q}) \cdot \underline{N}$ or $\underline{\mathcal{L}}'(\mathbf{q}) = \underline{L}'(\mathbf{q}) \cdot \underline{N}$ and

$$\mathbf{w}_{\mathbf{q}} = \left(\alpha_{\mathbf{q}}^{(1)}, \ldots, \alpha_{\mathbf{q}}^{(M)}, \alpha_{-\mathbf{q}}^{(1)\dagger}, \ldots, \alpha_{-\mathbf{q}}^{(M)\dagger}\right). \tag{4.27}$$

The new boson operators $\alpha_{\mathbf{q}}^{(n)}$ and $\alpha_{\mathbf{q}}^{(n)\dagger}$ obey canonical commutation relations as do the new vectors $\mathbf{w_q}$:

$$[\mathbf{w_q}, \mathbf{w}_{\mathbf{q}'}^{\dagger}] = \underline{N}\delta_{\mathbf{q},\mathbf{q}'}, \tag{4.28}$$

$$[\mathbf{w_q}, \mathbf{w}_{\mathbf{q}'}] = [\mathbf{w}_{\mathbf{q}}^{\dagger}, \mathbf{w}_{\mathbf{q}'}^{\dagger}] = 0. \tag{4.29}$$

The $2M$-dimensional matrix $\underline{\mathcal{L}}'(\mathbf{q})$ is diagonal with real eigenvalues $\epsilon_n(\mathbf{q}) = \hbar\omega_n(\mathbf{q})/2 > 0$ $(n = 1, \ldots, M)$ and $\epsilon_n(\mathbf{q}) = -\hbar\omega_n(-\mathbf{q})/2 < 0$ $(n = M+1, \ldots, 2M)$. So for each $\mathbf{q}$, there are M positive and M negative eigenvalues. The commutation relations yield

$$\mathcal{H}_2 = \hbar\sum_{n=1}^{M}\sum_{\mathbf{q}}{}' \,\omega_n(\mathbf{q})\left\{\alpha_{\mathbf{q}}^{(n)\dagger}\alpha_{\mathbf{q}}^{(n)} + \frac{1}{2}\right\}, \tag{4.30}$$

which is identical to the result for a harmonic oscillator when $\omega_n(\mathbf{q})$ is the frequency for mode n with wavevector $\mathbf{q}$. It also follows that $\alpha_{\mathbf{q}}^{(n)\dagger}$ is the creation operator for a SW with energy $\omega_n(\mathbf{q})$ and momentum $\mathbf{q}$ and $\alpha_{\mathbf{q}}^{(n)}$ is the corresponding annihilation operator:

$$\alpha_{\mathbf{q}}^{(n)\dagger}\,|0\rangle = |n, \mathbf{q}\rangle, \tag{4.31}$$

$$\alpha_{\mathbf{q}}^{(n)}\,|n, \mathbf{q}\rangle = |0\rangle, \tag{4.32}$$

where $|0\rangle$ is the ground state and $|n, \mathbf{q}\rangle$ is an excited state containing a single SW with momentum $\mathbf{q}$.

To see the origin of the negative eigenvalues in $\underline{\mathcal{L}}'(\mathbf{q})$, take $\mathcal{H}_2$ in the diagonal form of equation (4.26) and evaluate

$$i\hbar\frac{d\mathbf{w_q}}{dt} = -[\mathcal{H}_2, \mathbf{w_q}] = \underline{L}'(\mathbf{q})\cdot\underline{N}\cdot\mathbf{w_q} = \underline{\mathcal{L}}'(\mathbf{q})\cdot\mathbf{w_q}. \tag{4.33}$$

So the boson commutation relations produce the $\underline{N}$ matrix and the corresponding negative eigenvalues of $\underline{\mathcal{L}}'(\mathbf{q})$.

Vector operators $\mathbf{w_q}$ and $\mathbf{v_q}$ are related by $\mathbf{w_q} = \underline{X}(\mathbf{q})\cdot\mathbf{v_q}$ or $\mathbf{v_q} = \underline{X}^{-1}(\mathbf{q})\cdot\mathbf{w_q}$, where the $2M$-dimensional matrix $\underline{X}$ is normalized by

$$\underline{X}(\mathbf{q})\cdot\underline{N}\cdot\underline{X}(\mathbf{q})^{\dagger} = \underline{N}. \tag{4.34}$$

For fixed $\mathbf{q}$,

$$\sum_j\left\{\mathcal{L}_{ij}(\mathbf{q}) - \delta_{ij}\,\epsilon_n(\mathbf{q})\right\}X_{nj}(\mathbf{q})^* = 0, \tag{4.35}$$

where $\underline{X}(\mathbf{q})^*$ is the complex conjugate of $\underline{X}(\mathbf{q})$. Despite the normalization condition given above, each row of $\underline{X}(\mathbf{q})$ (corresponding to a given eigenvalue n in equation (4.35)) is only defined up to some complex factor $\exp(i\kappa_n)$. But the phases κ_n have no physical effects.

The relation $\mathbf{v_q} = \underline{X}^{-1}(\mathbf{q}) \cdot \mathbf{w_q}$ implies that (exercise 4.2)

$$X^{-1}_{s,\,l}(-\mathbf{q})^* = X^{-1}_{s+M,\,l+M}(\mathbf{q}), \tag{4.36}$$

$$X^{-1}_{s,\,l+M}(-\mathbf{q})^* = X^{-1}_{s+M,\,l}(\mathbf{q}), \tag{4.37}$$

where s and l lie between 1 and M. These conditions indicate that only the phase factors κ_l with $1 \leqslant l \leqslant M$ are independent since the l and $l + M$ rows of $\underline{X}(\mathbf{q})$ are related.

4.3 Spin–spin correlation function and INS

Neutron scattering measures the spin–spin correlation function

$$S_{\alpha\beta}(\mathbf{q},\omega) = \frac{1}{2\pi N}\sum_{i,j}\int dt\, e^{-i\omega t}e^{-i\mathbf{q}\cdot(\mathbf{R}_i-\mathbf{R}_j)}\langle S_{i\alpha}(0)S_{j\beta}(t)\rangle, \tag{4.38}$$

which was introduced in equation (2.5) and satisfies the symmetry relation $S_{\alpha\beta}(\mathbf{q},\omega) = S_{\beta\alpha}(\mathbf{q},\omega)^*$. We now create a bridge between the scattering cross section and the eigenvalues and eigenvectors of $\underline{\mathcal{L}}$. First use the unitary matrices $\underline{U}_r$ to transform the spins into the local boson operators $a_{\mathbf{q}}^{(r)}$. Then use $\underline{X}(\mathbf{q})$ to transform the local operators $a_{\mathbf{q}}^{(r)}$ into the diagonalized operators $\alpha_{\mathbf{q}}^{(n)}$. By inserting a complete set of states at $T = 0$, we find

$$\begin{aligned} S_{\alpha\beta}(\mathbf{q},\omega) &= \frac{S}{4\pi M}\sum_{r,s=1}^{M}\int dt\, e^{-i\omega t}\Big\langle\Big\{V^-_{r\alpha}a_{\mathbf{q}}^{(r)} + V^+_{r\alpha}a_{-\mathbf{q}}^{(r)\dagger}\Big\} \\ &\quad\times e^{i\mathcal{H}t/\hbar}\Big\{V^-_{s\beta}\,a_{-\mathbf{q}}^{(s)} + V^+_{s\beta}\,a_{\mathbf{q}}^{(s)\dagger}\Big\}e^{-i\mathcal{H}t/\hbar}\Big\rangle \\ &= \frac{S}{4\pi M}\sum_{n=1}^{M}\sum_{r,s=1}^{M}\int dt\, e^{i(\omega_n(\mathbf{q})-\omega)t}\Big\{V^-_{r\alpha}\,X^{-1}_{r,\,n}(\mathbf{q}) + V^+_{r\alpha}X^{-1}_{r+M,\,n}(\mathbf{q})\Big\} \\ &\quad\times\Big\{V^-_{s\beta}X^{-1}_{s,\,n+M}(-\mathbf{q}) + V^+_{s\beta}X^{-1}_{s+M,\,n+M}(-\mathbf{q})\Big\} \\ &\quad\times\langle 0|\alpha_{\mathbf{q}}^{(n)}|n,\,\mathbf{q}\rangle\langle n,\,\mathbf{q}|\alpha_{\mathbf{q}}^{(n)\dagger}|0\rangle, \end{aligned} \tag{4.39}$$

$$V^{\pm}_{r\alpha} = (U_r^{-1})_{\alpha x} \pm i(U_r^{-1})_{\alpha y} = (U_r)_{x\alpha} \pm i(U_r)_{y\alpha}. \tag{4.40}$$

Finally, use the symmetry relations equations (4.36) and (4.37) for $\underline{X}(\mathbf{q})$ to write

$$\begin{aligned} S_{\alpha\beta}(\mathbf{q},\omega) &= \frac{S}{2M}\sum_{n=1}^{M}\sum_{r,s=1}^{M}W^{(n)}_{r\alpha}(\mathbf{q})W^{(n)}_{s\beta}(\mathbf{q})^*\,\delta(\omega-\omega_n(\mathbf{q})) \\ &\equiv \sum_n S^{(n)}_{\alpha\beta}(\mathbf{q})\,\delta(\omega-\omega_n(\mathbf{q})), \end{aligned} \tag{4.41}$$

$$S_{\alpha\beta}^{(n)}(\mathbf{q}) = S_{\beta\alpha}^{(n)}(\mathbf{q})^* = \frac{S}{2M}\sum_{r,s=1}^{M} W_{r\alpha}^{(n)}(\mathbf{q})W_{s\beta}^{(n)}(\mathbf{q})^*, \tag{4.42}$$

$$W_{r\alpha}^{(n)}(\mathbf{q}) = V_{r\alpha}^{-}\, X_{r,\,n}^{-1}(\mathbf{q}) + V_{r\alpha}^{+}\, X_{r+M,\,n}^{-1}(\mathbf{q}), \tag{4.43}$$

where n now runs from 1 to M, corresponding to positive eigenvalues $\epsilon_n(\mathbf{q})$. The coefficient $S_{\alpha\beta}^{(n)}(\mathbf{q})$ in equation (4.41) gives the strength or spectral weight of each SW frequency $\omega_n(\mathbf{q})$. According to equation (4.42), the phase κ_n of each column of $\underline{X}^{-1}$ has no physical effect.

The spin–spin correlation function can be generalized to nonzero temperatures by inserting the thermal factor $\exp(-\beta\mathcal{H})/Z$ where

$$Z = \mathrm{Tr}\,[\exp(-\beta\mathcal{H})] \tag{4.44}$$

is the partition function and Tr is the trace. As you will show in exercise 4.5, SW theory implies that

$$\begin{aligned} S_{\alpha\beta}(\mathbf{q},\omega) &= \frac{S}{2M}[n_{\mathrm{B}}(\omega) + 1]\sum_{n=1}^{M}\sum_{r,s=1}^{M}\Big\{W_{r\alpha}^{(n)}(\mathbf{q})W_{s\beta}^{(n)}(\mathbf{q})^*\,\delta(\omega - \omega_n(\mathbf{q})) \\ &\quad - W_{r\alpha}^{(n)}(-\mathbf{q})^* W_{s\beta}^{(n)}(-\mathbf{q})\,\delta(\omega + \omega_n(-\mathbf{q}))\Big\} \\ &= [n_{\mathrm{B}}(\omega) + 1]\sum_{n=1}^{M}\Big\{S_{\alpha\beta}^{(n)}(\mathbf{q})\,\delta(\omega - \omega_n(\mathbf{q})) \\ &\quad - S_{\beta\alpha}^{(n)}(-\mathbf{q})\,\delta(\omega + \omega_n(-\mathbf{q}))\Big\}, \end{aligned} \tag{4.45}$$

where

$$n_{\mathrm{B}}(\omega) = \frac{1}{\exp(\hbar\omega/k_{\mathrm{B}}T) - 1} \tag{4.46}$$

is the Boson distribution function. The detailed balance condition is given by (exercise 4.6)

$$S_{\alpha\beta}(-\mathbf{q},-\omega) = e^{-\hbar\omega/k_{\mathrm{B}}T}S_{\beta\alpha}(\mathbf{q},\omega) = e^{-\hbar\omega/k_{\mathrm{B}}T}S_{\alpha\beta}(\mathbf{q},\omega)^*, \tag{4.47}$$

which was also introduced in chapter 2. The fluctuation–dissipation theorem [6], which relates the scattering function $S_{\alpha\beta}(\mathbf{q},\omega)$ to the spin susceptibility $\chi_{\alpha\beta}^{\mathrm{ss}}(\mathbf{q},\omega)$, is derived in appendix 4.E. While these expressions are useful to predict the relative strengths of the creation and annihilation peaks in $S_{\alpha\alpha}(\mathbf{q},\omega)$, they do *not* generalize the SW frequencies themselves to nonzero temperatures. Because the $T > 0$ SW modes involve states with more than one spin flip [7], the temperature dependence of the SW frequencies depends on higher-order $1/S$ terms in the HP expansion.

Unlike the spectral weight $S^{(n)}_{\alpha\beta}(\mathbf{q})$, the SW frequencies $\omega_n(\mathbf{q})$ are periodic functions of the ordering wavevector $\mathbf{Q}$ and of the reciprocal-lattice vectors $\boldsymbol{\tau}$:

$$\omega_n(\mathbf{q} + \mathbf{Q}) = \omega_n(\mathbf{q} + \boldsymbol{\tau}) = \omega_n(\mathbf{q}). \tag{4.48}$$

For a Bravais lattice with one magnetic ion in each structural unit cell, the spectral weight is a periodic function of the reciprocal lattice vectors $\boldsymbol{\tau}$:

$$S^{(n)}_{\alpha\beta}(\mathbf{q} + \boldsymbol{\tau}) = S^{(n)}_{\alpha\beta}(\mathbf{q}). \tag{4.49}$$

An example of the different periods of $\omega_n(\mathbf{q})$ and $S^{(n)}_{\alpha\beta}(\mathbf{q})$ is given in section 5.3. Recall that the measured cross section contains the square of the magnetic form factor $f(\boldsymbol{\kappa} = \mathbf{q} + \boldsymbol{\tau})$, which is never a periodic function of $\boldsymbol{\tau}$.

Including the kinematical constraint that requires neutrons only couple to spin fluctuations perpendicular to their change in momentum $\boldsymbol{\kappa}$, we obtain

$$S(\boldsymbol{\kappa},\omega) = \sum_{\alpha,\beta}\left\{\delta_{\alpha\beta} - \frac{\kappa_\alpha\kappa_\beta}{\kappa^2}\right\} S_{\alpha\beta}(\boldsymbol{\kappa},\omega) = \sum_n S^{(n)}(\boldsymbol{\kappa})\,\delta(\omega - \omega_n(\boldsymbol{\kappa})), \tag{4.50}$$

$$S^{(n)}(\boldsymbol{\kappa}) = \sum_{\alpha,\beta}\left\{\delta_{\alpha\beta} - \frac{\kappa_\alpha\kappa_\beta}{\kappa^2}\right\} S^{(n)}_{\alpha\beta}(\boldsymbol{\kappa}). \tag{4.51}$$

Since $S^{(n)}_{\alpha\beta}(\boldsymbol{\kappa})^* = S^{(n)}_{\beta\alpha}(\boldsymbol{\kappa})$, $S(\boldsymbol{\kappa},\omega)$, and $S^{(n)}(\boldsymbol{\kappa})$ are always real.

Because $S^{(n)}_{\alpha\beta}(\boldsymbol{\kappa})$ is Hermitian in spin space, one can always find a set of orthogonal spin axes where it is diagonal (just like one can find the principle axes that diagonalize the moment-of-inertia tensor I_{ij}). Using those principle axes,

$$S^{(n)}(\boldsymbol{\kappa}) = \sum_{\alpha}\left\{1 - \frac{{\kappa_\alpha}^2}{\kappa^2}\right\} S^{(n)}_{\alpha\alpha}(\boldsymbol{\kappa}). \tag{4.52}$$

If the net moment lies along one of the principle axes but the two orthogonal axes are not principal axes, then the off-diagonal matrix elements involving the other two axes will be nonzero but complex and odd with $S^{(n)}_{\alpha\beta}(\boldsymbol{\kappa}) = -S^{(n)}_{\beta\alpha}(\boldsymbol{\kappa})$ $(\alpha \neq \beta)$. So those off-diagonal terms would not contribute to $S(\boldsymbol{\kappa},\omega)$. If the magnetic moment does not lie along a high symmetry direction and you cannot be bothered to find the principle axes that diagonalize $S^{(n)}_{\alpha\beta}(\boldsymbol{\kappa})$, then all bets are off and you will have to evaluate both the diagonal and off-diagonal matrix elements of $S^{(n)}_{\alpha\beta}(\boldsymbol{\kappa})$.

As touched upon in chapter 2, various sum rules constrain the integral of ω^n $S_{\alpha\beta}(\mathbf{q},\omega)$ over $\mathbf{q}$ within the first BZ and over all frequencies ω [8]. We shall just mention the 'mother-of-all sum rules' with $n = 0$ (exercise 4.7):

$$\frac{1}{N}\sum_{\alpha,\mathbf{q}} \int d\omega\, S_{\alpha\alpha}(\mathbf{q},\omega) = S(S + 1). \tag{4.53}$$

The order S^2 part of the right-hand side comes from classical magnetic order. For example, a FM with spins aligned along **z** has

$$\langle S_{i\alpha}(t)\rangle = S\delta_{\alpha z}, \tag{4.54}$$

$$\langle S_{i\alpha}(0)S_{j\beta}(t)\rangle = S^2\delta_{\alpha z}\delta_{\beta z} + \vartheta(S), \tag{4.55}$$

and the static contribution

$$\frac{1}{2\pi N^2}\sum_{\mathbf{q}}\sum_{i,j}\int dt\, e^{-i\omega t}e^{-i\mathbf{q}\cdot(\mathbf{R}_j-\mathbf{R}_i)}\langle S_{i\alpha}(0)S_{j\beta}(t)\rangle = S^2\delta_{\alpha z}\,\delta_{\beta z}\,\delta(\omega), \tag{4.56}$$

which contributes to the elastic neutron-scattering cross section. The order S part of the right-hand side of equation (4.53) comes from the SWs (exercise 5.7).

The INS intensity is the sum over delta functions because linear SW theory (to second order in $1/\sqrt{S}$) does not include interactions between the SWs. Therefore, the SWs are undamped. To turn the sum over delta functions in equation (4.50) into a smooth function of frequency and wavevector, we have to integrate over a resolution function $R(\boldsymbol{\kappa},\omega)$ that gives the instrumental error distribution in $\{\boldsymbol{\kappa},\omega\}$ space for any measurement. Strictly speaking, $R(\boldsymbol{\kappa},\omega)$ is different for every experiment even on the same instrument! Evaluating the instrumental resolution function can involve the Monte-Carlo simulation of over 10^{11} neutron packets propagating down the incident beamline, through the material and the instrument, and finally hitting a detector [9]. Typically, the resulting $R(\boldsymbol{\kappa},\omega)$ is a four-dimensional ellipsoid.

If the instrumental resolution function is not available, as is often the case, we recommend simply integrating over a Gaussian:

$$\sum_n \delta(\omega-\omega_n(\boldsymbol{\kappa}))S^{(n)}(\boldsymbol{\kappa}) \rightarrow A\sum_n\sum_{\mathbf{k}} e^{-(\omega-\omega_n(\boldsymbol{\kappa}))^2/\Delta_\omega^2}e^{-|\mathbf{k}-\boldsymbol{\kappa}|^2/\Delta_\kappa^2}S^{(n)}(\mathbf{k}), \tag{4.57}$$

where A is a normalization constant and Δ_ω and Δ_κ are the resolution widths in frequency and wavevector taken from experiment. For cold neutrons, Δ_ω is about 0.1 meV and Δ_κ is about a hundredth of an inverse Å. Of course, you can also include extrinsic damping due to impurities by setting Δ_κ to some value above the instrumental resolution.

What about powders? It seems a shame after developing this beautiful formalism to throw away so much information. But to treat powders with randomly oriented particles, $S(\boldsymbol{\kappa},\omega)$ is integrated over all orientations of $\boldsymbol{\kappa}$:

$$S(\kappa,\omega) = \frac{1}{4\pi}\int d\Omega_\kappa\, S(\boldsymbol{\kappa},\omega), \tag{4.58}$$

which depends only on the magnitude $\kappa = |\boldsymbol{\kappa}|$. Since individual SW excitations are not usually seen, the SW spectrum of a powder is more difficult to analyze than the spectrum of a single crystal.

4.4 THz spectroscopy

Because photons are so fast (it is hard to beat the speed of light!), the SW probed by absorption of a photon at THz frequencies has wavevector $\mathbf{q}$ very close to 0. Although THz spectroscopy cannot measure the SW dispersion, it compensates for that deficiency in several respects. First, the energy resolution of optical absorption is much higher than for neutron scattering. Whereas most neutron-scattering measurements are lucky to achieve an energy resolution of 0.1 meV, the energy resolution of optical spectroscopy is about 0.01 meV. Second, the wavevector resolution of optical spectroscopy is also far greater than for neutron scattering. Whereas typical neutron-scattering measurements have a wavevector resolution of about 10^{-2} Å^{-1}, optical spectroscopy has a wavevector resolution of $\Delta q = \nu/c \approx 3 \times 10^{-10}$ Å^{-1} for $\nu = 1$ THz (4.137 meV). As we shall see, this wavevector resolution is very useful to separate the contributions of SWs with wavevectors close to each other.

4.4.1 Optical absorption and susceptibilities

In THz spectroscopy, either the photon is completely absorbed to produce a SW or (at nonzero temperature) the SW is completely absorbed to produce a photon. At zero temperature, the absorption of THz light is given by

$$\alpha(\omega) = \frac{2\omega}{c}\,\mathrm{Im}\,\mathcal{N}(\omega) \tag{4.59}$$

where [10, 11]

$$\begin{aligned}\mathcal{N}(\omega) &= \sqrt{(\epsilon^b + \chi^{\mathrm{ee}}(\omega))(1 + \chi^{\mathrm{mm}}(\omega))} \pm \frac{1}{2}\{\chi^{\mathrm{me}}(\omega) + \chi^{\mathrm{em}}(\omega)\}\\ &\approx \sqrt{\epsilon^b} + \frac{1}{2\sqrt{\epsilon^b}}\chi^{\mathrm{ee}}(\omega) + \frac{\sqrt{\epsilon^b}}{2}\chi^{\mathrm{mm}}(\omega) \pm \frac{1}{2}\{\chi^{\mathrm{me}}(\omega) + \chi^{\mathrm{em}}(\omega)\}\end{aligned} \tag{4.60}$$

is the complex refractive index for a linearly polarized beam, $\chi^{\mathrm{ee}}(\omega)$, $\chi^{\mathrm{mm}}(\omega)$, $\chi^{\mathrm{me}}(\omega)$, and $\chi^{\mathrm{em}}(\omega)$ are the dielectric, magnetic, magnetoelectric, and electromagnetic susceptibilities describing the dynamical response of the spin system. The background dielectric constant ϵ^b is associated with charge excitations at energies above the SW spectrum and the background permeability μ^b is usually set to 1. For comparison with the results in section 3.3, recall that $\nu = \omega/2\pi$ is the frequency in Hz.

Both the dielectric constant ϵ^b and the dielectric susceptibility $\chi^{\mathrm{ee}}(\omega)$ depend on the orientation $\mathbf{e}$ of the THz electric field $\mathbf{E}^{\omega}$.[1] The magnetic susceptibility $\chi^{\mathrm{mm}}(\omega)$ depends on the orientation $\mathbf{h}$ of the THz magnetic field $\mathbf{H}^{\omega}$. The magnetoelectric and electromagnetic susceptibilities $\chi^{\mathrm{me}}(\omega)$ and $\chi^{\mathrm{em}}(\omega)$ depend on both. Of course, $\mathbf{e}\perp\mathbf{h}$ and $\mathbf{e} \times \mathbf{h}$ is the direction of light propagation.

Our previous expression for $\mathcal{N}(\omega)$ in section 3.3 neglected the cross term $\pm(\chi^{\mathrm{me}}(\omega) + \chi^{\mathrm{em}}(\omega))/2$. Since the sign of these terms depends on the direction of

[1] In terms of the dielectric tensor $\underline{\epsilon}^b$, $\epsilon^b(\mathbf{e}) = \mathbf{e}\cdot\underline{\epsilon}^b\cdot\mathbf{e} = \sum_{\alpha,\beta}\epsilon^b_{\alpha\beta}e_\alpha e_\beta$ is a scalar.

light propagation, it cancels out when two opposite propagation directions are averaged. Following Miyahara and Furukawa [11], the susceptibility tensors at $T = 0$ are given by

$$\chi^{\mathrm{mm}}(\omega) = \frac{\mu_0}{\hbar V} \sum_{n=1}^{M} \frac{\langle 0|\mathbf{h}\cdot\mathbf{M}|n, \mathbf{q}=0\rangle\langle n, \mathbf{q}=0|\mathbf{h}\cdot\mathbf{M}|0\rangle}{\omega_n(0) - \omega - i\varepsilon}, \tag{4.61}$$

$$\chi^{\mathrm{ee}}(\omega) = \frac{1}{\hbar\epsilon_0 V} \sum_{n=1}^{M} \frac{\langle 0|\mathbf{e}\cdot\mathbf{P}|n, \mathbf{q}=0\rangle\langle n, \mathbf{q}=0|\mathbf{e}\cdot\mathbf{P}|0\rangle}{\omega_n(0) - \omega - i\varepsilon}, \tag{4.62}$$

$$\chi^{\mathrm{em}}(\omega) = \sqrt{\frac{\mu_0}{\epsilon_0}} \frac{1}{\hbar V} \sum_{n=1}^{M} \frac{\langle 0|\mathbf{e}\cdot\mathbf{P}|n, \mathbf{q}=0\rangle\langle n, \mathbf{q}=0|\mathbf{h}\cdot\mathbf{M}|0\rangle}{\omega_n(0) - \omega - i\varepsilon}, \tag{4.63}$$

$$\chi^{\mathrm{me}}(\omega) = \sqrt{\frac{\mu_0}{\epsilon_0}} \frac{1}{\hbar V} \sum_{n=1}^{M} \frac{\langle 0|\mathbf{h}\cdot\mathbf{M}|n, \mathbf{q}=0\rangle\langle n, \mathbf{q}=0|\mathbf{e}\cdot\mathbf{P}|0\rangle}{\omega_n(0) - \omega - i\varepsilon}, \tag{4.64}$$

where $\varepsilon = 0^+$ and the speed of light in vacuum $c = 1/\sqrt{\mu_0\epsilon_0}$. While $\mathbf{M}$ is the magnetization operator, $\mathbf{P}$ is the electric polarization operator. Hence, the coupling of photons to SWs depends on the matrix elements $\langle n, \mathbf{q}=0|\mathbf{P}|0\rangle$ and $\langle n, \mathbf{q}=0|\mathbf{M}|0\rangle$ where $|n, \mathbf{q}=0\rangle$ is an excited state containing a single SW at $\mathbf{q}=0$ with frequency $\omega_n(0)$. Only $\mathbf{q}=0$ eigenstates contribute to the sum over eigenstates because $\mathbf{M}$ and $\mathbf{P}$ separately only involve $\mathbf{q}=0$ creation and annihilation operators, as shown explicitly for $\mathbf{M}$ below.

The $T = 0$ optical absorption can be written compactly as

$$\alpha^{\pm}(\omega) = \omega \sum_n (B_n \pm A_n)\, \delta(\omega - \omega_n(0)), \tag{4.65}$$

$$A_n = \frac{1}{N} \frac{X}{\mathcal{V}\mu_{\mathrm{B}}} \mathrm{Re}\{\langle 0|\mathbf{e}\cdot\mathbf{P}|n, 0\rangle\langle n, 0|\mathbf{h}\cdot\mathbf{M}|0\rangle\}, \tag{4.66}$$

$$B_n = \frac{1}{N}\left\{\frac{Y_1}{\mathcal{V}^2}|\langle 0|\mathbf{e}\cdot\mathbf{P}|n, 0\rangle|^2 + \frac{Y_2}{\mu_{\mathrm{B}}^2}|\langle 0|\mathbf{h}\cdot\mathbf{M}|n, 0\rangle|^2\right\}, \tag{4.67}$$

where $\mathcal{V} = V/N$ is the volume per spin so that $\mathbf{P}/\mathcal{V}$ has units of nC cm^{-2}. With polarization units factored into X and Y_1, the matrix elements $\langle 0|\mathbf{e}\cdot\mathbf{P}|n, 0\rangle/\mathcal{V}$ and $\langle 0|\mathbf{M}\cdot\mathbf{h}|n, 0\rangle/\mu_{\mathrm{B}}$ are dimensionless. The $\pm$ sign refers to light propagating in the $\pm\mathbf{k}$ directions.

It can then be shown that [11, 12]

$$X = \frac{2\pi\mu_{\mathrm{B}}}{\hbar c}\sqrt{\frac{\mu_0}{\epsilon_0}}\,\mathrm{nC\,cm^{-2}}, \tag{4.68}$$

$$Y_1 = \frac{\pi \mathcal{V}}{\hbar c \epsilon_0 \sqrt{\epsilon^b}} \ \mathrm{nC^2\,cm^{-4}}, \tag{4.69}$$

$$Y_2 = \frac{\pi \mu_B^2 \mu_0 \sqrt{\epsilon^b}}{\hbar c \mathcal{V}}, \tag{4.70}$$

which are related by $X = 2\sqrt{Y_1 Y_2}$. If the A_n and B_n terms cancel each other, the material is transparent for one propagation direction of light with frequency $\omega_n(0)$ [13].

4.4.2 Magnetic absorption

We now examine the matrix elements of the magnetization $\mathbf{M} = 2\mu_B \sum_i \mathbf{S}_i$ required to obtain the magnetic optical absorption. Transforming into the local reference frame, $\mathbf{S}_i$ can be expanded to order $\sqrt{S}$ as

$$\begin{aligned}
S_{i\alpha} &= \sum_\beta (U_i^{-1})_{\alpha\beta} \bar{S}_{i\beta} = \langle S_{i\alpha}\rangle + \sqrt{\frac{S}{2}} \left\{ V_{i\alpha}^- a_i^{(i)} + V_{i\alpha}^+ a_i^{(i)\dagger} \right\} + \vartheta(S^0) \\
&= \langle S_{i\alpha}\rangle + \sqrt{\frac{S}{2N_u}} \sum_{\mathbf{q}} e^{i\mathbf{q}\cdot\mathbf{R}_i} \left\{ V_{i\alpha}^- a_{\mathbf{q}}^{(r)} + V_{r\alpha}^+ a_{-\mathbf{q}}^{(r)\dagger} \right\} + \vartheta(S^0) \\
&= \langle S_{i\alpha}\rangle + \sqrt{\frac{SM}{2N}} \sum_{m=1}^{M} \sum_{\mathbf{q}} e^{i\mathbf{q}\cdot\mathbf{R}_i} \Big\{ \left[V_{r\alpha}^- X_{r,\,m}^{-1}(\mathbf{q}) + V_{r\alpha}^+ X_{r+M,\,m}^{-1}(\mathbf{q}) \right] \alpha_{\mathbf{q}}^{(m)} \\
&\quad + \left[V_{r\alpha}^+ X_{r+M,\,m+M}^{-1}(\mathbf{q}) + V_{r\alpha}^- X_{r,\,m+M}^{-1}(\mathbf{q}) \right] \alpha_{-\mathbf{q}}^{(m)\dagger} \Big\} + \vartheta(S^0),
\end{aligned} \tag{4.71}$$

where site i lies on sublattice r. Neglected terms of order S^0 are second-order in the creation and annihilation operators and so do not contribute to the matrix elements coupling the ground state to an excited SW state.

Summing over all sites, we obtain the matrix element for the magnetization (exercise 4.9):

$$\langle 0|M_\alpha|n,\, \mathbf{q} = 0\rangle = \sqrt{2SN_u}\, \mu_B \sum_{r=1}^{M} W_{r\alpha}^{(n)}(0), \tag{4.72}$$

which only involves the $\mathbf{q} = 0$ component of $W_{r\alpha}^{(n)}(\mathbf{q})$. Consequently,

$$\begin{aligned}
\alpha_m(\omega) &= \omega \frac{Y_2}{N\mu_B^2} \sum_{n=1}^{M} |\langle n,\, 0|\mathbf{M}\cdot\mathbf{h}|0\rangle|^2\, \delta(\omega - \omega_n(0)) \\
&= 2S\omega \frac{Y_2}{M} \sum_{\alpha,\beta} h_\alpha h_\beta \sum_{n=1}^{M} \sum_{r,s=1}^{M} W_{r\alpha}^{(n)}(0)^* W_{s\beta}^{(n)}(0)\, \delta(\omega - \omega_n(0)) \\
&= 2S\omega \frac{Y_2}{M} \sum_{n=1}^{M} \left| \sum_{r=1}^{M} \sum_\alpha h_\alpha W_{r\alpha}^{(n)}(0) \right|^2 \delta(\omega - \omega_n(0))
\end{aligned} \tag{4.73}$$

is the purely magnetic contribution to the optical absorption at $T = 0$.

Starting with the general expression

$$\alpha_{\rm m}(\omega) = \frac{\omega\mu_0\sqrt{\epsilon_b}}{2\hbar c V}\int dt\, e^{-i\omega t}\langle \mathbf{h}\cdot\mathbf{M}(0)\,\mathbf{h}\cdot\mathbf{M}(t)\rangle, \tag{4.74}$$

the magnetic optical absorption can be generalized to nonzero temperatures using the same method as for the spin–spin correlation function. Within SW theory (exercise 4.10)

$$\begin{aligned}\alpha_{\rm m}(\omega) &= 2S\frac{Y_2}{M}\,\omega[n_{\rm B}(\omega)+1]\sum_{\alpha,\beta}h_\alpha h_\beta\sum_{n=1}^{M}\sum_{r,s=1}^{M}\\ &\times\left\{W_{r\alpha}^{(n)}(0)W_{s\beta}^{(n)}(0)^*\,\delta(\omega-\omega_n(0)) - W_{r\alpha}^{(n)}(0)^*W_{s\beta}^{(n)}(0)\,\delta(\omega+\omega_n(0))\right\}\\ &= 2S\frac{Y_2}{M}\,\omega[n_{\rm B}(\omega)+1]\sum_{n=1}^{M}\left|\sum_{r=1}^{M}\sum_{\alpha}h_\alpha W_{r\alpha}^{(n)}(0)\right|^2\\ &\times\{\delta(\omega-\omega_n(0)) - \delta(\omega+\omega_n(0))\}.\end{aligned} \tag{4.75}$$

The $\delta(\omega + \omega_n(0))$ term now corresponds to the annihilation of a SW with the creation of a photon! Since optical absorption measures the difference between the number of photons going in and the number that come out, the measured magnetic absorption for $\omega > 0$ is (exercise 4.11)

$$\begin{aligned}\alpha_{\rm m}^{\rm meas}(\omega) &= \alpha_{\rm m}(\omega) + \alpha_{\rm m}(-\omega)\\ &= 2S\frac{Y_2}{M}\,\omega\sum_{n=1}^{M}\left|\sum_{r=1}^{M}\sum_{\alpha}h_\alpha W_{r\alpha}^{(n)}(0)\right|^2\delta(\omega-\omega_n(0)),\end{aligned} \tag{4.76}$$

independent of temperature!

As proven in appendix 4.E, the $\mathbf{q} = 0$ spin–spin correlation function

$$S_{\alpha\beta}(\mathbf{q}=0,\omega) = \frac{1}{8\pi\mu_{\rm B}^2 N}\int dt\, e^{-i\omega t}\langle M_\alpha(0)M_\beta(t)\rangle \tag{4.77}$$

is related to the purely magnetic contribution to the optical absorption by

$$\begin{aligned}\alpha_{\rm m}(\omega) &= \frac{\sqrt{\epsilon_{\rm b}}}{c}\,\omega[n_{\rm B}(\omega)+1]\,{\rm Im}\,\chi^{\rm mm}(\omega) = 4Y_2\,\omega\sum_{\alpha,\beta}h_\alpha h_\beta\, S_{\alpha\beta}(\mathbf{q}=0,\omega)\\ &= 4Y_2\,\omega[n_{\rm B}(\omega)+1]\sum_{\alpha,\beta}h_\alpha h_\beta\,{\rm Im}\,\chi_{\alpha\beta}^{\rm ss}(\mathbf{q}=0,\omega),\end{aligned} \tag{4.78}$$

which is not based on SW theory. Consequently, the optical and $\mathbf{q} = 0$ spin susceptibilities are generally related by

$$\chi^{\rm mm}(\omega) = \frac{4\pi\mu_0\mu_{\rm B}^2}{\hbar V}\sum_{\alpha,\beta}h_\alpha h_\beta\,\chi_{\alpha\beta}^{\rm ss}(\mathbf{q}=0,\omega), \tag{4.79}$$

as shown in exercise 4.12.

4.4.3 Polarization matrix elements and the SC polarization

Polarization matrix elements depend on the precise form for the polarization operator **P**, which may be associated with the spin-current (SC) [14, 15], *p–d* hybridization [16], or exchange striction [17]. Prior to evaluating the matrix elements, **P** must be expanded to linear order in the operators $\alpha_{\mathbf{q}}^{(n)}$ and $\alpha_{\mathbf{q}}^{(n)\dagger}$ that diagonalize the Hamiltonian.

To demonstrate this procedure, consider the SC polarization [14] associated with the inverse DM interaction [15]. For a one-dimensional system with lattice constant a, the SC polarization can be written

$$\mathbf{P} = \lambda \sum_i \mathbf{e}_{i,\, i+1} \times (\mathbf{S}_i \times \mathbf{S}_{i+1}), \tag{4.80}$$

where $\mathbf{e}_{ij} = \mathbf{R}_j - \mathbf{R}_i$. Now expand **P** to linear order in the creation and annihilation operators:

$$\begin{aligned}\mathbf{P} = \langle \mathbf{P} \rangle + \lambda a \sqrt{\frac{N_u S}{2}} \sum_{r=1}^{M} \Big\{ \Delta S_{rx} \Big[\mathbf{z}\Big(V_{rz}^{-} a_{\mathbf{q}=0}^{(r)} + V_{rz}^{+} a_{\mathbf{q}=0}^{(r)\dagger} \Big) \\ + \mathbf{y}\Big(V_{ry}^{-} a_{\mathbf{q}=0}^{(r)} + V_{ry}^{+} a_{\mathbf{q}=0}^{(r)\dagger} \Big) \Big] - \mathbf{y} \Delta S_{ry} \Big(V_{rx}^{-} a_{\mathbf{q}=0}^{(r)} + V_{rx}^{+} a_{\mathbf{q}=0}^{(r)\dagger} \Big) \\ - \mathbf{z} \Delta S_{rz} \Big(V_{rx}^{-} a_{\mathbf{q}=0}^{(r)} + V_{rx}^{+} a_{\mathbf{q}=0}^{(r)\dagger} \Big) \Big\} + \vartheta(S),\end{aligned} \tag{4.81}$$

where $\Delta S_{r\alpha} = S_{r+1,\, \alpha} - S_{r-1,\, \alpha}$. Notice that the constant term $\langle \mathbf{P} \rangle$ is of order S^2 while the creation and annihilation terms are of order $S^{3/2}$. As for the magnetization operator **M**, only $\mathbf{q} = 0$ operators contribute to the electric polarization operator.

It follows that the matrix elements of **P** are (exercise 4.13)

$$\langle n,\, 0 | P_x | 0 \rangle = 0, \tag{4.82}$$

$$\langle n,\, 0 | P_y | 0 \rangle = \lambda a \sqrt{\frac{S N_u}{2}} \sum_{r=1}^{M} \Big\{ \Delta S_{rx} W_{ry}^{(n)}(0)^* - \Delta S_{ry} W_{rx}^{(n)}(0)^* \Big\}, \tag{4.83}$$

$$\langle n,\, 0 | P_z | 0 \rangle = \lambda a \sqrt{\frac{S N_u}{2}} \sum_{r=1}^{M} \Big\{ \Delta S_{rx} W_{rz}^{(n)}(0)^* - \Delta S_{rz} W_{rx}^{(n)}(0)^* \Big\}. \tag{4.84}$$

Like the matrix element of the magnetization in equation (4.72), the polarization matrix elements scale like $\sqrt{N}$ (exercise 4.14).

We now apply this result to a simple cycloid propagating along **x** in the xz plane and given by the classical spin state

$$\mathbf{S}_i = S(\sin(\mathbf{Q} \cdot \mathbf{R}_i),\, 0,\, \cos(\mathbf{Q} \cdot \mathbf{R}_i)), \tag{4.85}$$

with period Ma so that $\mathbf{Q} = (2\pi/Ma)\mathbf{x}$. The static polarization

$$\langle \mathbf{P} \rangle = \lambda a N S^2 \, \mathbf{z} \sin\left(\frac{2\pi}{M}\right) \tag{4.86}$$

lies along $\mathbf{z}$, perpendicular to both $\mathbf{e}_{i,\,i+1} = a\mathbf{x}$ and the helicity $\mathbf{S}_i \times \mathbf{S}_{i+1} \parallel \mathbf{y}$. Therefore (exercise 4.15),

$$\langle n,\, 0|P_x|0\rangle = \langle n,\, 0|P_z|0\rangle = 0, \tag{4.87}$$

$$\begin{aligned}\langle n,\, 0|P_y|0\rangle = &- i\lambda a\sqrt{2N_u}\, S^{3/2} \sin\left(\frac{2\pi}{M}\right)\sum_{r=1}^{M} \cos\left(\frac{2\pi r}{M}\right) \\ &\times \left\{X^{-1}_{r+M,\,n}(0)^* - X^{-1}_{r,\,n}(0)^*\right\}.\end{aligned} \tag{4.88}$$

So for a simple cycloid in the xz plane, $\mathbf{Q}$ along $\mathbf{x}$, $\langle \mathbf{P} \rangle$ along $\mathbf{z}$, and $\langle n,\, 0|\mathbf{P}|0\rangle$ along $\mathbf{y}$ are mutually perpendicular!

The $\pm A_n$ term in $\alpha^{\pm}(\omega)$ is linear in the polarization matrix element and produces non-reciprocal directional dichroism [13], which means that the absorption of light changes with reversal of the direction of light propagation (or with reversal of the magnetization in an external magnetic field). The change in absorption for linearly polarized light with components $\mathbf{h}$ and $\mathbf{e}$ is

$$\Delta\alpha(\omega) = \alpha^{+}(\omega) - \alpha^{-}(\omega) = \frac{2\omega}{\hbar c}\,\mathrm{Im}\,\{\chi^{\mathrm{me}}(\omega) + \chi^{\mathrm{em}}(\omega)\}. \tag{4.89}$$

Because the SW mode must be both electric- and magnetic-dipole active (i.e. an electromagnon), materials that exhibit this effect are multiferroic.

4.5 SW amplitudes

What is the spin disturbance associated with a SW? To answer that important question, suppose that a perturbing potential $V(\mathbf{R},\, t)$ is turned on for a short time while it acts on the ground state $|0\rangle$. The frequency ω and wavevector $\mathbf{q}$ of the potential are tuned to match the energy $\omega_n(\mathbf{q})$ of a SW state with wavevector $\mathbf{q}$. At time $t = 0$, the potential is turned off and the new spin state evolves in time as [18]

$$|n,\, \mathbf{q},\, t\rangle' = c_0\,|0\rangle + c_n e^{-i\omega_n t}\,|n,\, \mathbf{q}\rangle, \tag{4.90}$$

which mixes the ground state $|0\rangle$ and the SW state $|n,\, \mathbf{q}\rangle$. Hence, the expectation value of the spin at site r evolves as

$$\begin{aligned}{}'\langle n,\, \mathbf{q},\, t|\, S_{r\alpha}\,|n,\, \mathbf{q},\, t\rangle' = &|c_0|^2\langle 0|\, S_{r\alpha}\,|0\rangle + |c_n|^2\langle n,\, \mathbf{q}|\, S_{r\alpha}\,|n,\, \mathbf{q}\rangle \\ &+ c_0^* c_n e^{-i\omega_n t}\langle 0|\, S_{r\alpha}\,|n,\, \mathbf{q}\rangle + c_0 c_n^* e^{i\omega_n t}\langle n,\, \mathbf{q}|S_{r\alpha}|0\rangle.\end{aligned} \tag{4.91}$$

Since the first two terms are stationary, the oscillating spin distortion produced by

$$\Delta\mathbf{S}_r^{(n)}(\mathbf{q}, t) = 2\sqrt{N}\ \mathrm{Re}\{\mathrm{e}^{-i\omega_n t}\delta\mathbf{S}_r(n, \mathbf{q})\}, \tag{4.92}$$

$$\delta S_{r\alpha}(n, \mathbf{q}) = \langle 0|S_{r\alpha}|n, \mathbf{q}\rangle = \sqrt{\frac{S}{2N_u}}\,\mathrm{e}^{i\mathbf{q}\cdot\mathbf{R}_r}W_{r\alpha}^{(n)}(\mathbf{q}), \tag{4.93}$$

which ignores a constant prefactor and incorporates its phase into $W_{r\alpha}^{(n)}(\mathbf{q})$. Both the SW frequency $\omega_n(\mathbf{q})$ and the spin disturbance $\delta\mathbf{S}_r(n, \mathbf{q})$ at site r are the same at $\mathbf{q} = 0$ and $\mathbf{q} = \mathbf{Q}$. Because $W_{r\alpha}^{(n)}(\mathbf{q})$ depends on position r, the spatial (r) dependence of $\Delta S_{r\alpha}^{(n)}(\mathbf{q}, t)$ is different than $\exp(i\mathbf{q}\cdot\mathbf{R}_r)$. In section 6.4.3, we shall use equation (4.93) to find the SW amplitudes associated with the modes of a cycloid produced by the DM interaction.

From equations (4.42) and (4.78), the INS intensity $S_{\alpha\beta}^{(n)}(\mathbf{Q})$ for mode n at $\mathbf{Q}$ is proportional to

$$\sum_{r,s=1}^{M} \mathrm{e}^{-i\mathbf{Q}\cdot(\mathbf{R}_r-\mathbf{R}_s)}\delta S_{r\alpha}(n, \mathbf{Q})\ \delta S_{s\beta}(n, \mathbf{Q})^* \tag{4.94}$$

while the magnetic contribution to the optical absorption for mode n is proportional to

$$\omega_n \left|\sum_{r=1}^{M} \mathbf{h}\cdot\delta\mathbf{S}_r(n, \mathbf{q} = 0)\right|^2. \tag{4.95}$$

As mentioned above, $\delta\mathbf{S}_r(n, \mathbf{q} = 0) = \delta\mathbf{S}_r(n, \mathbf{Q})$.

4.6 General considerations

The most time-consuming task in evaluating the SW spectrum of any material is diagonalizing $\underline{\mathcal{L}}$. For systems with large magnetic unit cells, such as cycloidal materials like $BiFeO_3$, the $2M \times 2M$ matrix can easily be $10^3 \times 10^3$ in size. Because the matrix must be diagonalized separately for every momentum $\mathbf{q}$, those tasks can fortunately be run in parallel. Matrix diagonalization can be expedited by transforming $\underline{\mathcal{L}}$ into a Hermitian matrix: numerical diagonalization techniques for Hermitian matrices are about 15% faster than for non-Hermitian matrices. That method is discussed in appendix A.

How can you be sure that the SW frequencies and intensities are correct? Although there is no way to be certain, there are definite indications that things have gone horribly wrong. Choosing the wrong spin state will produce imaginary SW solutions when the anisotropy is turned off. Choosing the wrong sign for the exchange or anisotropy interactions can also produce imaginary roots because the spin state no longer minimizes the energy.

There are also a few ways to check your solution for the spin dynamics. If the spin state is rotationally invariant about some axis, then a Goldstone mode with zero frequency must be associated with that rotation. If spin fluctuations are the same in the x and y directions, then $S_{xx}(\mathbf{q})$ and $S_{yy}(\mathbf{q})$ should be equal. After solving for the $\underline{X}$ matrix, check that the normalization and eigenvector conditions are satisfied.

Probably the best way to check your results for the spin state is to evaluate the first-order term $\mathcal{H}_1$ in the $1/S$ expansion of the Hamiltonian. As mentioned above, $\mathcal{H}_1$ involves the forces transverse to each spin, which must vanish in equilibrium. Each term in $\mathcal{H}_1$ is proportional to a single boson operator $a_{\mathbf{q}}^{(r)}$ or $a_{\mathbf{q}}^{(r)\dagger}$. Unless the Hamiltonian can create SWs out of thin air (highly unlikely), the terms multiplying every boson operator must vanish.

Checking that $\mathcal{H}_1$ vanishes is equivalent to checking that the classical force acting on each spin is parallel to that spin. This can be verified by minimizing the classical energy for the spin at site i in the mean field produced by all the other spins, by the anisotropies, and by the magnetic field. As described in appendix 4.B, the resulting orientation for $\mathbf{S}_i$ should agree with the one assumed in the spin state!

Several shortcuts for coding the exchange, anisotropy, and magnetic-field interactions are described in appendix 4.C. Finally, appendix 4.D briefly reviews the (H, K, L) wavevector notation for orthogonal and hexagonal systems.

4.7 Appendix 4.A: Symmetry and matrices

This appendix goes into a bit more detail about the $2M \times 2M$ $\underline{L}(\mathbf{q})$ matrix, which can be generally written as

$$\underline{L}(\mathbf{q}) = \begin{pmatrix} \underline{P}(\mathbf{q}) & \underline{Q}(\mathbf{q}) \\ \underline{Q}'(\mathbf{q}) & \underline{P}'(\mathbf{q}) \end{pmatrix}, \tag{4.96}$$

where $\underline{P}(\mathbf{q})$, $\underline{Q}(\mathbf{q})$, $\underline{P}'(\mathbf{q})$, and $\underline{Q}'(\mathbf{q})$ are $M \times M$ matrices. Because $\underline{L}(\mathbf{q})$ is Hermitian, it follows that

$$\underline{P}'(\mathbf{q}) = \underline{P}(-\mathbf{q})^*, \tag{4.97}$$

$$\underline{Q}'(\mathbf{q}) = \underline{Q}(-\mathbf{q})^*. \tag{4.98}$$

We prove the second relation and leave the first to the reader (exercise 4.17). By rearranging terms, notice that

$$\begin{aligned} \sum_{\mathbf{q}}{}' a_{\mathbf{q}}^{(r)\dagger} L_{r,\, s+M}(\mathbf{q}) a_{-\mathbf{q}}^{(s)\dagger} &= \sum_{\mathbf{q}}{}' a_{-\mathbf{q}}^{(r)\dagger} L_{r,\, s+M}(-\mathbf{q}) a_{\mathbf{q}}^{(s)\dagger} \\ &= \sum_{\mathbf{q}}{}' a_{\mathbf{q}}^{(s)\dagger} L_{s,\, r+M}(\mathbf{q}) a_{-\mathbf{q}}^{(r)\dagger}. \end{aligned} \tag{4.99}$$

Hence,

$$L_{s,\, r+M}(\mathbf{q}) = L_{r,\, s+M}(-\mathbf{q}) = L_{s+M,\, r}(-\mathbf{q})^*, \tag{4.100}$$

where the last equality follows from the Hermiticity of $\underline{L}(\mathbf{q})$. So equation (4.98) has been proven.

You might conclude from equations (4.97) and (4.98) that $\underline{P}'(\mathbf{q}) = \underline{P}(\mathbf{q})$ and $\underline{Q}'(\mathbf{q}) = \underline{Q}(\mathbf{q})$, since the complex conjugate just changes $\mathbf{q} \rightarrow -\mathbf{q}$. But that is only right under certain circumstances, such as when all the spins are confined to the xz plane with $\phi_n = 0$. Otherwise, the matrix elements of $\underline{L}$ may involve nontrivial imaginary parts produced by the transformation into the local reference frame of each spin. We will see this explicitly in appendix 4.C.

We now briefly review the Cholesky decomposition method, which transforms the problem of evaluating the eigenvectors and eigenvalues of the non-Hermitian matrix $\underline{\mathcal{L}} = \underline{L} \cdot \underline{N}$ into a Hermitian problem. For a mathematical proof of this method, see [19].

First apply the Cholesky decomposition to find an upper triangular matrix $\underline{K}$ ($K_{ij} = 0$ for $j < i$) that satisfies

$$\underline{L} = \underline{K}^\dagger \cdot \underline{K}, \tag{4.101}$$

which implies that $\underline{L}$ is Hermitian. The Cholesky decomposition also requires that $\underline{L}$ is positive definite. Because zero-frequency Goldstone modes can mess this up, you might want to add some very small positive numbers to the diagonal matrix elements of $\underline{L}$. Then solve for the eigenvalues and eigenvectors of

$$\underline{C} = \underline{K} \cdot \underline{N} \cdot \underline{K}^\dagger. \tag{4.102}$$

From the eigenvectors of $\underline{C}$, construct the matrix $\underline{Y}$ that diagonalizes $\underline{C}$ such that

$$\underline{\mathcal{L}}' = \underline{Y}^\dagger \cdot \underline{C} \cdot \underline{Y} \tag{4.103}$$

is diagonal with the first M eigenvalues $\omega_n/2$ positive and the last M eigenvalues $-\omega_n/2$ negative. As above, the diagonal matrix $\underline{L}' = \underline{\mathcal{L}}' \cdot \underline{N}$ contains only positive eigenvalues, the first M of which are $\omega_n/2$. Finally, evaluate $\underline{K}^{-1}$ to construct the matrix

$$\underline{X}^{-1} = \underline{K}^{-1} \cdot \underline{Y} \cdot \underline{L}'^{\,1/2}, \tag{4.104}$$

which completes the procedure. Because constructing $\underline{K}^{-1}$ is rather time consuming, this method only saves about 15% in time compared to diagonalizing the non-Hermitian matrix $\underline{\mathcal{L}}$.

4.8 Appendix 4.B: Classical check

There are two ways to check that the assumed spin state is locally stable. First, the SW frequencies should all be real. Second, each spin state should lie at a minimum of the classical energy. The first method is not foolproof because anisotropy or a magnetic field may lift the SW frequencies even when the spin state is unstable. So we turn to the second method.

Assume that all spins outside of site i are fixed in the assumed spin state. For any set of interactions, the energy at site i can be written in the form

$$E_i = -\mathbf{A}_i \cdot \mathbf{S}_i - K_i\, S_{iz}^2. \tag{4.105}$$

Exchange, DM interactions, and magnetic fields all contribute to the classical field $\mathbf{A}_i$. When rotated onto the local z spin axis, single-ion anisotropy contributes to K_i. We are interested in how the local spin

$$\mathbf{S}_i = S\big(\cos\phi_i \sin\theta_i,\ \sin\phi_i \sin\theta_i,\ \cos\theta_i\big) \tag{4.106}$$

at site i responds to the effective field $\mathbf{A}_i$ and local anisotropy K_i.

Minimizing $E_i(\phi_i, \theta_i)$ with respect to ϕ_i and θ_i gives

$$\frac{\partial E_i}{\partial \phi_i} = -S \sin\theta_i\big(A_{ix} \cos\phi_i - A_{iy} \sin\phi_i\big) = 0, \tag{4.107}$$

$$\begin{aligned}\frac{\partial E_i}{\partial \theta_i} = &- S \cos\theta_i\big(A_{ix} \cos\phi_i + A_{iy} \sin\phi_i\big) \\ &+ SA_{iz} \sin\theta_i + K_i S^2 \sin 2\theta_i = 0.\end{aligned} \tag{4.108}$$

Inserting the solution for ϕ_i from equation (4.107) (assuming that $\sin\theta_i \neq 0$) into equation (4.108) gives a single condition for θ_i .

Plugging the assumed values for $\{\phi_i, \theta_i\}$ into equations (4.107) and (4.108) should give zero. In other words,

$$\left\langle \frac{|\nabla_i E_i|^2}{E_i^2} \right\rangle = \frac{1}{M} \sum_{r=1}^{M} \frac{1}{E_r^2} \left\{ \left(\frac{\partial E_r}{\partial \theta_r}\right)^2 + \frac{1}{\sin^2\theta_r} \left(\frac{\partial E_r}{\partial \phi_r}\right)^2 \right\} \tag{4.109}$$

should be small.

Alternatively, the energy $E_i(\phi_i, \theta_i)$ can be converted into a function of only θ_i using equation (4.107) for ϕ_i and then numerically minimized as a function of θ_i . The calculated value for $\mathbf{S}_i'$ should be very close to the assumed value $\mathbf{S}_i$ at every site. Typically,

$$\left\langle \frac{\delta S_i^2}{S^2} \right\rangle = \frac{1}{MS^2} \sum_{r=1}^{M} |\, \mathbf{S}_r' - \mathbf{S}_r \,|^2 \tag{4.110}$$

should be less than about 10^{-6}.

While the first check is satisfied even if the spin maximizes the energy, the second check guarantees that the energy is at least a local minimum. These two checks are easy enough that they should be performed for any assumed spin state

4.9 Appendix 4.C: Shortcuts

This appendix provides some general guidelines and shortcuts to construct $\underline{L}(\mathbf{q})$. Some of these results were previously obtained by Haraldsen and Fishman [20].

The first rule is to always symmetrize terms in the Hamiltonian so they contribute equally to all four quadrants of $\underline{L}(\mathbf{q})$. For example, the SW Hamiltonian of an AF on a square lattice with $M = 2$ sublattices is

$$\mathcal{H}_2 = -4JS \sum_{\mathbf{q}}{}' \left\{ a_{\mathbf{q}}^{(1)\dagger} a_{\mathbf{q}}^{(1)} + a_{\mathbf{q}}^{(2)\dagger} a_{\mathbf{q}}^{(2)} - \Gamma_{\mathbf{q}} \left(a_{\mathbf{q}}^{(1)} a_{-\mathbf{q}}^{(2)} + a_{\mathbf{q}}^{(1)\dagger} a_{-\mathbf{q}}^{(2)\dagger} \right) \right\}, \tag{4.111}$$

$$\Gamma_{\mathbf{q}} = \frac{1}{2} \left\{ \cos(q_x a) + \cos(q_y a) \right\}. \tag{4.112}$$

Symmetrize $\mathcal{H}_2$ as

$$\begin{aligned} \mathcal{H}_2 = &- 2JS \sum_{\mathbf{q}}{}' \left\{ a_{\mathbf{q}}^{(1)\dagger} a_{\mathbf{q}}^{(1)} + a_{\mathbf{q}}^{(2)\dagger} a_{\mathbf{q}}^{(2)} - \Gamma_{\mathbf{q}} \left(a_{\mathbf{q}}^{(1)} a_{-\mathbf{q}}^{(2)} + a_{\mathbf{q}}^{(1)\dagger} a_{-\mathbf{q}}^{(2)\dagger} \right) \right\} \\ &- 2JS \sum_{\mathbf{q}}{}' \left\{ a_{-\mathbf{q}}^{(1)\dagger} a_{-\mathbf{q}}^{(1)} + a_{-\mathbf{q}}^{(2)\dagger} a_{-\mathbf{q}}^{(2)} - \Gamma_{\mathbf{q}} \left(a_{-\mathbf{q}}^{(1)} a_{\mathbf{q}}^{(2)} + a_{-\mathbf{q}}^{(1)\dagger} a_{\mathbf{q}}^{(2)\dagger} \right) \right\}, \end{aligned} \tag{4.113}$$

which uses $\Gamma_{\mathbf{q}} = \Gamma_{-\mathbf{q}}$. This is sometimes called the 'spread it around' rule. It is then clear that

$$L_{11} = L_{22} = L_{33} = L_{44} = -2JS, \tag{4.114}$$

$$L_{14} = L_{23} = L_{41} = L_{32} = 2JS\Gamma_{\mathbf{q}} \tag{4.115}$$

with factors of $2JS$ rather than $4JS$ (see exercise 4.21).

Do not worry about the constant terms of order ϑ (S^0) that come from rearranging the creation and annihilation operator in $\mathcal{H}_2$. Those terms contribute $1/S^2$ corrections to the energy $E_0 = \langle \mathcal{H} \rangle$ and are only important when studying quantum fluctuations.

Since exchange is the most common magnetic interaction, it pays to develop some shortcuts for evaluating the contributions of those terms to $\underline{L}(\mathbf{q})$. Exchange couplings come in two types: those that couple spins within a given sublattice and those that couple spins on different sublattices. Each set may contain several distinct exchange couplings. Exchange coupling u within sublattice r is denoted as $J_{rr}^{(u)}$. Between sublattices r and s, it is denoted as $J_{rs}^{(u)} = J_{sr}^{(u)}$.

For example, consider the two AF-coupled chains with $M = 2$ sublattices in figure 4.2. The series of sublattice 1 and 2 chains continues indefinitely in the $\pm y$ directions. In the proposed notation, exchange couplings within a given sublattice are $J_1 = J_{11}^{(1)}$,

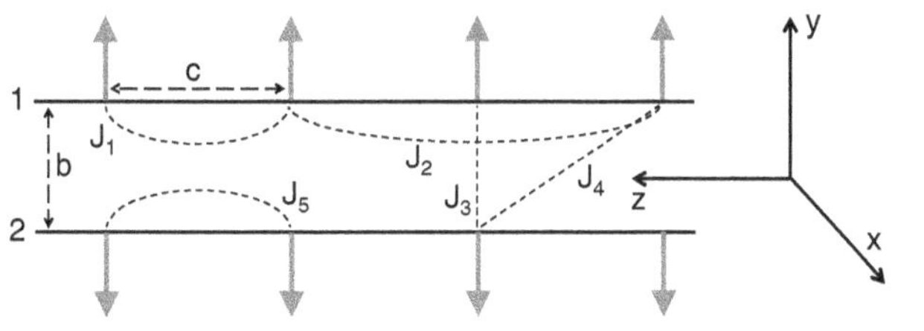

Figure 4.2. Coupled chains with exchange interactions indicated. Chains 1 and 2 repeat from $y = +\infty$ to $y = -\infty$.

$J_2 = J_{11}^{(2)}$, and $J_5 = J_{22}^{(1)}$. Interactions between spins on different sublattices are $J_3 = J_{12}^{(1)}$ and $J_4 = J_{12}^{(2)}$.

The number of neighbors associated with $J_{rs}^{(u)}$ is designated as $z_{rs}^{(u)} = z_{sr}^{(u)}$. For the example in figure 4.2, $z_{11}^{(1)} = z_{22}^{(1)} = 2$ since there are nearest neighbors on either side of each spin on each chain connected by $J_1 = J_{11}^{(1)}$ and $J_5 = J_{22}^{(1)}$, respectively. We also have $z_{11}^{(2)} = 2$ because there are two next-nearest neighbors for each spin within each chain connected by $J_2 = J_{11}^{(2)}$. For interactions that couple different sublattices, $z_{12}^{(1)} = 2$ because each spin on chain 1 couples 2 spins on either side in chain 2 through $J_3 = J_{12}^{(1)}$ and $z_{12}^{(2)} = 4$ because each spin on chain 1 couples 4 spins on on either side in chain 2 through $J_4 = J_{12}^{(2)}$ (which couples sites separated by $(0, b, \pm c)$ and $(0, -b, \pm c)$).

Exchange between spins on sublattices r and s involves the dot product

$$\begin{aligned}\mathbf{S}_r \cdot \mathbf{S}_s &= \sum_{\alpha,\beta,\gamma} (U_r^{-1})_{\alpha\beta}\bar{S}_{r\beta}(U_s^{-1})_{\alpha\gamma}\bar{S}_{s\gamma} = \sum_{\alpha,\beta,\gamma} \bar{S}_{r\beta}(U_r)_{\beta\alpha}(U_s^{-1})_{\alpha\gamma}\bar{S}_{s\gamma} \\ &= \bar{S}_r \cdot \underline{F}(r, s) \cdot \bar{S}_s,\end{aligned} \tag{4.116}$$

where

$$\underline{F}(r, s) = \underline{U}_r \cdot \underline{U}_s^{-1} \tag{4.117}$$

$$F_{\alpha\beta}(r, s) = \sum_{\gamma}(U_r)_{\alpha\gamma}(U_s^{-1})_{\gamma\beta} = \sum_{\gamma}(U_r)_{\alpha\gamma}(U_s)_{\beta\gamma}. \tag{4.118}$$

Note that $\underline{F}(r, r) = \underline{I}$, where $\underline{I}$ is the three-dimensional unit matrix.

For a system with only exchange interactions, $\underline{L}$ contains (exercise 4.22) diagonal components ($r = s$)

$$\begin{aligned}L_{rr} = L_{r+M,\, r+M} &= \frac{1}{2}\sum_u S_r z_{rr}^{(u)} J_{rr}^{(u)}\left\{1 - \Gamma_{rr}^{(u)}(\mathbf{q})\right\} \\ &+ \frac{1}{2}\sum_u \sum_{s\neq r} S_s z_{rs}^{(u)} J_{rs}^{(u)} F_{zz}(r, s)\end{aligned} \tag{4.119}$$

and off-diagonal components ($r \neq s$)

$$L_{rs} = -\frac{1}{4}\sum_u \sqrt{S_r S_s}\, z_{rs}^{(u)} J_{rs}^{(u)} \Gamma_{rs}^{(u)}(\mathbf{q})^* \, G_1(r, s), \tag{4.120}$$

$$L_{r+M,\, s+M} = -\frac{1}{4}\sum_u \sqrt{S_r S_s}\, z_{rs}^{(u)} J_{rs}^{(u)} \Gamma_{rs}^{(u)}(\mathbf{q})^* \, G_1(r, s)^*, \tag{4.121}$$

$$L_{r,\, s+M} = -\frac{1}{4}\sum_u \sqrt{S_r S_s}\, z_{rs}^{(u)} J_{rs}^{(u)} \Gamma_{rs}^{(u)}(\mathbf{q})^* \, G_2(r, s)^*, \tag{4.122}$$

$$L_{r+M,\,s} = -\frac{1}{4}\sum_{u}\sqrt{S_r S_s}\,z_{rs}^{(u)}J_{rs}^{(u)}\Gamma_{rs}^{(u)}(\mathbf{q})^*\;G_2(r,\,s), \tag{4.123}$$

where we consider the general case of different spins $S_r \neq S_s$ on sublattices r and $s \neq r$. We have defined

$$\Gamma_{rs}^{(u)}(\mathbf{q}) = \frac{1}{z_{rs}^{(u)}}\sum_{\mathbf{d}_{rs}^{(u)}} e^{-i\mathbf{q}\cdot\mathbf{d}_{rs}^{(u)}}, \tag{4.124}$$

where $\mathbf{d}_{rs}^{(u)} = \mathbf{R}_s - \mathbf{R}_r$ is one of the $z_{rs}^{(u)}$ different vectors that couple sublattice r to sublattice s through $J_{rs}^{(u)}$. Note that

$$\Gamma_{rs}^{(u)}(\mathbf{q})^* = \Gamma_{rs}^{(u)}(-\mathbf{q}) = \Gamma_{sr}^{(u)}(\mathbf{q}) \tag{4.125}$$

and $\Gamma_{rs}^{(u)}(\mathbf{q}=0) = 1$. For the AF-coupled chains in figure 4.2, $\mathbf{d}_{11}^{(1)} = \mathbf{d}_{22}^{(1)} = \pm c\mathbf{z}$, $\mathbf{d}_{11}^{(2)} = \pm 2c\mathbf{z}$, $\mathbf{d}_{12}^{(1)} = \pm b\mathbf{y}$, and $\mathbf{d}_{12}^{(2)} = b\mathbf{y} \pm c\mathbf{z}$ or $-b\mathbf{y} \pm c\mathbf{z}$.

With spins on sublattice r given by

$$\mathbf{S}_r = S_r\big(\cos\phi_r\sin\theta_r,\,\sin\phi_r\sin\theta_r,\,\cos\theta_r\big), \tag{4.126}$$

we define the functions (exercise 4.24)

$$F_{zz}(r,\,s) = \sin\theta_r\sin\theta_s\cos(\phi_r-\phi_s) + \cos\theta_r\cos\theta_s, \tag{4.127}$$

$$\begin{aligned} G_1(r,\,s) &= F_{xx}(r,\,s) + F_{yy}(r,\,s) - i\big\{F_{xy}(r,\,s) - F_{yx}(r,\,s)\big\} \\ &= (\cos\theta_r\cos\theta_s + 1)\cos(\phi_r-\phi_s) + \sin\theta_r\sin\theta_s \\ &\quad - i\sin(\phi_r-\phi_s)\{\cos\theta_r + \cos\theta_s\}, \end{aligned} \tag{4.128}$$

$$\begin{aligned} G_2(r,\,s) &= F_{xx}(r,\,s) - F_{yy}(r,\,s) - i\big\{F_{xy}(r,\,s) + F_{yx}(r,\,s)\big\} \\ &= (\cos\theta_r\cos\theta_s - 1)\cos(\phi_r-\phi_s) + \sin\theta_r\sin\theta_s \\ &\quad - i\sin(\phi_r-\phi_s)\{\cos\theta_r - \cos\theta_s\}. \end{aligned} \tag{4.129}$$

It is easy to see that $F_{zz}(r,\,s) = \langle\mathbf{S}_r\rangle\cdot\langle\mathbf{S}_s\rangle/S_sS_r$ is just the cosine of the angle between classical spins $\mathbf{S}_r$ and $\mathbf{S}_s$. Exercise 4.25 shows how these relations change with the choice of more general unitary matrices $\underline{U}_r$ and $\underline{U}_s$. Regardless of the choice of unitary matrix, these functions obey the symmetry relations

$$F_{zz}(r,\,s) = F_{zz}(s,\,r) = F_{zz}(r,\,s)^*, \tag{4.130}$$

$$G_1(r,\,s) = G_1(s,\,r)^*, \tag{4.131}$$

$$G_2(r,\,s) = G_2(s,\,r). \tag{4.132}$$

Hence, both $G_1(r,\,s)$ and $G_2(r,\,s)$ are complex while $F_{zz}(r,\,s)$ is real. For $r = s$, $\underline{F}(r,\,r) = \underline{I}$, $G_1(r,\,r) = 2$, and $G_2(r,\,r) = 0$.

Now apply these expressions to a simple AF on a square lattice with spin S. Since $\phi_1 = \phi_2 = 0$, $\theta_1 = 0$ and $\theta_2 = \pi$, we find $F_{zz}(1, 2) = F_{zz}(2, 1) = -1$, $G_1(1,2) = 0$, and $G_2(1,2) = -2$. Because $z_{12} = 4$ and $\mathbf{d}_{12} = \pm a\mathbf{x}$ and $\pm a\mathbf{y}$,

$$L_{11} = L_{22} = L_{33} = L_{44} = -2JS, \tag{4.133}$$

$$L_{12} = L_{21} = L_{34} = L_{43} = 0, \tag{4.134}$$

$$L_{14} = L_{23} = L_{41} = L_{32} = 2JS\Gamma_{\mathbf{q}}, \tag{4.135}$$

which uses $\Gamma_{12}(\mathbf{q}) = \Gamma_{\mathbf{q}}$. These are the same relations we will obtain in section 5.5.

The second-most common term found in any Hamiltonian is anisotropy. For single-ion (meaning it only involves single sites) anisotropy $-K\sum_i(\mathbf{m} \cdot \mathbf{S}_r)^2$ along the unit vector

$$\mathbf{m} = (\sin\eta\cos\delta,\ \sin\eta\sin\delta,\ \cos\eta), \tag{4.136}$$

we find

$$\mathbf{m} \cdot \mathbf{S}_r = A_x\bar{S}_{rx} + A_y\bar{S}_{ry} + A_z\bar{S}_{rz}, \tag{4.137}$$

$$A_x = \sin\eta\cos\theta_r\cos(\delta - \phi_r) - \cos\eta\sin\theta_r, \tag{4.138}$$

$$A_y = \sin\eta\sin(\delta - \phi_r), \tag{4.139}$$

$$A_z = \sin\eta\sin\theta_r\cos(\delta - \phi_r) + \cos\eta\cos\theta_r. \tag{4.140}$$

These terms change $\underline{L}$ by (exercise 4.26)

$$L_{rr} \to L_{rr} - \frac{KS_r}{2}\left(A_x^2 + A_y^2 - 2A_z^2\right), \tag{4.141}$$

$$L_{r+M,\,r+M} \to L_{r+M,\,r+M} - \frac{KS_r}{2}\left(A_x^2 + A_y^2 - 2A_z^2\right), \tag{4.142}$$

$$L_{r,\,r+M} \to L_{r,\,r+M} - \frac{KS_r}{2}\left(A_x + iA_y\right)^2, \tag{4.143}$$

$$L_{r+M,\,r} \to L_{r+M,\,r} - \frac{KS_r}{2}\left(A_x - iA_y\right)^2. \tag{4.144}$$

When $\mathbf{m} = \mathbf{z}$, $A_x = -\sin\theta_r$, $A_y = 0$, and $A_z = \cos\theta_r$.

A magnetic field $\mathbf{B} = B\mathbf{m}$ with energy $-2\mu_B B\sum_i \mathbf{m} \cdot \mathbf{S}_r$ changes $\underline{L}$ by

$$L_{rr} \to L_{rr} + \mu_B B A_z, \tag{4.145}$$

$$L_{r+M,\,r+M} \to L_{r+M,\,r+M} + \mu_B B A_z. \tag{4.146}$$

As above, $A_z = \cos\theta_r$ when $\mathbf{m} = \mathbf{z}$.These expressions are readily generalized when the g-tensor is anisotropic.

Finally, consider a DM interaction coupling spins on sites i and j by

$$\frac{1}{2}\sum_{i,j}\mathbf{D}_{ij}\cdot(\mathbf{S}_i \times \mathbf{S}_j), \tag{4.147}$$

where the 1/2 avoids double counting and $\mathbf{D}_{ij} = -\mathbf{D}_{ji}$ is antisymmetric. Used to model the canting of an AF chain in section 6.2, this interaction changes $\underline{L}$ by (exercise 4.27)

$$L_{rr} \to L_{rr} - \sum_{s\neq r}\frac{S_s}{2}\sum_{\alpha,\beta,\gamma}\sum_{u} z_{rs}^{(u)}D_{rs\alpha}^{(u)}\,\epsilon_{\alpha\beta\gamma}(U_r^{-1})_{\beta z}(U_s^{-1})_{\gamma z}, \tag{4.148}$$

$$L_{r+M,\,r+M} \to L_{r+M,\,r+M} - \sum_{s\neq r}\frac{S_s}{2}\sum_{\alpha,\beta,\gamma}\sum_{u} z_{rs}^{(u)}D_{rs\alpha}^{(u)}\,\epsilon_{\alpha\beta\gamma}(U_r^{-1})_{\beta z}(U_s^{-1})_{\gamma z}, \tag{4.149}$$

and for $r \neq s$,

$$L_{rs} \to L_{rs} + \frac{\sqrt{S_rS_s}}{4}\sum_{\alpha,\beta,\gamma}\sum_{u} z_{rs}^{(u)}D_{rs\alpha}^{(u)}\,\epsilon_{\alpha\beta\gamma}V_{r\beta}^{+}V_{s\gamma}^{-}\,e^{i\mathbf{d}_{rs}\cdot\mathbf{q}}, \tag{4.150}$$

$$L_{r+M,\,s+M} \to L_{r+M,\,s+M} + \frac{\sqrt{S_rS_s}}{4}\sum_{\alpha,\beta,\gamma}\sum_{u} z_{rs}^{(u)}D_{rs\alpha}^{(u)}\,\epsilon_{\alpha\beta\gamma}V_{r\beta}^{-}V_{s\gamma}^{+}\,e^{i\mathbf{d}_{rs}\cdot\mathbf{q}}, \tag{4.151}$$

$$L_{r,\,s+M} \to L_{r,\,s+M} + \frac{\sqrt{S_rS_s}}{4}\sum_{\alpha,\beta,\gamma}\sum_{u} z_{rs}^{(u)}D_{rs\alpha}^{(u)}\,\epsilon_{\alpha\beta\gamma}V_{r\beta}^{+}V_{s\gamma}^{+}\,e^{i\mathbf{d}_{rs}\cdot\mathbf{q}}, \tag{4.152}$$

$$L_{r+M,\,s} \to L_{r+M,\,s} + \frac{\sqrt{S_rS_s}}{4}\sum_{\alpha,\beta,\gamma}\sum_{u} z_{rs}^{(u)}D_{rs\alpha}^{(u)}\,\epsilon_{\alpha\beta\gamma}V_{r\beta}^{-}V_{s\gamma}^{-}\,e^{i\mathbf{d}_{rs}\cdot\mathbf{q}}, \tag{4.153}$$

where $\epsilon_{\alpha\beta\gamma}$ is the antisymmetric tensor.

Exercise 4.28 shows that all shortcuts meet the requirement that $\underline{L}$ is Hermitian.

4.10 Appendix 4.D: Orthogonal and hexagonal notations

We briefly review the orthogonal (H, K, L) and hexagonal $(H, K, L)_h$ wavevector notations commonly used to present INS results. For an orthogonal system with structural-lattice vectors

$$\mathbf{a}_1 = a(1, 0, 0) \tag{4.154}$$

$$\mathbf{a}_2 = b(0, 1, 0) \tag{4.155}$$

$$\mathbf{a}_3 = c(0, 0, 1), \tag{4.156}$$

the corresponding reciprocal-lattice vectors are

$$\mathbf{b}_1 = \frac{2\pi}{a}(1, 0, 0), \tag{4.157}$$

$$\mathbf{b}_2 = \frac{2\pi}{b}(0, 1, 0), \tag{4.158}$$

$$\mathbf{b}_3 = \frac{2\pi}{c}(0, 0, 1). \tag{4.159}$$

As required, these two sets of vectors satisfy $\mathbf{a}_i \cdot \mathbf{b}_j = 2\pi\delta_{ij}$. A wavevector given in this basis by $\mathbf{q} = (H, K, L)$ can be expanded as

$$\mathbf{q} = H\mathbf{b}_1 + K\mathbf{b}_2 + L\mathbf{b}_3 = 2\pi\left(\frac{H}{a}\,\mathbf{x} + \frac{K}{b}\,\mathbf{y} + \frac{L}{c}\,\mathbf{z}\right). \tag{4.160}$$

For a cubic system,

$$\mathbf{q} = \frac{2\pi}{a}(H\,\mathbf{x} + K\,\mathbf{y} + L\,\mathbf{z}). \tag{4.161}$$

In either case, the wavevector (H, K, L) is said to be in reciprocal-lattice units (rlu).

Hexagonal notation has been the source of endless confusion to both INS experimentalists and analysts, present company included. Consider a system of hexagonal planes separated by $c\mathbf{z}$, where c can differ from the triangular lattice constant a. The structural-lattice vectors are then

$$\mathbf{a}_1 = \frac{a}{2}(1, \sqrt{3}, 0), \tag{4.162}$$

$$\mathbf{a}_2 = \frac{a}{2}(1, -\sqrt{3}, 0), \tag{4.163}$$

$$\mathbf{a}_3 = c(0, 0, 1). \tag{4.164}$$

Correspondingly, the reciprocal-lattice vectors are

$$\mathbf{b}_1 = \frac{2\pi}{3a}(3, \sqrt{3}, 0), \tag{4.165}$$

$$\mathbf{b}_2 = \frac{2\pi}{3a}(3, -\sqrt{3}, 0), \tag{4.166}$$

$$\mathbf{b}_3 = \frac{2\pi}{c}(0, 0, 1). \tag{4.167}$$

These two sets of vectors also satisfy $\mathbf{a}_i \cdot \mathbf{b}_j = 2\pi\delta_{ij}$. In the hexagonal basis, $\mathbf{q} = (H, K, L)_h$ is expanded as

$$\mathbf{q} = H\mathbf{b}_1 + K\mathbf{b}_2 + L\mathbf{b}_3. \tag{4.168}$$

If a wavevector is given in a tetragonal ($a = b$) basis as

$$\mathbf{q} = (A, B, C) = 2\pi\left(\frac{A}{a}\mathbf{x} + \frac{B}{a}\mathbf{y} + \frac{C}{c}\mathbf{z}\right), \tag{4.169}$$

then (exercise 4.29)

$$H = \frac{1}{2}(A + \sqrt{3}B), \tag{4.170}$$

$$K = \frac{1}{2}(A - \sqrt{3}B), \tag{4.171}$$

$$L = C. \tag{4.172}$$

This correspondence will come in handy in section 5.4, when we study a FM on a honeycomb lattice.

4.11 Appendix 4.E: Spin susceptibility

We now establish the fluctuation–dissipation theorem [6], which relates the spin–spin correlation function $S_{\alpha\beta}(\mathbf{q},\omega)$ to the spin susceptibility $\chi^{\mathrm{ss}}_{\alpha\beta}(\mathbf{q},\omega)$. This derivation is quite general and is not based on SW theory. Start with the linear-response expression

$$\begin{aligned}\chi^{\mathrm{ss}}_{\alpha\gamma}(\mathbf{q},\omega) &= \frac{i}{\pi\hbar N}\sum_{i,j}\int_{-\infty}^{\infty} dt\ \theta(-t)e^{-i\omega t}e^{-i\mathbf{q}\cdot(\mathbf{R}_i-\mathbf{R}_j)}\langle[S_{i\alpha}(0), S_{j\gamma}(t)]\rangle \\ &= \frac{i}{\pi\hbar N}\sum_{i,j}\int_{-\infty}^{0} dt\ e^{-i(\omega+i\varepsilon)t}e^{-i\mathbf{q}\cdot(\mathbf{R}_i-\mathbf{R}_j)}\langle[S_{i\alpha}(0), S_{j\gamma}(t)]\rangle,\end{aligned} \tag{4.173}$$

which provides the response of the α component of the spin to the γ component of a time- and space-dependent magnetic field with wavevector $\mathbf{q}$ and frequency ω. It satisfies the symmetry relation $\chi^{\mathrm{ss}}_{\alpha\gamma}(\mathbf{q},\omega) = \chi^{\mathrm{ss}}_{\alpha\gamma}(-\mathbf{q},-\omega)^*$ (exercise 4.30).

Using the cyclic property of the trace, $\mathrm{Tr}(AB) = \mathrm{Tr}(BA)$, we find that

$$\begin{aligned}
&\sum_{i,j} \int_{-\infty}^{\infty} dt\, e^{-i\omega t} e^{-i\mathbf{q}\cdot(\mathbf{R}_i-\mathbf{R}_j)} \langle S_{j\gamma}(t) S_{i\alpha}(0)\rangle \\
&= \frac{1}{Z} \sum_{i,j} \int_{-\infty}^{\infty} dt\, e^{-i\omega t} e^{-i\mathbf{q}\cdot(\mathbf{R}_i-\mathbf{R}_j)}\, \mathrm{Tr}\left\{ e^{-\beta\mathcal{H}} e^{i\mathcal{H}t/\hbar}\, S_{j\gamma} e^{-i\mathcal{H}t/\hbar} S_{i\alpha} \right\} \\
&= \frac{1}{Z} \sum_{i,j} \int_{-\infty}^{\infty} dt\, e^{-i\omega t} e^{-i\mathbf{q}\cdot(\mathbf{R}_i-\mathbf{R}_j)}\, \mathrm{Tr}\left\{ e^{-\beta\mathcal{H}} S_{i\alpha} e^{i\mathcal{H}(t+i\hbar\beta)/\hbar} S_{j\gamma} e^{-i\mathcal{H}(t+i\hbar\beta)/\hbar} \right\} \\
&= \frac{e^{-\hbar\beta\omega}}{Z} \sum_{i,j} \int_{-\infty}^{\infty} dt\, e^{-i\omega t} e^{-i\mathbf{q}\cdot(\mathbf{R}_i-\mathbf{R}_j)}\, \mathrm{Tr}\left\{ e^{-\beta\mathcal{H}} S_{i\alpha}\, e^{i\mathcal{H}t/\hbar} S_{j\gamma} e^{-i\mathcal{H}t/\hbar} \right\} \\
&= e^{-\hbar\beta\omega} \sum_{i,j} \int_{-\infty}^{\infty} dt\, e^{-i\omega t} e^{-i\mathbf{q}\cdot(\mathbf{R}_i-\mathbf{R}_j)} \langle S_{i\alpha}(0) S_{j\gamma}(t)\rangle .
\end{aligned} \tag{4.174}$$

Therefore,

$$\begin{aligned}
F_{\alpha\gamma}(\mathbf{q},\omega) &\equiv \frac{i}{\pi\hbar N} \sum_{i,j} \int_{-\infty}^{\infty} dt\, e^{-i\omega t} e^{-i\mathbf{q}\cdot(\mathbf{R}_i-\mathbf{R}_j)} \langle [S_{i\alpha}(0),\, S_{j\gamma}(t)]\rangle \\
&= \frac{i}{\pi\hbar N}(1 - e^{-\hbar\beta\omega}) \sum_{i,j} \int_{-\infty}^{\infty} dt\, e^{-i\omega t} e^{-i\mathbf{q}\cdot(\mathbf{R}_i-\mathbf{R}_j)} \langle S_{i\alpha}(0) S_{j\gamma}(t)\rangle \\
&= \frac{2i}{\hbar}(1 - e^{-\hbar\beta\omega}) S_{\alpha\gamma}(\mathbf{q},\omega).
\end{aligned} \tag{4.175}$$

With the symmetrized spin–spin correlation function defined as

$$S_{\alpha\gamma}(\mathbf{q},\omega)_s = \frac{1}{2\pi N} \sum_{i,j} \int_{-\infty}^{\infty} dt\, e^{-i\omega t} e^{-i\mathbf{q}\cdot(\mathbf{R}_i-\mathbf{R}_j)} \langle S_{i\alpha}(0) S_{j\gamma}(t) + S_{j\gamma}(t) S_{i\alpha}(0)\rangle , \tag{4.176}$$

we obtain the familiar relationships [6]

$$S_{\alpha\gamma}(\mathbf{q},\omega) = \frac{\hbar}{2i}[n_{\mathrm{B}}(\omega) + 1] F_{\alpha\gamma}(\mathbf{q},\omega), \tag{4.177}$$

$$S_{\alpha\gamma}(\mathbf{q},\omega)_s = \frac{\hbar}{2i} \coth(\hbar\beta\omega/2)\, F_{\alpha\gamma}(\mathbf{q},\omega). \tag{4.178}$$

To connect $F_{\alpha\gamma}(\mathbf{q},\omega)$ and $\chi^{\mathrm{ss}}_{\alpha\gamma}(\mathbf{q},\omega)$, use

$$\begin{aligned}
\int_{-\infty}^{\infty} dt\, e^{-i\omega t} \langle [S_{i\alpha}(0),\, S_{j\gamma}(t)]\rangle &= \int_{0}^{\infty} dt\, e^{-i\omega t} \langle [S_{i\alpha}(0),\, S_{j\gamma}(t)]\rangle \\
&\quad + \int_{-\infty}^{0} dt\, e^{-i\omega t} \langle [S_{i\alpha}(0),\, S_{j\gamma}(t)]\rangle \\
&= \int_{-\infty}^{0} dt\, e^{i\omega t}\, \langle [S_{i\alpha}(0),\, S_{j\gamma}(-t)]\rangle \\
&\quad + \int_{-\infty}^{0} dt\, e^{-i\omega t} \langle [S_{i\alpha}(0),\, S_{j\gamma}(t)]\rangle .
\end{aligned} \tag{4.179}$$

As demonstrated in exercise 4.31,

$$F_{\alpha\gamma}(\mathbf{q},\omega) = \chi^{\mathrm{ss}}_{\alpha\gamma}(\mathbf{q},\omega) - \chi^{\mathrm{ss}}_{\gamma\alpha}(\mathbf{q},\omega)^* \tag{4.180}$$

and

$$S_{\alpha\gamma}(\mathbf{q},\omega) = \frac{\hbar}{2i}[n_{\mathrm{B}}(\omega) + 1]\left\{\chi^{\mathrm{ss}}_{\alpha\gamma}(\mathbf{q},\omega) - \chi^{\mathrm{ss}}_{\gamma\alpha}(\mathbf{q},\omega)^*\right\}, \tag{4.181}$$

$$S_{\alpha\gamma}(\mathbf{q},\omega)_s = \frac{\hbar}{2i}\coth(\beta\omega/2)\left\{\chi^{\mathrm{ss}}_{\alpha\gamma}(\mathbf{q},\omega) - \chi^{\mathrm{ss}}_{\gamma\alpha}(\mathbf{q},\omega)^*\right\}. \tag{4.182}$$

Both expressions are statements of the fluctuation–dissipation theorem. Comparing equation (4.181) with equation (4.45), we see that $F_{\alpha\gamma}(\mathbf{q},\omega)$ does not depend on temperature within SW theory.

Generally, equations (4.181) and (4.182) relate complex quantities on each side. But for $\alpha = \gamma$, both sides are real and

$$S_{\alpha\alpha}(\mathbf{q},\omega) = \hbar[n_{\mathrm{B}}(\omega) + 1]\,\mathrm{Im}\,\chi^{\mathrm{ss}}_{\alpha\alpha}(\mathbf{q},\omega), \tag{4.183}$$

$$S_{\alpha\alpha}(\mathbf{q},\omega)_s = \hbar\coth(\hbar\beta\omega/2)\,\mathrm{Im}\,\chi^{\mathrm{ss}}_{\alpha\alpha}(\mathbf{q},\omega). \tag{4.184}$$

Also consider the sum

$$\begin{aligned}\sum_{\alpha,\gamma} h_\alpha h_\gamma\, S_{\alpha\gamma}(\mathbf{q},\omega) &= \frac{\hbar}{2i}[n_{\mathrm{B}}(\omega) + 1]\sum_{\alpha,\gamma} h_\alpha h_\gamma\left\{\chi^{\mathrm{ss}}_{\alpha\gamma}(\mathbf{q},\omega) - \chi^{\mathrm{ss}}_{\gamma\alpha}(\mathbf{q},\omega)^*\right\}\\ &= \hbar[n_{\mathrm{B}}(\omega) + 1]\sum_{\alpha,\gamma} h_\alpha h_\gamma\,\mathrm{Im}\,\chi^{\mathrm{ss}}_{\alpha\gamma}(\mathbf{q},\omega),\end{aligned} \tag{4.185}$$

where the left-hand side is real due to the symmetry relation $S_{\alpha\gamma}(\mathbf{q},\omega) = S_{\gamma\alpha}(\mathbf{q},\omega)^*$. This result was used in equation (4.78).

This method can also be used to connect the magnetic optical absorption

$$\alpha_{\mathrm{m}}(\omega) = \frac{\omega\mu_0\sqrt{\epsilon_b}}{2\hbar c V}\int dt\, e^{-i\omega t}\langle \mathbf{h}\cdot\mathbf{M}(0)\,\mathbf{h}\cdot\mathbf{M}(t)\rangle \tag{4.186}$$

to the optical magnetic susceptibility

$$\chi^{\mathrm{mm}}(\omega) = \frac{i\mu_0}{\hbar V}\int_{-\infty}^{\infty} dt\,\theta(-t)e^{-i\omega t}\langle[\mathbf{h}\cdot\mathbf{M}(0),\,\mathbf{h}\cdot\mathbf{M}(t)]\rangle, \tag{4.187}$$

with the general result

$$\alpha_{\mathrm{m}}(\omega) = \frac{\sqrt{\epsilon_{\mathrm{b}}}}{c}\,\omega[n_{\mathrm{B}}(\omega) + 1]\,\mathrm{Im}\,\chi^{\mathrm{mm}}(\omega) \tag{4.188}$$

that was again used in equation (4.78). Comparing equation (4.188) with equation (4.75), we see that within SW theory, $\mathrm{Im}\,\chi^{\mathrm{mm}}(\omega)$ is the sum over two delta functions at $\pm\omega_n(\mathbf{q} = 0)$ and does not depend on temperature.

4.12 Exercises

1. Verify the commutation relations of equation (4.12).
2. Prove equations (4.36) and (4.37) for the symmetry relations of $\underline{X}^{-1}(\mathbf{q})$.
3. Use the identities for $\underline{X}$ given by equations (4.36) and (4.37) to show that

$$W_{ra}^{(n)}(\mathbf{q}) = W_{ra}^{(n+M)}(-\mathbf{q})^*. \tag{4.189}$$

*4. Why do the $a_i^\dagger a_i$ and $a_j^\dagger a_j$ terms in equation (4.38) not contribute to equation (4.39) when $\omega > 0$?

*5. Use SW theory to prove equation (4.45) for the spin–spin correlation function at nonzero temperatures. Include the transition between a state with $m \geqslant 1$ SWs of frequency ω_n to one with $m \pm 1$ SWs of that frequency.

6. Show that $S_{\alpha\beta}(\mathbf{q},\omega)$ obeys the detailed balance condition of equation (4.47).
7. Derive the 'mother-of-all sum rules' in equation (4.53).
8. Show that optical absorption $\alpha^{\pm}(\omega)$ given by equation (4.65) has units of cm^{-1} so that $\alpha^{\pm}(\omega)/\omega$ is dimensionless.
9. Derive equation (4.72) for the magnetic-moment matrix element.

*10. Prove equation (4.75) for $\alpha_{\rm m}(\omega)$ at nonzero temperatures.

11. Derive equation (4.76) for the measured magnetic optical absorption.
12. Derive equation (4.79) relating the spin and optical susceptibilities. Hint: Once you relate the imaginary parts, the Kramers–Kronig relation [6] can be used to relate the real parts.

*13. Prove equations (4.82)–(4.84) for the matrix elements of $\mathbf{P}$.

14. Show that each matrix element of $\underline{X}$ and $\underline{X}^{-1}$ scales like $1/\sqrt{M}$. What does that imply about the scaling of $S_{\alpha\alpha}^{(n)}(\mathbf{q})$, $W_{ra}^{(n)}(\mathbf{q})$, $\Delta S_{ra}(n, \mathbf{q}, t)$, and $\langle n, \mathbf{q}|M_\alpha|0\rangle$ with N_u and M? How about the magnetic optical absorption strength $\alpha_{\rm m}(\omega)$?

*15. Derive equations (4.87)–(4.88) using $V_{ra}^{\pm}$ for the simple cycloid.

16. How can a system exhibit spin fluctuations at site r along $\mathbf{y}$ for mode n with $\mathbf{q} = \mathbf{Q}$ through equation (4.92) but also have $S_{yy}^{(n)}(\mathbf{Q}) = 0$?
17. Prove equation (4.94): $\underline{P}'(\mathbf{q}) = \underline{P}(-\mathbf{q})^*$.
18. Use the normalization condition, equation (4.34), for $\underline{X}$ and equation (4.104) to show that $\underline{Y}^\dagger \cdot \underline{Y} = \underline{I}$.
19. What is the relation between $\chi^{\rm me}(\omega)$ and $\chi^{\rm em}(\omega)$?
20. If polarization per unit volume has dimensions of $\mu\text{C cm}^{-2}$, what are the dimensions of the coupling constant λ?
21. What is the analogue of the detailed balance condition (see exercise 4.6) for the magnetic absorption in optical spectroscopy?
22. By explicitly evaluating the exchange contributions to $\mathcal{H}$, prove equations (4.119)–(4.123). Show that those expressions are consistent with equations (4.97)–(4.98).
23. Derive the classical field $\mathbf{A}_i$ in equation (4.105) for a DM interaction that couples neighboring spins in one dimension. Take the spins $\mathbf{S}_i = S(\sin\theta_i, 0, \cos\theta_i)$ to lie in the xz plane.

*24. Use the unitary matrix $\underline{U}_i$ given by equation (4.2) to derive equations (4.127)–(4.129).

*25. The most general unitary rotation matrix is

$$\underline{U} = \begin{pmatrix} \cos\theta\cos\phi\cos\alpha - \sin\phi\sin\alpha & \cos\theta\sin\phi\cos\alpha + \cos\phi\sin\alpha & -\sin\theta\cos\alpha \\ -\sin\phi\cos\alpha - \cos\theta\cos\phi\sin\alpha & \cos\phi\cos\alpha - \cos\theta\sin\phi\sin\alpha & \sin\theta\sin\alpha \\ \sin\theta\cos\phi & \sin\theta\sin\phi & \cos\theta \end{pmatrix}, \tag{4.190}$$

where the additional angle α rotates the local reference frame about $\bar{\mathbf{z}}$ at every site. First prove that $\underline{U}$ is unitary. Then use this rotation matrix to show that equation (4.127) for $F_{zz}(r, s)$ is unchanged but that

$$G_1(r, s) = \{\cos\theta_r\cos\theta_s\cos(\phi_r - \phi_s) + \sin\theta_r\sin\theta_s + \cos(\phi_r - \phi_s) - i\sin(\phi_r - \phi_s)[\cos\theta_r + \cos\theta_s]\}e^{i(\alpha_r-\alpha_s)}, \tag{4.191}$$

$$G_2(r, s) = \{\cos\theta_r\cos\theta_s\cos(\phi_r - \phi_s) + \sin\theta_r\sin\theta_s - \cos(\phi_r - \phi_s) - i\sin(\phi_r - \phi_s)[\cos\theta_r - \cos\theta_s]\}e^{i(\alpha_r+\alpha_s)}. \tag{4.192}$$

Maple or Mathematica might be helpful here! Are the symmetry relations for $G_1(r, s)$ and $G_2(r, s)$ given by equations (4.131) and (4.132) changed by these phase factors?

26. Prove equations (4.141)–(4.144) for the effect of easy-axis anisotropy along **m**. How do the changes in $\underline{L}(\mathbf{q})$ simplify when $\mathbf{m} = \mathbf{x}$? When $\mathbf{m} = \mathbf{y}$?

*27. Prove equations (4.148)–(4.153) for the effect of the DM interaction on $\underline{L}(\mathbf{q})$.

28. Show that the shortcuts in equations (4.119)–(4.123) for exchange satisfy the requirement that $\underline{L}(\mathbf{q})$ is Hermitian. Then show that this requirement is met for the anisotropy terms in equations (4.141)–(4.144) and the DM terms in equations (4.148)–(4.153).

29. Prove equations (4.165)–(4.167) for a hexagonal system. Then use those relations to prove equations (4.170)–(4.172).

30. Use equation (4.173) to show that the spin susceptibility satisfies the symmetry relation $\chi^{\mathrm{ss}}_{\alpha\beta}(\mathbf{q},\omega) = \chi^{\mathrm{ss}}_{\alpha\beta}(-\mathbf{q},-\omega)^*$.

31. Use equation (4.179) to prove equation (4.180) for $F_{\alpha\gamma}(\mathbf{q},\omega)$.

References

[1] Walker L R and Walstedt R E 1977 Computer model of metallic spin-glasses *Phys. Rev. Lett.* **38** 514–8

[2] Walker L R and Walstedt R E 1980 Computer model of metallic spin-glasses *Phys. Rev.* B **22** 3816–42

[3] Tóth S and Lake B 2015 Linear spin wave theory for single-Q incommensurate magnetic structures *J. Phys.: Condens. Matter* **27** 166002
[4] Holstein T and Primakoff H 1940 Field dependence of the intrinsic domain magnetization of a ferromagnet *Phys. Rev.* **58** 1098–13
[5] Kübler J, Hock K-H, Sticht J and Williams A R 1988 Density functional theory of non-collinear magnetism *J. Phys. F: Met. Phys.* **18** 469
[6] White R H 1970 *Quantum Theory of Magnetism* (New York: Springer)
[7] Mattis D C 1965 *Theory of Magnetism* (New York: Harper and Row)
[8] Balcar E and Lovesey S W 1989 *Theory of Magnetic Neutron and Photon Scattering* (Oxford: Clarendon)
[9] Hahn S E, Podlesnyak A A, Ehlers G, Granroth G E, Fishman R S, Kolesnikov A I, Pomjakushina E and Conder K 2014 Inelastic neutron scattering studies of $YFeO_3$ *Phys. Rev.* B **89** 014420
[10] Miyahara S and Furukawa N 2011 Theory of magnetoelectric resonance in two-dimensional $S = 3/2$ antiferromagnet $Ba_2CoGe_2O_7$ via spin-dependent metal-ligand hybridization mechanism *J. Phys. Soc. Jpn.* **80** 073708
[11] Miyahara S and Furukawa N 2014 Theory of magneto-optical effects in helical multiferroic materials via toroidal magnon excitation *Phys. Rev.* B **89** 195145
[12] Fishman R S, Lee J H, Bordács S, Kézsmárki I and Nagel U 2015 Spin-induced polarizations and nonreciprocal directional dichroism of the room-temperature multiferroic $BiFeO_3$ *Phys. Rev.* B **92** 094422
[13] Kézsmárki I *et al* 2014 One-way transparency of four-coloured spin-wave excitations in multiferroic materials *Nat. Commun.* **5** 3203
[14] Katsura H, Nagaosa N and Balatsky A V 2005 Spin current and magnetoelectric effect in noncollinear magnets *Phys. Rev. Lett.* **95** 057205
[15] Sergienko I A and Dagotto E 2006 Role of the Dzyaloshinskii–Moriya interaction in multiferroic perovskites *Phys. Rev.* B **73** 094434
[16] Arima T-H 2007 Ferroelectricity induced by proper-screw type magnetic order *J. Phys. Soc. Jpn.* **76** 073702
[17] Tokura Y, Seki S and Nagaosa N 2014 Multiferroics of spin origin *Rep. Prog. Phys.* **77** 076501
[18] Cohen-Tannoudji C, Diu B and Laloë F 2005 *Quantum Mechanics* (New York: Wiley)
[19] Colpa J H P 1978 Diagonalization of the quadratic boson Hamiltonian *Phys. A: Stat. Mech. Appl.* **93** 327–53
[20] Haraldsen J T and Fishman R S 2009 Spin rotation technique for non-collinear magnetic systems: application to the generalized Villain model *J. Phys.: Condens. Matter* **21** 216001

IOP Concise Physics

Spin-Wave Theory and its Applications to Neutron Scattering and THz Spectroscopy

Randy S Fishman, Jaime A Fernandez-Baca and Toomas Rõõm

Chapter 5

Model collinear magnets

It is not enough to calculate just the eigenvalues, the eigenvectors are also important! Try to imagine a pizza with no cheese!

—Unknown Italian mathematician

How you pluck the string is just as important as which string you pluck.—

'The Zen Guide to Guitar Mastery'

5.1 Introduction

This and the following chapter use SW theory to evaluate the inelastic spectra and optical absorption of some model systems. This chapter treats CL magnets; the next treats NC magnets.

A list of some materials with CL spin states is given in table 5.1. Many transition-metal FMs like Ni and Fe are itinerant rather than localized. Because they cannot be described by the Heisenberg model, we do not give a value for their localized spins S. Whenever possible, we have cited references containing INS or optical measurements. As you can see from the dates of those references, most CL magnets were discovered over 40 years ago! Due to strong spin–orbit coupling in rare-earth compounds including Gd and EuX, the localized 'spin' is actually J rather than S.

We will study five examples of CL spin states. Starting with a simple FM on a square lattice, we then treat a FM chain with alternating exchange interactions and a FM on the non-Bravais honeycomb lattice. The latter two FM examples demonstrate the importance of evaluating the spectral weight of the SW modes along with their frequencies. We then derive the inelastic spectra of an AF on a square lattice in

doi:10.1088/978-1-64327-114-9ch5

Table 5.1. Some materials with CL spin states.

Material	Magnetic ion	Spin	Type	References
Ni	Ni	—	FM	[1, 2]
Fe	Fe	—	FM	[3]
Gd	Gd	7/2	FM	[4]
MnSb	Mn^{3+}	2	FM	[5, 6]
CrX_3 (X = Cl#, Br*, I*)	Cr^{3+}	3/2	*FM, #AF	[7, 8]
EuX (X = O*, S*, Se#, Te#)	Eu^{2+}	7/2	*FM, #AF	[9–11]
MnF_2	Mn^{2+}	5/2	AF	[12]
MnO	Mn^{2+}	5/2	AF	[13]
MnX (X = O, S, Se)	Mn^{2+}	5/2	AF	[14, 15]
Fe_2O_3	Fe^{3+}	5/2	AF	[13, 16]
$FeCl_2$	Fe^{2+}	2	AF	[17, 18]
FeF_2	Fe^{2+}	2	AF	[19]

a magnetic field. Finally, we numerically evaluate the powder spectra of a FM and a G-type AF on a cubic lattice.

Unlike most NC magnets, the CL systems discussed in this chapter can be solved analytically. Except for the powder spectra, we provide analytic solutions for all cases studied here. Although any system with more than two sublattices will almost surely be solved numerically, it is useful to compare numerical and analytic solutions whenever the latter are available.

For any CL magnet with spins aligned along $\pm\mathbf{z}$, the SW intensity can be simplified when no interaction (such as easy-plane anisotropy) breaks the symmetry between the x and y directions. Then for mode n, $S_{xx}^{(n)}(\mathbf{q}) = S_{yy}^{(n)}(\mathbf{q})$, $S_{zz}^{(n)}(\mathbf{q}) = 0$, and

$$S^{(n)}(\mathbf{q}) = \sum_{\alpha,\beta}\left\{\delta_{\alpha\beta} - \frac{q_\alpha q_\beta}{q^2}\right\}S_{\alpha\beta}^{(n)}(\mathbf{q}) = S_{xx}^{(n)}(\mathbf{q})\left\{1 + \frac{q_z^{\,2}}{q^2}\right\}, \tag{5.1}$$

which only depends on q_z and never vanishes. This relation was already introduced in equation (2.7). Unless otherwise stated, we will assume that $\boldsymbol{\kappa} = \mathbf{q}$ so that the wavevector is restricted to the first BZ.

5.2 A FM in a magnetic field

As pictured in figure 5.1, consider a FM on a square lattice with spins aligned along $\mathbf{z}$ and coupled by the nearest-neighbor exchange $J > 0$. We also add easy-axis anisotropy $K > 0$ and a magnetic field $\mathbf{B}$, both along $\mathbf{z}$. Of course, there is only $M = 1$ spin per magnetic unit cell. The Hamiltonian for this model is:

$$\mathcal{H} = -J\sum_{\langle i,j\rangle}\mathbf{S}_i\cdot\mathbf{S}_j - K\sum_i S_{iz}^{\,2} - 2\mu_B B\sum_i S_{iz}, \tag{5.2}$$

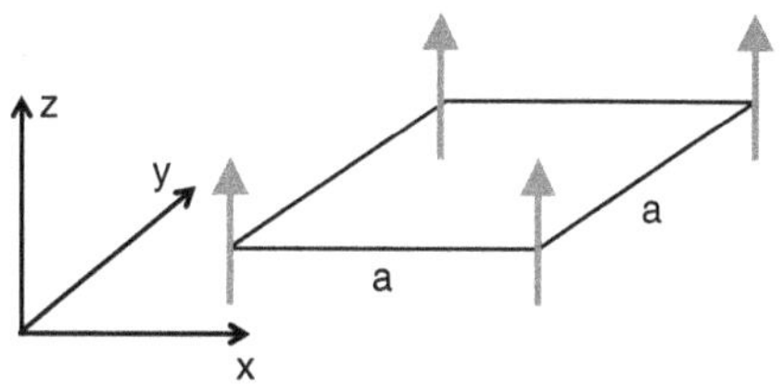

Figure 5.1. The ground state of a FM on a square lattice.

where the symbol $\langle i, j\rangle$ indicates that sites i and j are nearest neighbors and that each pair of nearest neighbors is only counted once. As elsewhere in this chapter and the next, we take the g-tensor to be isotropic with diagonal components of 2. Expanded in the boson operators a_i and $a_i^\dagger$, $\mathcal{H}$ becomes

$$\mathcal{H} = E_0 - JS\sum_{\langle i,j\rangle}\left\{a_i a_j^\dagger + a_i^\dagger a_j - a_i^\dagger a_i - a_j^\dagger a_j\right\} + 2KS\sum_i a_i^\dagger a_i + 2\mu_B B\sum_i a_i^\dagger a_i, \tag{5.3}$$

where

$$E_0 = -N\left\{2JS^2 + KS^2 + 2\mu_B BS\right\} \tag{5.4}$$

is the classical energy. Of course, each site on the square lattice has $z = 4$ nearest neighbors. To avoid double counting, each pair of spins is counted only once.

Without the field $\mathbf{B}$, E_0 would be of order S^2. Because the magnetic field linearly couples to the spins, it introduces a term of order SB. It is common to define a scaled field $B' = \mu_B B/S$ with units of energy. If B' is considered to be of order S^0, then $E_0 \sim S^2$. Moreover, the second-order term $\mathcal{H}_2$ is then of order S, down by two orders of $1/\sqrt{S}$ from $E_0 \sim S^2$, even with the magnetic field.

For a FM, the magnetic field can be simply added to the easy-axis anisotropy K in $\mathcal{H}_2$ to produce an effective anisotropy of

$$K' = K + \frac{\mu_B B}{S} = K + B' > 0. \tag{5.5}$$

As we shall see, that is *not* the case for an AF. Fourier transformed, $\mathcal{H}_2$ becomes

$$\mathcal{H}_2 = 4JS\sum_{\mathbf{q}} a_{\mathbf{q}}^\dagger a_{\mathbf{q}}\,(1 - \Gamma_{\mathbf{q}}) + 2K'S\sum_{\mathbf{q}} a_{\mathbf{q}}^\dagger a_{\mathbf{q}}, \tag{5.6}$$

where

$$\Gamma_{\mathbf{q}} = \frac{1}{2}\left\{\cos(q_x a) + \cos(q_y a)\right\} \tag{5.7}$$

is defined so that $\Gamma_{\mathbf{q}} = 1$ when $\mathbf{q} = 0$.

For future reference, $\Gamma_{\mathbf{q}}$ is generally defined as

$$\Gamma_{\mathbf{q}} = \frac{1}{z}\sum_{\mathbf{d}} e^{-i\mathbf{q}\cdot\mathbf{d}}, \tag{5.8}$$

where $\mathbf{d}$ is one of z different nearest-neighbor vectors. Once again, $\Gamma_{\mathbf{q}} = 1$ when $\mathbf{q} = 0$. For any lattice with inversion symmetry, $\Gamma_{\mathbf{q}}$ is real due to the opposite neighbors at $\pm\mathbf{d}$. But for a non-Bravais lattice like the honeycomb studied below, $\Gamma_{\mathbf{q}}$ can be imaginary when $\mathbf{q} \neq 0$.

It is then easy to show that $\underline{L}$ and $\underline{\mathcal{L}}$ are

$$\underline{L} = S\begin{pmatrix} 2J(1 - \Gamma_{\mathbf{q}}) + K' & 0 \\ 0 & 2J(1 - \Gamma_{\mathbf{q}}) + K' \end{pmatrix}, \tag{5.9}$$

$$\underline{\mathcal{L}} = \underline{L} \cdot \underline{N} = S\begin{pmatrix} 2J(1 - \Gamma_{\mathbf{q}}) + K' & 0 \\ 0 & -2J(1 - \Gamma_{\mathbf{q}}) - K' \end{pmatrix}. \tag{5.10}$$

Since $\underline{\mathcal{L}}$ has eigenvalues $\epsilon_n(\mathbf{q}) = \pm\hbar\omega(\mathbf{q})/2$, the SW frequency for a FM is

$$\hbar\omega(\mathbf{q}) = 4JS(1 - \Gamma_{\mathbf{q}}) + 2K'S. \tag{5.11}$$

In the absence of easy-axis anisotropy, the spins can rotate in any direction with no cost in energy. Consequently, $\omega(\mathbf{q})$ is a Goldstone mode with $\omega(\mathbf{q} = 0) = 0$, implying that a $\mathbf{q} = 0$ (i.e. uniform) fluctuation of the spins away from the z axis transforms the system into an equivalent spin state.

Because the easy-axis anisotropy K and the magnetic field $\mathbf{B}$ favor the spins along $\mathbf{z}$, they break that rotational invariance and $\hbar\omega(\mathbf{q} = 0) = 2K'S > 0$. A nonzero SW frequency at the ordering wavevector $\mathbf{Q}$ (in this case, $\mathbf{Q} = 0$) is called the SW gap Δ. For a FM, the easy-axis anisotropy and magnetic field simply lift the SW spectrum at every wavevector $\mathbf{q}$ by the energy $\Delta = 2(KS + \mu_B B)$.

Notice that the dispersion of a FM is quadratic in q near $\mathbf{Q} = 0$. The limit

$$D = \lim_{\mathbf{q}\to 0} \frac{\hbar\omega(\mathbf{q})}{q^2}. \tag{5.12}$$

is called the SW stiffness. Of course, D is undefined when $\Delta > 0$.

We now use the formalism developed in section 4.3 to evaluate the strength of the SW mode. The eigenvectors of $\underline{\mathcal{L}}$ are

$$X^*_{1j} = c_1{}^*\begin{pmatrix} 1 \\ 0 \end{pmatrix}, \tag{5.13}$$

$$X^*_{2j} = c_1\begin{pmatrix} 0 \\ 1 \end{pmatrix}, \tag{5.14}$$

which use the symmetry relations for $\underline{X}$. Hence,

$$\underline{X} = \begin{pmatrix} c_1 & 0 \\ 0 & c_1{}^* \end{pmatrix}. \tag{5.15}$$

The normalization condition $\underline{X} \cdot \underline{N} \cdot \underline{X}^{\dagger} = \underline{N}$ implies that $|c_1|^2 = 1$. Since

$$\underline{X}^{-1} = \begin{pmatrix} 1/c_1 & 0 \\ 0 & 1/c_1{}^* \end{pmatrix}, \tag{5.16}$$

we find that $W_{1z}^{(1)}(\mathbf{q}) = 0$ and

$$W_{1x}^{(1)}(\mathbf{q}) = iW_{1y}^{(1)}(\mathbf{q}) = \frac{1}{c_1}. \tag{5.17}$$

So the diagonal components of the spin–spin correlation function $S_{\alpha\beta}^{(1)}(\mathbf{q})$ are

$$S_{xx}^{(1)}(\mathbf{q}) = S_{yy}^{(1)}(\mathbf{q}) = \frac{S}{2|c_1|^2} = \frac{S}{2}, \tag{5.18}$$

and $S_{zz}^{(1)}(\mathbf{q}) = 0$. The off-diagonal components of $S_{\alpha\beta}^{(1)}(\mathbf{q})$ vanish.

These results imply that the SW strength of a FM is independent of $\mathbf{q}$. The xx and yy components of $S_{\alpha\beta}^{(1)}(\mathbf{q})$ are the same because spin fluctuations perpendicular to the spin orientation $\mathbf{z}$ along $\mathbf{x}$ and $\mathbf{y}$ have the same strength. Spin fluctuations along $\mathbf{z}$ are prohibited within SW theory because the magnitude of the spin is assumed to be fixed.

Figure 5.2 plots the SW frequency and strength along $(H, 0)$. In reciprocal lattice units (rlu), the 3D wavevector (H, K, L) corresponds to $\mathbf{q} = (2\pi/a)(H, K, L)$ (see appendix D of chapter 4). For our 2D model, the frequencies have no L dependence.

We use Gaussian resolution functions with widths $\hbar\Delta\omega = 0.1$ meV and $\Delta q = 0.04\pi/a$ (0.02 rlu). Unless otherwise stated, these resolution functions will be used for the other examples in this chapter and in the next. Since the resolution functions were not strictly normalized by the number of $\mathbf{q}$ or ω points under the

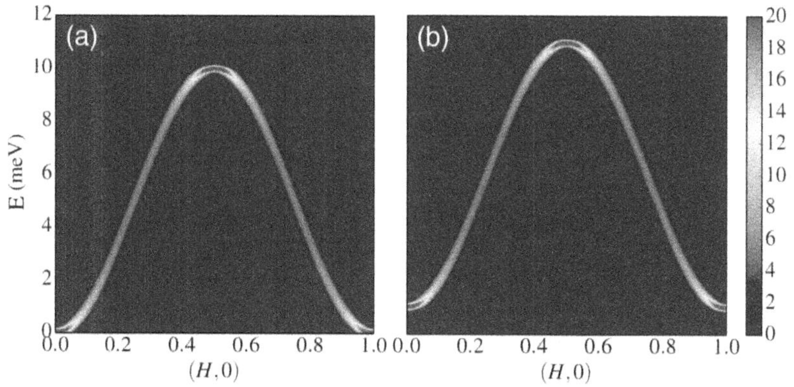

Figure 5.2. The inelastic spectrum of a FM on a square lattice with $S = 5/2$, $J = 1$ meV and either (a) $K' = 0$ or (b) 0.2 meV along $(H, 0)$.

Gaussians, the color scales in figures 5.2(a) and (b) give the relative but not the absolute intensities. Notice that

$$S^{(1)}(\mathbf{q}) = \sum_{\alpha,\beta}\left\{\delta_{\alpha\beta} - \frac{q_\alpha q_\beta}{q^2}\right\}S^{(1)}_{\alpha\beta}(\mathbf{q}) = \frac{S}{2} \tag{5.19}$$

for $q_y = 0$ or wavevector index $K = 0$. Although $S^{(1)}(\mathbf{q})$ is independent of $\mathbf{q}$, the inelastic intensities vary along $(H, 0)$ due to the integral over the resolution function. The SW strength is greatest when the dispersion $\omega(\mathbf{q})$ is flat because several $\mathbf{q}$ points then contribute to the intensity.

For the parameters used in figure 5.2(b), $\Delta = 2K'S = 1$ meV. As anticipated above, the SW branch simply rises uniformly with no relative change in intensity when $K' > 0$.

5.3 A FM chain with alternating exchange interactions

This example demonstrates how the number of observable SW branches may depend on model parameters. It also shows that the spectral weight $S^{(n)}_{\alpha\beta}(\mathbf{q})$ has a different wavevector periodicity than the SW frequencies $\omega^{(n)}(\mathbf{q})$.

Take a chain along $\mathbf{x}$ with all spins aligned along $\mathbf{z}$ by the alternating FM exchange interactions $J_1 > 0$ and $J_2 > 0$, as pictured in figure 5.3. For $J_2 \neq J_1$, there are $M = 2$ distinct spins: spins on sublattice 1 have J_1 to the right and J_2 to the left while spins on sublattice 2 have J_2 to the right and J_1 to the left. For $M = 2$, there should be two SW branches at every q point. But when $J_1 = J_2$, there can be only one SW branch per q point because the two sublattices are then equivalent. So how do the two SW branches collapse into one?

The Hamiltonian for this simple model is:

$$\mathcal{H} = -J_1\sum_n \mathbf{S}_{2n-1}\cdot\mathbf{S}_{2n} - J_2\sum_n \mathbf{S}_{2n}\cdot\mathbf{S}_{2n+1}, \tag{5.20}$$

where J_1 couples odd sites (sublattice 1) to even sites (sublattice 2) on the right while J_2 couples odd sites to even states on the left. The Fourier transform of $\mathcal{H}_2$ is

$$\begin{aligned}\mathcal{H}_2 = J_1S\sum_{\mathbf{q}}\left\{a_{\mathbf{q}}^{\dagger}a_{\mathbf{q}} + b_{\mathbf{q}}^{\dagger}b_{\mathbf{q}} - a_{\mathbf{q}}^{\dagger}b_{\mathbf{q}}e^{iqa} - a_{\mathbf{q}}b_{\mathbf{q}}^{\dagger}e^{-iqa}\right\}\\ + J_2S\sum_{\mathbf{q}}\left\{a_{\mathbf{q}}^{\dagger}a_{\mathbf{q}} + b_{\mathbf{q}}^{\dagger}b_{\mathbf{q}} - b_{\mathbf{q}}^{\dagger}a_{\mathbf{q}}e^{iqa} - b_{\mathbf{q}}a_{\mathbf{q}}^{\dagger}e^{-iqa}\right\},\end{aligned} \tag{5.21}$$

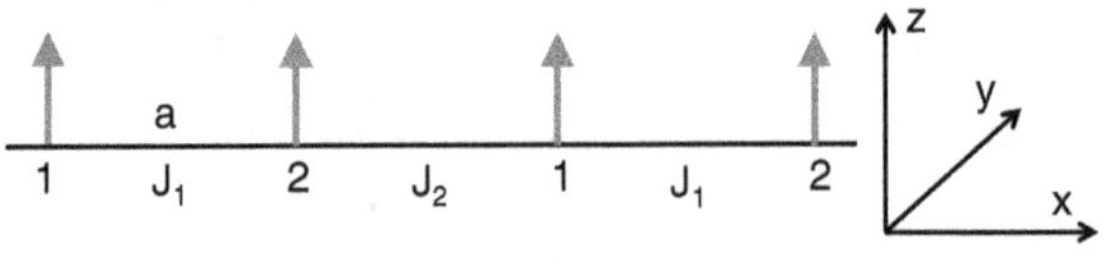

Figure 5.3. A FM chain with alternating exchange constants J_1 and J_2.

where $a_{\mathbf{q}}$ and $b_{\mathbf{q}}$ are operators on sublattices 1 and 2, respectively. It follows that

$$\underline{L} = \frac{S}{2}\begin{pmatrix} J_1 + J_2 & -J_1\xi - J_2\xi^* & 0 & 0 \\ -J_1\xi^* - J_2\xi & J_1 + J_2 & 0 & 0 \\ 0 & 0 & J_1 + J_2 & -J_1\xi - J_2\xi^* \\ 0 & 0 & -J_1\xi^* - J_2\xi & J_1 + J_2 \end{pmatrix}, \tag{5.22}$$

where $\xi = \exp(iqa)$.

Why does $\underline{L}$ now involve complex terms $\exp(\pm iqa)$ when we previously encountered $\cos(q_\alpha a)$ for the FM on a square lattice? On a square lattice, each spin couples to four neighboring spins on the same sublattice with the same exchange constant J. Hence, $\underline{L}$ involved $\exp(iq_x a) + \exp(-iq_x a) = 2\cos(q_x a)$ and $\exp(iq_y a) + \exp(-iq_y a) = 2\cos(q_y a)$. In the current example, sublattice 1 couples to sublattice 2 on the right with J_1 and to sublattice 2 on the left with J_2. Hence, $\exp(iqa)$ and $\exp(-iqa)$ appear separately.

Now $\underline{\mathcal{L}} = \underline{L} \cdot \underline{N}$ can be compactly written as

$$\underline{\mathcal{L}} = \frac{(J_1 + J_2)S}{2}\begin{pmatrix} 1 & -\Upsilon_{\mathbf{q}}^* & 0 & 0 \\ -\Upsilon_{\mathbf{q}} & 1 & 0 & 0 \\ 0 & 0 & -1 & \Upsilon_{\mathbf{q}}^* \\ 0 & 0 & \Upsilon_{\mathbf{q}} & -1 \end{pmatrix}, \tag{5.23}$$

where

$$\Upsilon_{\mathbf{q}} = \frac{J_1 e^{-iqa} + J_2 e^{iqa}}{J_1 + J_2} \tag{5.24}$$

is complex. The eigenfrequencies of $\underline{\mathcal{L}}$ are

$$\epsilon_{1,2} = \frac{(J_1 + J_2)S}{2}(1 \pm |\Upsilon_{\mathbf{q}}|) \geqslant 0, \tag{5.25}$$

$$\epsilon_{3,4} = -\frac{(J_1 + J_2)S}{2}(1 \pm |\Upsilon_{\mathbf{q}}|) \leqslant 0, \tag{5.26}$$

with corresponding SW frequencies $\hbar\omega_{1,2}(\mathbf{q}) = 2\epsilon_{1,2}(\mathbf{q}) \geqslant 0$.

The eigenvectors of $\underline{\mathcal{L}}$ give the inverse $\underline{X}$ matrix (exercise 5.9)

$$\underline{X}^{-1} = \frac{1}{2}\begin{pmatrix} -1/(|\Upsilon_{\mathbf{q}}|c_1) & 1/(|\Upsilon_{\mathbf{q}}|c_2) & 0 & 0 \\ 1/(\Upsilon_{\mathbf{q}}^* c_1) & 1/(\Upsilon_{\mathbf{q}}^* c_2) & 0 & 0 \\ 0 & 0 & -1/(|\Upsilon_{\mathbf{q}}|c_1^*) & 1/(|\Upsilon_{\mathbf{q}}|c_2^*) \\ 0 & 0 & 1/(\Upsilon_{\mathbf{q}}^* c_1^*) & 1/(\Upsilon_{\mathbf{q}}^* c_2^*) \end{pmatrix}. \tag{5.27}$$

The normalization condition $\underline{X} \cdot \underline{N} \cdot \underline{X}^{-1} = \underline{N}$ then requires that $|c_1|^2 = |c_2|^2 = 1/(2|\Upsilon_{\mathbf{q}}|^2)$.

For spins aligned along **z**, $W_{rz}^{(n)}(\mathbf{q}) = 0$ and

$$W_{1x}^{(1)}(\mathbf{q}) = iW_{1y}^{(1)}(\mathbf{q}) = -\frac{1}{2|\Upsilon_{\mathbf{q}}|c_1}, \tag{5.28}$$

$$W_{2x}^{(1)}(\mathbf{q}) = iW_{2y}^{(1)}(\mathbf{q}) = \frac{1}{2\Upsilon_{\mathbf{q}}^{*}c_1}, \tag{5.29}$$

$$W_{1x}^{(2)}(\mathbf{q}) = iW_{1y}^{(2)}(\mathbf{q}) = \frac{1}{2|\Upsilon_{\mathbf{q}}|c_2}, \tag{5.30}$$

$$W_{2x}^{(2)}(\mathbf{q}) = iW_{2y}^{(2)}(\mathbf{q}) = \frac{1}{2\Upsilon_{\mathbf{q}}^{*}c_2}. \tag{5.31}$$

It follows that $S_{zz}^{(n)}(\mathbf{q}) = 0$ and

$$S_{xx}^{(1)}(\mathbf{q}) = S_{yy}^{(1)}(\mathbf{q}) = \frac{S}{4}\left\{1 - \frac{\mathrm{Re}(\Upsilon_{\mathbf{q}})}{|\Upsilon_{\mathbf{q}}|}\right\}, \tag{5.32}$$

$$S_{xx}^{(2)}(\mathbf{q}) = S_{yy}^{(2)}(\mathbf{q}) = \frac{S}{4}\left\{1 + \frac{\mathrm{Re}(\Upsilon_{\mathbf{q}})}{|\Upsilon_{\mathbf{q}}|}\right\}. \tag{5.33}$$

Summing over the spin degrees of freedom, we obtain

$$\begin{aligned} S^{(n)}(\mathbf{q}) &= \sum_{\alpha,\beta}\left\{\delta_{\alpha\beta} - \frac{q_\alpha q_\beta}{q^2}\right\}S_{\alpha\beta}^{(n)}(\mathbf{q}) = S_{yy}^{(n)}(\mathbf{q}) + S_{zz}^{(n)}(\mathbf{q}) \\ &= \frac{S}{4}\left\{1 \mp \frac{\mathrm{Re}(\Upsilon_{\mathbf{q}})}{|\Upsilon_{\mathbf{q}}|}\right\}, \end{aligned} \tag{5.34}$$

for modes 1 (−) and 2 (+).

Modes 1 and 2 cross when $\Upsilon_{\mathbf{q}} = 0$ or

$$|\, J_1 e^{-iqa} + J_2 e^{iqa} \,| = \left\{J_1^2 + J_2^2 + 2J_1J_2\cos(2qa)\right\}^{1/2} = 0. \tag{5.35}$$

Since the minimum value for the left-hand side is $|J_1 - J_2|$, the SW modes do not cross when $J_1 \neq J_2$. If $J_1 = J_2$, modes 1 and 2 cross when $q = \pi/2a$ or $3\pi/2a$ ($H = 0.25$ or 0.75). But in that case, $\Upsilon_{\mathbf{q}} = \cos(qa)$ is real so that $S_{\alpha\alpha}^{(1)}(\mathbf{q}) = 0$ and only mode 2 with $\hbar\omega_2 = 2J_1S(1 - \Upsilon_{\mathbf{q}})$ is observable for any q.

The inelastic spectra for this model are plotted in figure 5.4. As expected, the chain has only one observable SW branch when $J_1 = J_2$. Now keep $J_1 = 1$ meV fixed as J_2 is reduced. For $J_2 = 0.9$ meV, SW gaps appear at $H = 0.25$ and 0.75. These gaps are larger for $J_2 = 0.75$ and 0.5 meV with two SW branches clearly seen in figures

5.4(c) and (d). As $J_2 \to 0$, the two SW branches become flat and dispersionless with frequencies $\hbar\omega = 0$ and 5 meV.

This example demonstrates how two SW branches collapse into one as the spectral weight of one branch disappears. Starting with $M = 2$ sublattices for $J_1 \neq J_2$, there are always two non-degenerate SW branches even as $J_1 \to J_2$. However, the spectral weight of the upper branch disappears near $H = 0$ and 1 unless $J_1 \neq J_2$. To emphasize this point, we have sketched the SW branch with little or no intensity for a given H by the dashed curves.

In this example, the spectral weight $S^{(n)}_{\alpha\beta}(\mathbf{q})$ has a different wavevector periodicity than the SW frequencies $\omega_n(\mathbf{q})$. When $J_1 \neq J_2$, the periodicity of the magnetic lattice is $2a$ so the ordering wavevector is $Q = 2\pi/2a = \pi/a$ or $H = 0.5$. Consequently, the SW frequencies have the periodicity $\Delta H = 0.5$ so that $\omega_n(H + 0.5) = \omega_n(H)$ as in figures 5.4(b)–(d). But the spectral weights have periodicity $\Delta H = 1$ so that $S^{(n)}_{\alpha\beta}(H + 1) = S^{(n)}_{\alpha\beta}(H)$.

5.4 A FM on a honeycomb lattice

In a Bravais lattice, all lattice sites are *structurally* identical [20]. So the world looks the same from any site. In a non-Bravais lattice, that is no longer the case. Two examples of non-Bravais lattices are the diamond and honeycomb lattices. The latter

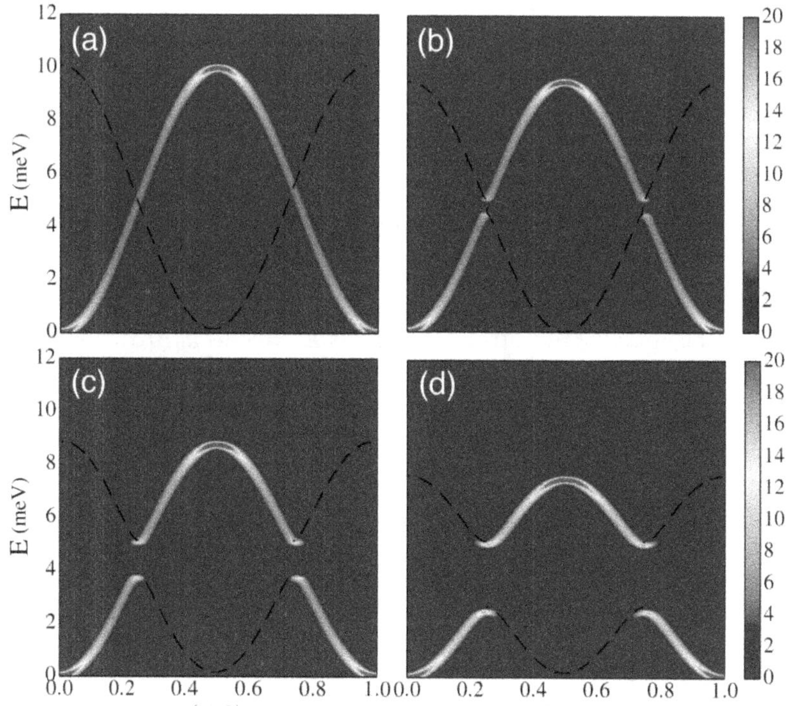

Figure 5.4. Inelastic spectra for a FM chain with alternating exchange interactions $S = 5/2$, $J_1 = 1$ meV and (a) $J_2 = 1$ meV, (b) 0.9 meV, (c) 0.75 meV, or (d) 0.5 meV. The black dashed curves indicate the SW branch with little or no intensity at a given H.

is sketched in figure 5.5, which indicates the two non-equivalent sites 1 and 2. Site 1 sees another site directly to the left but none to the right; site 2 sees another site directly to the right but none to the left. Recent interest [21, 22] in FM honeycombs like CrX_3 (X = Br or I) [7] was motivated by the possibility of 'Dirac' cones in the SW spectrum.

Actually, the previous section also provided an example of a non-Bravais lattice. When $J_1 \neq J_2$, sites 1 and 2 are magnetically inequivalent. The lattice will then dimerize by shifting neighboring sites together or further apart depending on whether they are bonded by J_1 or J_2. So sites 1 and 2 will become structurally inequivalent (exercise 5.10).

We now evaluate the SW spectrum for the honeycomb lattice assuming that only one FM exchange $J > 0$ couples nearest neighbors on different sublattices. Take the spins to lie along **x**, which is one of the hexagonal axes. Including easy-plane anisotropy $K < 0$ and a magnetic field **B** along **z**, the Hamiltonian can be written as

$$\mathcal{H} = -J\sum_{\langle i,j\rangle} \mathbf{S}_i \cdot \mathbf{S}_j - K\sum_i S_{iz}{}^2 - 2\mu_B B\sum_i S_{iz}. \tag{5.36}$$

With $M = 2$, we obtain (exercise 5.11)

$$\underline{L} = \frac{S}{2}\begin{pmatrix} 3J - K' & -3J\Gamma_{\mathbf{q}}{}^* & -K' & 0 \\ -3J\Gamma_{\mathbf{q}} & 3J - K' & 0 & -K' \\ -K' & 0 & 3J - K' & -3J\Gamma_{\mathbf{q}}{}^* \\ 0 & -K' & -3J\Gamma_{\mathbf{q}} & 3J - K' \end{pmatrix}, \tag{5.37}$$

where

$$\Gamma_{\mathbf{q}} = \frac{1}{z}\sum_{\mathbf{d}} e^{-i\mathbf{q}\cdot\mathbf{d}} = \frac{1}{3}\left\{ e^{iq_x a} + e^{-i(q_x + \sqrt{3}q_y)a/2} + e^{-i(q_x - \sqrt{3}q_y)a/2} \right\} \tag{5.38}$$

is a complex quantity when $\mathbf{q} \neq 0$, as suggested above.

As earlier, the effective anisotropy along **z** is $K' = K + \mu_B B/S$. For $K < 0$, easy-plane anisotropy and the magnetic field compete with each other. For $|K| > \mu_B B/S$,

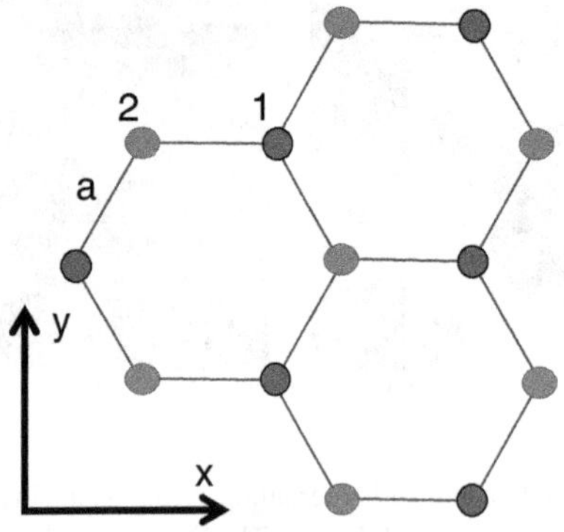

Figure 5.5. A honeycomb lattice showing inequivalent sites 1 (blue) and 2 (red).

the easy-plane anisotropy wins and the spins will lie in the xy plane. For $\mu_B B > |K|S$, the magnetic field wins and the spins will lie along the z axis.

To obtain a simple analytic solution, take $K' = 0$. Then,

$$\underline{\mathcal{L}} = \underline{L} \cdot \underline{N} = \frac{3JS}{2}\begin{pmatrix} 1 & -\Gamma_{\mathbf{q}}^* & 0 & 0 \\ -\Gamma_{\mathbf{q}} & 1 & 0 & 0 \\ 0 & 0 & -1 & -\Gamma_{\mathbf{q}}^* \\ 0 & 0 & -\Gamma_{\mathbf{q}} & -1 \end{pmatrix}. \tag{5.39}$$

Diagonalizing $\underline{\mathcal{L}}$, we obtain the eigenvalues

$$\epsilon_{1,2}(\mathbf{q}) = \frac{3JS}{2}(1 \pm |\Gamma_{\mathbf{q}}|) \geqslant 0, \tag{5.40}$$

$$\epsilon_{3,4}(\mathbf{q}) = -\frac{3JS}{2}(1 \pm |\Gamma_{\mathbf{q}}|) \leqslant 0. \tag{5.41}$$

Mode 2 is a Goldstone mode with $\hbar\omega_2(\mathbf{q} = 0) = 0$ whereas $\hbar\omega_1(\mathbf{q} = 0) = 6JS$. The condition for modes 1 and 2 to cross is $\Gamma_{\mathbf{q}} = 0$. This condition is met at six points in the first BZ: $(\pm 1/3, \sqrt{3}/9)$, $(\pm 1/3, -\sqrt{3}/9)$, and $(0, \pm 2\sqrt{3}/9)$, all in units of $2\pi/a$.

Notice that equation (5.39) for $\underline{\mathcal{L}}$ is formally identical to equation (5.23) for the FM chain with alternating exchange! Following the derivation in the previous example but with spins aligned along $\mathbf{x}$ rather than $\mathbf{z}$, we obtain $S_{xx}^{(n)}(\mathbf{q}) = 0$ and

$$S_{yy}^{(1)}(\mathbf{q}) = S_{zz}^{(1)}(\mathbf{q}) = \frac{S}{4}\left\{1 - \frac{\mathrm{Re}(\Gamma_{\mathbf{q}})}{|\Gamma_{\mathbf{q}}|}\right\}, \tag{5.42}$$

$$S_{yy}^{(2)}(\mathbf{q}) = S_{zz}^{(2)}(\mathbf{q}) = \frac{S}{4}\left\{1 + \frac{\mathrm{Re}(\Gamma_{\mathbf{q}})}{|\Gamma_{\mathbf{q}}|}\right\}. \tag{5.43}$$

Although these strengths are undefined at the mode crossing $\Gamma_{\mathbf{q}} = 0$, the sums

$$S_{yy}^{(1)}(\mathbf{q}) + S_{yy}^{(2)}(\mathbf{q}) = S_{zz}^{(1)}(\mathbf{q}) + S_{zz}^{(2)}(\mathbf{q}) = \frac{S}{2} \tag{5.44}$$

are well defined for all $\mathbf{q}$.

One of the motivations for studying the honeycomb lattice is that the SW spectrum contains 'Dirac' cones [21, 22] that also appear in the electronic band structure of graphene [23]. These are symmetrical cones where two SW branches cross at the six wavevectors given above. If these $\mathbf{q}$ points are approached from $\mathbf{q} = 0$, then $\Gamma_{\mathbf{q}}$ is real so only mode 2 has nonzero intensity and the cone will be invisible. The best way to see the 'Dirac' cone is to approach these points along $(H, \sqrt{3}/9)$ or $(1/3, K)$. The inelastic spectra are then plotted in figure 5.6 with the same frequency and intensity scales for both paths.

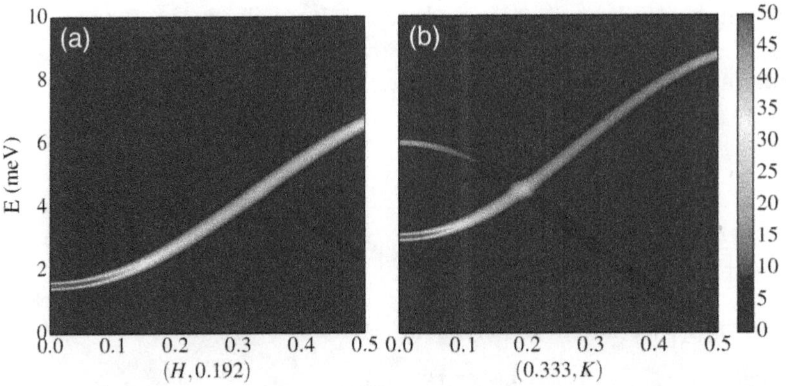

Figure 5.6. The inelastic spectrum of a FM on a honeycomb lattice with $J = 1$ meV and $S = 3/2$ along either (a) $(H, \sqrt{3}/9)$ or (b) $(1/3, K)$.

Even along those two paths, mode 2 still dominates the SW spectrum. Along $(H, \sqrt{3}/9)$, mode 2 has an intensity about 14 times higher than mode 1 near the mode crossing. Along $(1/3, K)$, the ratio of intensities at the mode crossing is 3. This example again demonstrates the importance of evaluating the SW intensities as well as their frequencies. Although interesting, the 'Dirac' cones may be difficult to observe in the FM Cr trihalides [7].

The essential difference between this example and the FM chain with alternating exchange is that $\Gamma_{\mathbf{q}}$ can vanish for the honeycomb lattice but $\Upsilon_{\mathbf{q}}$ cannot vanish for the FM chain. Consequently, the two modes of the FM chain are gapped whereas those of the honeycomb lattice cross at a 'Dirac' point.

5.5 An AF in a magnetic field

We now consider an AF on the square lattice plotted in figure 5.7. In addition to a magnetic field **B** along **z**, we also add easy-axis anisotropy K along **z** and easy-plane anisotropy E that favors the spins to lie in the yz plane. So the Hamiltonian is

$$\mathcal{H} = -J\sum_{\langle i,j\rangle} \mathbf{S}_i \cdot \mathbf{S}_j - K\sum_i S_{iz}{}^2 - E\sum_i S_{ix}{}^2 - 2\mu_B B\sum_i S_{iz}, \tag{5.45}$$

where $J < 0$, $K > 0$, and $E < 0$. Due to the presence of both easy-axis and easy-plane anisotropies, z is the easiest axis for spin alignment, x is the hardest axis, and y falls in between.

There are $M = 2$ spins in the unit cell. For spins aligned along **z** in sublattice 1, the local reference frame is given by $\bar{S}_{1x} = S_{1x}$, $\bar{S}_{1y} = S_{1y}$, and $\bar{S}_{1z} = S_{1z}$. On sublattice 2 with spins aligned along $-$**z**, the local reference frame is given by $\bar{S}_{2x} = -S_{2x}$, $\bar{S}_{2y} = S_{2y}$, and $\bar{S}_{2z} = -S_{2z}$. We define the boson operators on sublattice 1 as a_i and those on sublattice 2 as b_i.

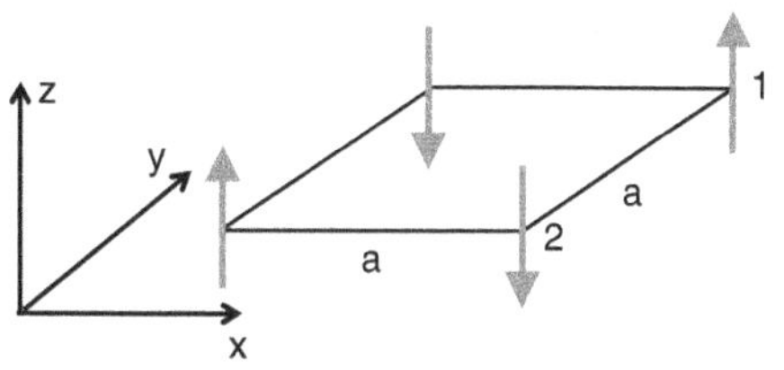

Figure 5.7. The ground state of an AF on a square lattice.

After Fourier transforming, $\mathcal{H}_2$ becomes

$$\begin{aligned}\mathcal{H}_2 = 4|J|S\sum_{\mathbf{q}}\left\{a_{\mathbf{q}}^{\dagger}a_{\mathbf{q}} + b_{\mathbf{q}}^{\dagger}b_{\mathbf{q}} - \Gamma_{\mathbf{q}}\left(a_{\mathbf{q}}b_{-\mathbf{q}} + a_{\mathbf{q}}^{\dagger}b_{-\mathbf{q}}^{\dagger}\right)\right\} \\ + 2\mu_{\mathrm{B}}B\sum_{\mathbf{q}}\left\{a_{\mathbf{q}}^{\dagger}a_{\mathbf{q}} - b_{\mathbf{q}}^{\dagger}b_{\mathbf{q}}\right\} + 2KS\sum_{\mathbf{q}}\left\{a_{\mathbf{q}}^{\dagger}a_{\mathbf{q}} + b_{\mathbf{q}}^{\dagger}b_{\mathbf{q}}\right\} \\ - \frac{ES}{2}\sum_{\mathbf{q}}\left\{a_{\mathbf{q}}a_{-\mathbf{q}} + a_{\mathbf{q}}^{\dagger}a_{-\mathbf{q}}^{\dagger} + 2a_{\mathbf{q}}^{\dagger}a_{\mathbf{q}} + b_{\mathbf{q}}b_{-\mathbf{q}} + b_{\mathbf{q}}^{\dagger}b_{-\mathbf{q}}^{\dagger} + 2b_{\mathbf{q}}^{\dagger}b_{\mathbf{q}}\right\}.\end{aligned} \tag{5.46}$$

It follows that

$$\underline{L} = \frac{S}{2}\begin{pmatrix} A_+ & 0 & -E & 4J\Gamma_{\mathbf{q}} \\ 0 & A_- & 4J\Gamma_{\mathbf{q}} & -E \\ -E & 4J\Gamma_{\mathbf{q}} & A_+ & 0 \\ 4J\Gamma_{\mathbf{q}} & -E & 0 & A_- \end{pmatrix}, \tag{5.47}$$

$$A_{\pm} = 4|J| + 2K - E \pm 2B', \tag{5.48}$$

where $\Gamma_{\mathbf{q}}$ was defined by equation (5.7). Unlike for a FM, the magnetic field *cannot* be simply added to the easy-axis anisotropy to produce an effective anisotropy K'. This difference occurs because the external field is not staggered like the AF moments.

Generally, the eigenvalues and eigenvectors of $\underline{\mathcal{L}} = \underline{L}\cdot\underline{N}$ must be solved numerically. But a simple analytic solution can be obtained when the easy-plane anisotropy E is set to zero. In that case, $a_{\mathbf{q}}^{\dagger}$ is decouples from both $b_{\mathbf{q}}$ and $a_{-\mathbf{q}}^{\dagger}$ while $b_{\mathbf{q}}^{\dagger}$ decouples from both $a_{\mathbf{q}}$ and $b_{-\mathbf{q}}^{\dagger}$. It is then straightforward to obtain the SW frequencies

$$\hbar\omega_{1,2}(\mathbf{q}) = 2S\left\{(2|J| + K)^2 - 4(J\Gamma_{\mathbf{q}})^2\right\}^{1/2} \pm 2\mu_{\mathrm{B}}B, \tag{5.49}$$

where $\hbar\omega_1(\mathbf{q})$ corresponds to the positive sign and $\hbar\omega_2(\mathbf{q})$ to the negative sign. For small K and $B = 0$, the SW gap at the AF ordering wavevector $\mathbf{Q} = (0.5, 0.5)$ is $\Delta \approx 4S\sqrt{|J|K}$.

When $K = 0$ and $B = 0$, the SW modes vanish linearly at $\mathbf{q} = \mathbf{Q}$. The SW velocity v is then defined by the limit

$$v = \lim_{\mathbf{q}\to\mathbf{Q}} \frac{\hbar\omega(\mathbf{q})}{|\mathbf{q}-\mathbf{Q}|} = \lim_{\mathbf{q}\to 0} \frac{\hbar\omega(\mathbf{q})}{|\mathbf{q}|}, \tag{5.50}$$

where the last equality follows from the relation $\omega(\mathbf{q}+\mathbf{Q}) = \omega(\mathbf{q})$. When $K \neq 0$, the AF dispersion becomes quadratic near $\mathbf{Q}$ (see figure 5.8(b)) and the SW velocity is not defined.

For $K > 0$, $E = 0$, and $B > 0$ (exercise 5.22),

$$\underline{X}^{-1} = \frac{1}{2|J|\hbar\omega_0(\mathbf{q})} \begin{pmatrix} F_+/(c_1\Gamma_\mathbf{q}) & 0 & 0 & -F_-/(c_2\Gamma_\mathbf{q}) \\ 0 & -2|J|/c_2^* & 2|J|/c_1^* & 0 \\ 0 & -F_-/(c_2^*\Gamma_\mathbf{q}) & F_+/(c_1^*\Gamma_\mathbf{q}) & 0 \\ 2|J|/c_1 & 0 & 0 & -2|J|/c_2^* \end{pmatrix}, \tag{5.51}$$

$$F_\pm = 2|J| + K \pm \hbar\omega_0(\mathbf{q})/2S \geqslant 0, \tag{5.52}$$

where $\omega_0(\mathbf{q}) = (\omega_1(\mathbf{q}) + \omega_2(\mathbf{q}))/2$ is the SW frequency at zero field. The normalization condition $\underline{X}^{-1} = \underline{N}\cdot\underline{X}^\dagger\cdot\underline{N}$ gives

$$|c_1|^2 = \frac{1}{2\hbar\omega_0(\mathbf{q})F_-}, \tag{5.53}$$

$$|c_2|^2 = \frac{1}{2\hbar\omega_0(\mathbf{q})F_+}. \tag{5.54}$$

The nonzero components of the spin–spin correlation function are (exercise 5.22)

$$S_{xx}^{(n)}(\mathbf{q}) = S_{yy}^{(n)}(\mathbf{q}) = \frac{S}{4}\left\{\frac{2|J|(1-\Gamma_\mathbf{q}) + K}{2|J|(1+\Gamma_\mathbf{q}) + K}\right\}^{1/2} \tag{5.55}$$

for both modes $n = 1$ and 2, independent of field! Unlike for a FM, the SW intensity of an AF depends on the wavevector $\mathbf{q}$. When $K = B = 0$, the intensity blows up as $\omega_0(\mathbf{q}) \to 0$ at $\mathbf{Q} = (0.5, 0.5)$ with $\Gamma_\mathbf{Q} = -1$ but vanishes at (0, 0) with $\Gamma_{\mathbf{q}=0} = 1$.

Figures 5.8(a) and (d) plot the SW intensity for an AF in zero field without easy-plane anisotropy E. Although the SW intensity in figure 5.8(a) blows up at $\mathbf{Q}$, the intensity scale cuts off at a maximum of 80. In figure 5.8(d), a SW gap of $\Delta \approx 3.2$ meV is produced by easy-axis anisotropy. As predicted by equation (5.55), the SW intensity $\sim S\sqrt{|J|/K}$ no longer blows up at $\mathbf{Q}$ in the presence of a SW gap. Notice that easy-axis anisotropy does not break the degeneracy of the SW modes.

Because there is only one SW mode in zero field when $E = 0$, it is sometimes assumed that a simple AF in zero field can be treated as a system with $M = 1$ spin per unit cell. However, easy-plane anisotropy breaks the degeneracy of the two SW modes. In the absence of easy-plane anisotropy, spin fluctuations along $\mathbf{x}$ and $\mathbf{y}$ are

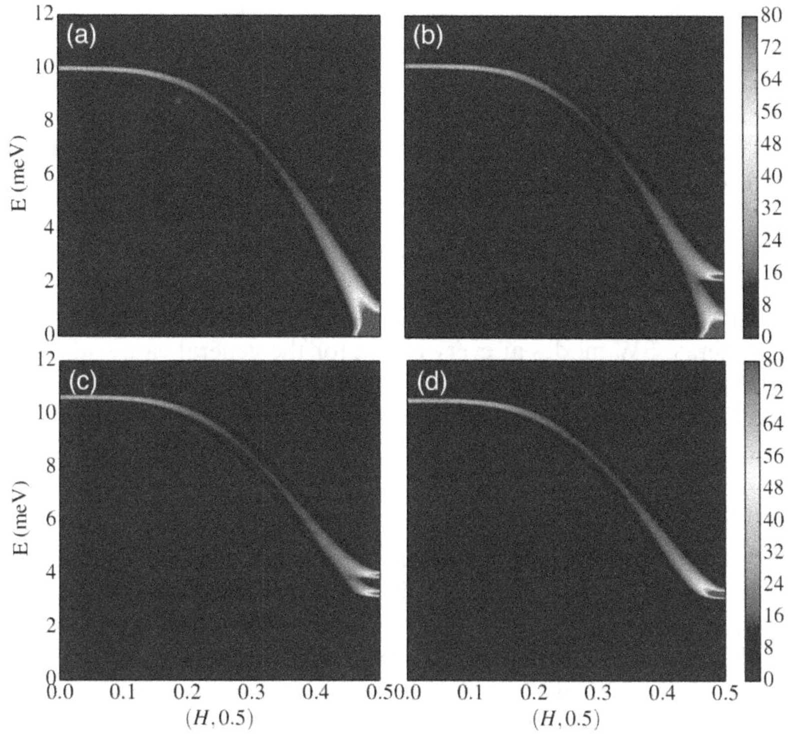

Figure 5.8. The inelastic spectrum of an AF on a square lattice in zero magnetic field with $S = 5/2$, $J = -1$, (a) $K = 0$, $E = 0$, (b) $K = 0$, $E = -0.05$, (c) $K = 0.1$, $E = -0.05$, and (d) $K = 0.1$, $E = 0$ along $(H, 0.5)$ and all parameters in units of meV.

equivalent. But for easy-plane anisotropy $E < 0$ in the yz plane and spins aligned along $\pm\mathbf{z}$, spin fluctuations are suppressed along $\mathbf{x}$ compared to spin fluctuations along $\mathbf{y}$.

To obtain the SW modes in the presence of easy-plane anisotropy, we numerically solve for the eigenvalues and eigenvectors of $\underline{\mathcal{L}}$. The resulting spectra are plotted in figures 5.8(c) and (d). As anticipated, easy-plane anisotropy breaks the degeneracy between the two SW modes in zero field. That splitting is largest at $\mathbf{Q}$. When $E < 0$, the upper SW branch contains spin fluctuations along $\mathbf{x}$ (only $S_{xx}(\mathbf{q})$ is nonzero) and the lower branch contains spin fluctuations along $\mathbf{y}$ (only $S_{yy}(\mathbf{q})$ is nonzero) because fluctuations along $\mathbf{x}$ are suppressed. When $E > 0$, spin fluctuations along $\mathbf{y}$ would be suppressed and the x and y fluctuations of the two branches would be reversed. In either case, the higher-energy SW branch is associated with the suppressed spin fluctuations. The intensity of the single degenerate SW mode when $E = 0$ is split between the two non-degenerate SW modes when $E \neq 0$.

For $K = 0$ in figure 5.8(b), the frequency of the lower branch still vanishes at $\mathbf{Q}$ because it is associated with rotations about the x axis and a uniform spin rotation with wavevector $\mathbf{Q}$ costs no energy. As expected, a SW gap develops when $K > 0$ in figure 5.8(d). The SW branches are both split and gapped in figure 5.8(c), which includes both easy-axis and easy-plane anisotropies.

In nonzero field B, the spins remain aligned along $\pm\mathbf{z}$ so long as the anisotropy energy is larger than the Zeeman energy. As the two SW branches split, each contains half the initial weight. For $B > 0$, the SW intensities do not change with field while the AF state persists.

For $B > B_{SF} \approx 2S\sqrt{|J|K}/\mu_B$, the AF phase is unstable and the spins 'flop' toward the xy plane while also tilting towards $\mathbf{z}$ to gain some Zeeman energy. The transition from the AF phase to the 'spin-flop' or canted AF phase is marked by the softening of the lower SW frequency at $\mathbf{q} = 0$ and $\mathbf{Q}$ (exercise 6.4).

There are many other instances when a unit cell with even M spins breaks into two identical subsystems with $M/2$ spins each. In those cases, there will be $M/2$ doubly-degenerate SW modes at every $\mathbf{q}$. But for the general case with M spins per unit cell, there are M non-degenerate SW branches at each $\mathbf{q}$. It is always important to evaluate the SW strengths because one or more of those SW modes may have no intensity, as in the $J_1 \to J_2$ limit of a FM chain with alternating exchange.

5.6 Powder spectra

As emphasized in chapters 2 and 3, single crystals are a scientist's best friend. If a single crystal is a bit too small for inelastic neutron-scattering, then several crystallites can sometimes be arranged by hand to form one large pseudo-crystal with enough cross section to generate a decent spectrum. However, even this heroic effort becomes impossible if the individual crystallites are too small. Then, the only way to study spin excitations is by measuring and fitting the inelastic spectrum of a powder consisting of small, randomly oriented particles.

To demonstrate how to evaluate powder spectra, we consider zero-field FM and AF states on the cubic lattices sketched in figure 5.9. The pictured AF configuration is called a G-type AF. By contrast, an A-type AF with alternating FM planes and a C-type AF with the same AF spin configuration on each plane are sketched in figure 5.11.

For a FM on a cubic lattice, the SW frequencies are

$$\hbar\omega(\mathbf{q}) = 6JS(1 - \Gamma_{\mathbf{q}}) + 2KS. \tag{5.56}$$

For an AF on a cubic lattice, they are

$$\hbar\omega(\mathbf{q}) = 2S\left\{(3|J| + K)^2 - (3|J|\Gamma_{\mathbf{q}})^2\right\}^{1/2}. \tag{5.57}$$

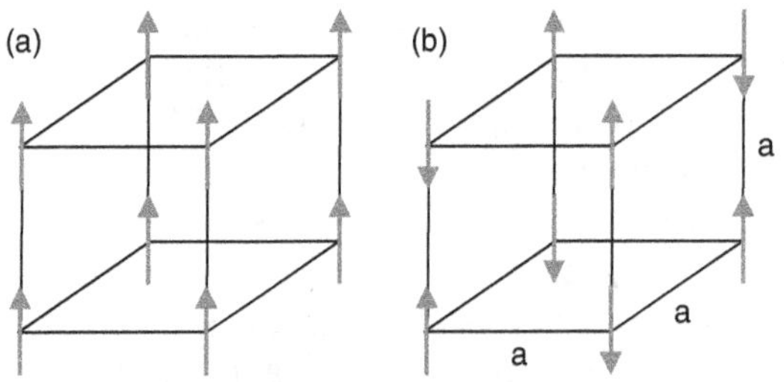

Figure 5.9. The (a) FM and (b) G-type AF spin configurations on a cubic lattice.

In both cases, $\Gamma_{\mathbf{q}}$ is given by equation (5.8). For the FM, the intensity is constant. But for the AF, it varies as

$$\left\{\frac{3|J|(1-\Gamma_{\mathbf{q}})+K}{3|J|(1+\Gamma_{\mathbf{q}})+K}\right\}^{1/2}.$$

As usual, the delta functions in $S_{\alpha\alpha}(\mathbf{q},\omega)$ are replaced by Gaussians with frequency widths of 0.1 meV.

Averaging the SW spectrum over all orientations of a crystal is equivalent to integrating over all orientations of $\mathbf{q}$. With spins aligned along $\mathbf{z}$, $S_{zz}(\mathbf{q},\omega)=0$ and $S_{xx}(\mathbf{q},\omega)=S_{yy}(\mathbf{q},\omega)$. Allowing

$$\mathbf{q}=q(\sin\theta\cos\phi,\ \sin\theta\sin\phi,\ \cos\theta) \tag{5.58}$$

to extend beyond the first BZ and taking advantage of the cubic symmetry, we find that

$$\begin{aligned} S(q,\omega) &= \frac{4}{\pi}\int_0^{\pi/4} d\phi \int_0^1 d(\cos\theta)\left\{2-\frac{{q_x}^2+{q_y}^2}{q^2}\right\}S_{xx}(\mathbf{q},\omega) \\ &= \frac{4}{\pi}\int_0^{\pi/4} d\phi \int_0^1 d(\cos\theta)(1+\cos^2\theta)S_{xx}(\mathbf{q},\omega), \end{aligned} \tag{5.59}$$

where the kinematical constraint $(\delta_{\alpha\beta}-q_\alpha q_\beta/q^2)$ introduces the factor $1+{q_z}^2/q^2=1+\cos^2\theta$ as in equation (5.1).

The resulting powder intensities $S(q,\omega)$ for a FM and AF with $K=0$ are plotted in figures 5.10(a) and (c). Even after averaging all orientations of $\mathbf{q}$, the powder spectrum still contains a great deal of information! For the FM, the first lobe of intensity touches the q axis at $qa/2\pi=1$, corresponding to the six FM Bragg peaks at $(\pm1, 0, 0)$, $(0, \pm1, 0)$ and $(0, 0, \pm1)$. The twelve ordering wavevectors of an AF at $(\pm1/2, 1/2, 1/2)$, $(1/2, \pm1/2, 1/2)$, and $(1/2, 1/2, \pm1/2)$ (along with the opposing wavevectors) all have a magnitude of $\sqrt{3}\pi/a$. Hence, the first spike of intensity for the AF powder spectrum nearly touches the q axis at $qa/2\pi=\sqrt{3}/2\approx0.87$.

For a FM, the largest SW frequency of $12JS=30$ meV is reached when $\Gamma_{\mathbf{q}}=-1$. The largest AF frequency of $6|J|S=15$ meV is reached when $\Gamma_{\mathbf{q}}=0$. Because there are many combinations of wavevectors that give $\Gamma_{\mathbf{q}}=0$, the powder intensity for an AF is quite large at its maximum frequency. By comparison, the FM intensity is largest in the middle of the band, near $\Gamma_{\mathbf{q}}=0$ or $6JS=15$ meV.

The SW density-of-states is defined by [24]

$$g(\omega)=\frac{1}{\hbar N}\sum_n\sum_{\mathbf{q}}{}' \delta(\omega-\omega_n(\mathbf{q})), \tag{5.60}$$

for $\mathbf{q}$ within the first BZ. Although easy to evaluate theoretically by replacing the delta function with a Lorentzian, the SW density-of-states cannot be directly measured because the scattering from a magnetically ordered system is coherent.

With $K=0.2$ meV, the powder spectra are plotted in figures 5.10(b) and (d). As expected, the FM spectrum is simply lifted by the SW gap $\Delta=2KS=1$ meV. For the

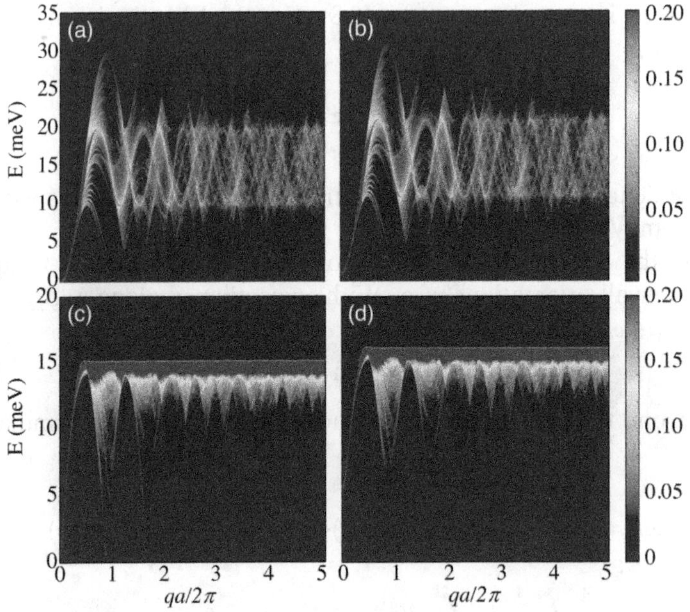

Figure 5.10. The inelastic powder-averaged spectrum on a cubic lattice with $S = 5/2$, for (a,b) a FM with $J = 1$ or (c,d) an AF with $J = -1$ meV, and (a,c) $K = 0$ or (b,d) 0.2 meV.

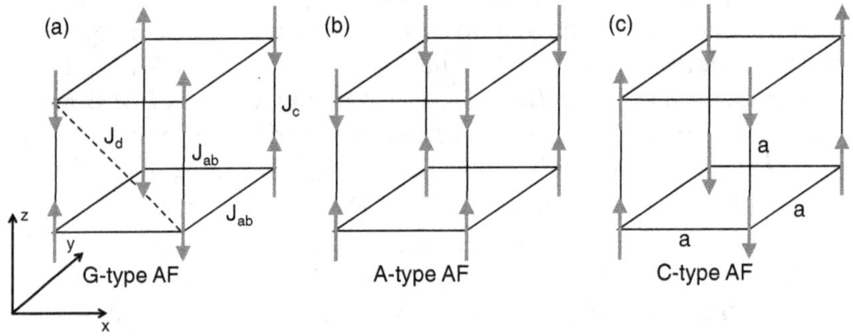

Figure 5.11. (a) G-type, (b) A-type, and (c) C-type AFs on a cubic lattice with three different exchange interactions. $J_{ab} < 0$ and $J_c < 0$ are assumed to be AF. J_d only acts between sites with $\Delta z = \pm a$ do does couple sites within the xy plane.

AF, the bottom of the spectrum is lifted by $\Delta \approx 2S\sqrt{6|J|K} \approx 5.5$ meV. Because the top of the AF band only rises by about 1 meV, the band of intensity narrows significantly with K. Although a great deal of information is lost in the powder averaging, it is still easy to distinguish a FM from an AF. In section 7.4, important information about the exchange parameters is extracted from the powder spectrum of a multiferroic metal-organic framework.

Powder spectra were used by McQueeney *et al* [24] to distinguish between G-, A-, and C-type AF order for three perovskite compounds. Their powder spectra indicated that $LaFeO_3$ is a G-type AF, $LaVO_3$ is a C-type AF, and $LaMnO_3$ is an A-type AF.

5.7 Exercises

1. Would the SW frequency for a FM change with the addition of easy-plane anisotropy E. Would the SW intensity change?
2. Evaluate the SW stiffness D for a FM on a square lattice from the expression $\hbar\omega(\mathbf{q}) = \Delta + Dq^2$ near $\mathbf{q} = 0$ for $K \neq 0$.
3. Use the measured SW stiffness for face-centered cubic Ni in figure 1.1 to estimate its nearest-neighbor exchange coupling J. Using the full expression for the SW frequency (not just the quadratic expansion), evaluate the predicted SW frequencies at wavevector 0.5 Å^{-1} in directions [1, 0, 0], [1, 1, 0], and [1, 1, 1]. How do these compare with the predicted value from the quadratic expansion?
4. For a FM with spins aligned along $\mathbf{z}$, would easy-axis anisotropy K and a magnetic field $\mathbf{B}$ compete with one another if the magnetic field was along $-\mathbf{z}$ instead of along $\mathbf{z}$?
5. Show that $\mathcal{H}_1$ vanishes for the FM on a square lattice.
6. Generalize the results for a FM on a square lattice by adding the next-nearest-neighbor coupling J'.
7. Show that the 'mother-of-all sum rules' stated in equation (4.53) is obeyed by the FM on a square lattice.
8. In the limit $J_2 \to 0$, the FM chain with alternating exchange develops a flat, zero-frequency mode. Physically, why does this SW branch appear?
9. Derive the inverse $\underline{X}^{-1}$ matrix of equation (5.27) for the FM chain with alternating exchange.
10. Reconsider the FM chain with alternating exchange of Section 5.3. Suppose that the lattice constants α_1 (associated with J_1) and α_2 (associated with J_2) are different with $\alpha_1 + \alpha_2 = 2\alpha$. What are the new SW frequencies and their intensities? If $\alpha_1 = n_1\alpha/m$ and $\alpha_2 = n_2\alpha/m$, where n_i and m are integers with $n_1 + n_2 = 2m$, what is the wavevector periodicity of the SW intensities?
11. Derive $\underline{L}$ of equation (5.37) for the FM on a honeycomb lattice with nearest-neighbor exchange.
12. Show that the 'Dirac' points of the honeycomb lattice lie at the corners of the first structural BZ.
13. Show that the ratio of intensities of modes 1 and 2 on the honeycomb lattice as either of the 'Dirac' points $(1/3, \pm\sqrt{3}/9)$ is approached from the x axis equals 1/3.
14. Evaluate the SW frequencies for an AF on a square lattice with easy-plane anisotropy E but $K = 0$ and $B = 0$. Does the Goldstone mode survive?
15. Can you think of some other interaction term that would break the degeneracy of the two SW modes for an AF on a square lattice in zero field?
16. Derive the AF SW velocity v of a square lattice using $\hbar\omega(\mathbf{q}) = v|\mathbf{q} - \mathbf{Q}|$ near $\mathbf{Q} = (1, 1)\pi/a$ when $K = E = 0$.

17. Evaluate the canted moment along **z** above B_{SF} for the AF on a square lattice in a magnetic field with easy-axis anisotropy. Derive B_{SF} from the balance between the anisotropy and Zeeman energies. Does the lower SW frequency $\omega(\mathbf{Q})$ of the AF precisely vanish as B_{SF} is approached from below?
18. For an AF on a square lattice at nonzero field $0 < B < B_{SF}$, compare the intensities of the upper and lower SW branches at **Q** when the easy-plane anisotropy $E < 0$.
19. To obtain $\underline{L}(\mathbf{q})$ for the AF state, we took the spins $r = 1$ and 2 to have $\cos\theta_r = \pm 1$, $\cos\phi_r = 1$, and $\sin\phi_r = 0$. Of course, the value of ϕ_r should not matter when the spins are aligned along $\pm$**z**. But what would happen to the E terms in equation (5.47) for $\underline{L}(\mathbf{q})$ if ϕ_r were arbitrary? Could this change have any effect on the inelastic spectra?
20. Using the expression for the AF SW frequencies on a cubic lattice, evaluate the width of the band in the powder spectrum (maximum minus the minimum frequency). Do the same for a FM.
*21. Consider the G-type AF in figure 5.11(a) with different exchanges $J_{ab} < 0$ in the ab plane and $J_c < 0$ along the c axis. Also add the diagonal interaction J_d between next-nearest neighbors on adjacent planes separated by $\Delta z = \pm a$ and easy-axis anisotropy K along **z**. Show that the G-type AF becomes unstable with respect to the A- or C-type AFs when $J_d < -|J_c|/4$ or $J_d < -|J_{ab}|/2$, respectively, regardless of K. When $K = 0$, show that the SW frequency of the G-type AF becomes imaginary when either of those conditions is satisfied.
*22. Derive equations (5.51) and (5.55) for the SW intensity of a square-lattice AF with easy-axis anisotropy $K > 0$ and in a magnetic field along **z** but $E = 0$.

References

[1] Mook H A, Nicklow R M, Thompson E D and Wilkinson M K 1969 Spin-wave spectrum of nickel metal *J. Appl. Phys.* **40** 1450–1

[2] Mook H A, Lynn J W and Nicklow R M 1973 Temperature dependence of the magnetic excitations in nickel *Phys. Rev. Lett.* **30** 556–9

[3] Mook H A and Nicklow R M 1973 Neutron scattering investigation of the magnetic excitations in iron *Phys. Rev.* B **7** 336–42

[4] Cable J W and Wollan E O 1968 Neutron diffraction study of the magnetic behavior of gadolinium *Phys. Rev.* **165** 733–4

[5] Reimers W, Hellner E, Treutmann W and Heger G 1982 Magnetic phase diagram of the system $Mn_{1-x}Cr_xSb$ ($0 \leqslant x \leqslant 1$ *J. Phys. C: Solid State Phys.* **15** 3597

[6] Radhakrishna P and Cable J W 1996 Inelastic-neutron-scattering studies of spin-wave excitations in the pnictides MnSb and CrSb *Phys. Rev.* B **54** 11940–3

[7] Samuelsen E J, Silberglitt R, Shirane G and Remeika J P 1971 Spin waves in ferromagnetic $CrBr_3$ studied by inelastic neutron scattering *Phys. Rev.* B **3** 157–66

[8] Narath A 1964 Nuclear magnetic resonance studies in antiferromagnetic $CrCl_3$ *J. Appl. Phys.* **35** 838

[9] Ray R K, Tsang J C, Dresselhaus M S, Aggarwal R L and Reed T B 1971 Raman scattering from europium chalcogenides *Phys. Lett.* A **37** 129–31

[10] Güntherodt G, Merlin R and Grünberg P 1979 Spin-disorder-induced Raman scattering from phonons in europium chalcogenides. I. Experiment *Phys. Rev.* B **20** 2834–49

[11] Suits J C and Argyle B E 1965 Magnetic Birefringence of EuSe *Phys. Rev. Lett.* **14** 687

[12] Stout J W and Adams H E 1942 Magnetism and the third law of thermodynamics. the heat capacity of manganous fluoride from 13 to 320 °K *J. Am. Chem. Soc.* **64** 1535–8

[13] Shull C G, Strauser W A and Wollan E O 1951 Neutron diffraction by paramagnetic and antiferromagnetic substances *Phys. Rev.* **83** 333–45

[14] Keffer F and O'Sullivan W 1957 Problem of spin arrangements in MnO and similar antiferromagnets *Phys. Rev.* **108** 637–44

[15] Lines M E and Jones E D 1965 Antiferromagnetism in the face-centered cubic lattice. II. Magnetic properties of MnO *Phys. Rev.* **139** A1313–27

[16] Kuiper P, Searle B G, Rudolf P, Tjeng and Chen C T 1993 x-ray magnetic dichroism of antiferromagnet Fe_2O_3: the orientation of magnetic moments observed by Fe 2p x-ray absorption spectroscopy *Phys. Rev. Lett.* **70** 1549–52

[17] Wilkinson M K, Cable J W, Wollan E O and Koehler W C 1959 Neutron diffraction investigations of the magnetic ordering in $FeBr_2$, $CoBr_2$, $FeCl_2$, and $CoCl_2$ *Phys. Rev.* **113** 497–507

[18] Jacobs I S and Lawrence P E 1967 Metamagnetic phase transitions and hysteresis in $FeCl_2$ *Phys. Rev.* **164** 866–78

[19] Guggenheim H J, Hutchings M T and Rainford B D 1968 Neutron-scattering determination of spin-wave dispersion relations in FeF_2 *J. Appl. Phys.* **39** 1120–1

[20] Kittel C 1986 *Introduction to Solid State Physics* 6th edn (New York: Wiley)

[21] Pershoguba S S, Banerjee S, Lashley J C, Park J, Ågren H, Aeppli G and Balatsky A V 2018 Dirac magnons in honeycomb ferromagnets *Phys. Rev.* X **8** 011010

[22] Boyko D, Balatsky A V and Haraldsen J T 2018 Evolution of magnetic Dirac bosons in a honeycomb lattice *Phys. Rev.* B **97** 014433

[23] Castro Neto A H, Guinea F, Peres N M R, Novoselov K S and Geim A K 2009 The electronic properties of graphene *Rev. Mod. Phys.* **81** 109–62

[24] McQueeney R J, Yan J-Q, Chang S and Ma J 2008 Determination of the exchange anisotropy in perovskite antiferromagnets using powder inelastic neutron scattering *Phys. Rev.* B **78** 184417

Chapter 6

Model non-collinear magnets

And he built a crooked house.

—Robert Henlein

To the artist, there is never anything ugly in nature.

—Auguste Rodin

6.1 Introduction

Unsurprisingly, materials with NC spin states are not as well understood as materials with CL states. Many materials with NC states have been the subject of recent interest due to their possible multiferroic behavior, meaning that their electrical and magnetic properties are coupled. The inelastic scattering and optical absorption spectra provide valuable information about the NC spin states of multiferroic materials.

NC spin states emerge from the competition between different exchange interactions or between exchange and DM interactions [1, 2]. The former includes cases where the geometric structure of the lattice does not allow all nearest-neighbor AF exchange interactions to be minimized. Said to be geometrically frustrated (GF), those crystal structures include stacked triangular layers, kagome lattices, and spinels [3]. Competing exchange (CE) on a non-frustrated, bipartite lattice involves the competition between FM and AF exchange interactions or between different AF interactions.

Because the DM interaction is typically an order of magnitude weaker than exchange interactions, the competition between the DM and exchange energies produces cycloids or helices with very long wavelengths or with wavevectors **Q** that

doi:10.1088/978-1-64327-114-9ch6

are almost commensurate. So these systems are almost FM or AF. In contrast, the difference between exchange interactions is typically of the same order as the individual exchange energies. Hence, CE produces cycloids or helices with short wavelengths or with wavevectors that are far from commensurate.

Table 6.1 provides details about a few NC compounds, most of which exhibit multiferroic behavior. This list includes both type-I and type-II multiferroics. In a type-I or 'proper' multiferroic, the electric polarization **P** develops at a higher temperature than the magnetic order. A small spin-induced change in the polarization then appears just below the magnetic transition temperature. In a type-II or 'improper' multiferroic, the electric polarization and magnetic order are coupled and develop at the same temperature [1, 2]. Recently, Benedek and Fennie [4] proposed an additional category called 'hybrid-improper' ferroelectrics like $Ca_3Mn_2O_7$, where the ferroelectric and magnetic properties are driven by a combination of two nonpolar rotational modes, which may not have the same ordering temperature.

Multiferroic behavior in a type-II multiferroic is caused by one of three different mechanisms: the inverse DM interaction, also called the SC mechanism [35, 36]; *p–d* orbital hybridization [37] produced by the hopping of electrons from the transition metal to the neighboring oxygen; or exchange striction [2] due to the dependence of the exchange interactions on an electric field. As already mentioned in section 4.4, the SC or inverse DM interaction produces an electric polarization $\mathbf{P} \propto (\mathbf{R}_j - \mathbf{R}_i) \times (\mathbf{S}_i \times \mathbf{S}_j)$ perpendicular to both the wavevector **Q** and the helicity $\mathbf{S}_i \times \mathbf{S}_j$. With **P** parallel to **Q**, *p–d* orbital hybridization appears in Ga- or Al-doped $CuFeO_2$ [13] and

Table 6.1. Some materials with NC spin states. $\mathbf{Q}_h$ is in hexagonal basis (see section 4.10, appendix 4.D), $\mathbf{Q}^*$ indicates an additional FM component. Spin is not listed for MnSi because it is itinerant.

Material	S	**Q**	State, MF	Origin	References
$BiFeO_3$	5/2	(0.5045,0.5, 0.4955),	Cycloid, I	DM	[5–8]
$YMnO_3$	2	$(0.33,0.33,0)_h$	Cycloid, I	GF	[9, 10]
$(NH_4)_2[FeCl_5(H_2O)]$	5/2	(0, 0, 0.23)	Cycloid, II	CE	[11, 12]
$CuFeO_2$	5/2	$(0.21, 0.21, 1.5)_h$	Helix, II	GF	[13, 14]
$MnWO_4$	2	(−0.21, 0.5, 0.46) (AF2)	Cycloid, II	CE	[15–17]
		(0.23, 0.5,−0.46) (AF5)	Cycloid, II	CE	[15, 16]
$TbMnO_3$	2	(0, 0.27, 1)	Cycloid, II	CE	[18–20]
$CaMn_7O_{12}$	2, 3/2	(0, 1, 0.96) (AFM I)	Helix, II	DM	[21]
$CuCrO_2$	3/2	$(0.33, 0.33, 0)_h$	Helix, II	GF	[22, 23]
$CoCr_2O_4$	3/2	$(0.63,0.63,0)_h$	Cycloid, II	GF	[24, 25]
$Ni_3V_2O_8$	1	(0,1, 0. 40) (C′)	Cycloid, II	CE	[26, 27]
$LiCu_2O_2$	1/2	(0.5, 0.17, 0)	Cycloid, II	CE	[28, 29]
$LiCuVO_4$	1/2	(0, 0.53, 0)	Cycloid, II	CE	[30]
$SrMnVO_4$	5/2	(0, 0, 0.5)*	Canted AF	DM	[31]
$MnCl_3$(bpy)	2	(0, 0.5, 0.5)	Canted AF	DM	[32, 33]
$Ba_2CuGe_2O_7$	1/2	(1.027, 1.027, 0)	Helix	DM	[34]
MnSi		(0.014, 0.014, 0.014)	Helix	DM	[38, 39]

$CuCrO_2$ [22]. Finally, exchange striction can induce an electric polarization in either CL or NC magnets and occurs in $YMnO_3$ [10].

Since the first two mechanisms require spin–orbit coupling, it is natural to wonder if spin–orbit coupling also distorts the NC spin states of type-II multiferroics [40]. However, GF and CE play the dominant roles in determining the spin states of many multiferroic materials even though they do not involve spin–orbit coupling. While **P** may be produced by the inverse DM interaction, the spin state itself could be controlled by CE. So it is important to distinguish between the mechanism that produces ferroelectricity and the mechanism that controls the NC state. Nevertheless, *both* the DM interaction and CE may play important roles in governing the spin states of some multiferroics like $MnWO_4$ [40].

As mentioned in section 4.4, the THz absorption in a multiferroic can be changed by switching the direction of light propagation. Called non-reciprocal directional dichroism, this effect was observed in $Ba_2CoGe_2O_7$ [41–43], $Ca_2CoSi_2O_7$, $Sr_2CoSi_2O_7$ [43], (Eu, Y)MnO_3 [44], $BiFeO_3$ [45], $Sm_{0.5}La_{0.5}Fe_3(BO_3)_4$ [46], $CaBaCo_4O_7$ [47], $MnWO_4$ [48] and CuB_2O_4 [49]. In some cases, a SW mode may have very low optical absorption for one propagation direction and high absorption for the other [43]. So at certain SW frequencies, a multiferroic material may behave like an optical diode [45]!

Due to the coupling $-\mathbf{E} \cdot \mathbf{P}$, a dc electric field **E** can alter the SW frequencies and optical absorption in a multiferroic material, which was observed by Raman scattering in $BiFeO_3$ [50]. As demonstrated by the change of the THz absorption in the rare-earth borates $NdFe_3(BO_3)_4$ and $SmFe_3(BO_3)_4$ [51] and the rotation of the THz polarization in $DyMnO_3$ [52], an electric field can control both magnetic excitations and domains.

This chapter begins by treating a canted AF chain. We then examine cycloids produced by either CE or the DM interaction. We also show how to treat a cycloid or helix that is incommensurate in 2D or 3D. Finally, we examine a helix produced by GF on a triangular lattice.

A problem with $M > 2$ sublattices or with $\underline{\mathcal{L}}$ bigger than 4×4 is best solved numerically, i.e. with a computer. There are several software packages that can be used to obtain the eigenvalues and eigenvectors of a non-Hermitian matrix. We favor the free packages available through the Netlib repository[1], which are written in FORTRAN. Those codes can also be used to perform a Cholesky decomposition, which allows you to evaluate the SW spectrum by diagonalizing the Hermitian matrix derived in appendix A of chapter 4. While FORTRAN has been superseded by more advanced programming languages, the netlib codes are easy to incorporate into sophisticated fitting programs.

6.2 An AF chain with alternating DM interactions

Table 6.1 indicates that the most common effect of the DM interaction is to cant the spins of an AF. We start by studying a chain of spins coupled by the AF exchange

[1] www.netlib.org

$J < 0$ and canted away from an easy-axis with $K > 0$ by an alternating DM interaction D. The ground state is sketched in figure 6.1. A slight generalization of this model was recently used [33] to study the $S = 2$ molecule-based magnet $MnCl_3$(bpy), where bpy is an organic cation that separates the chains and breaks inversion symmetry.

We will ignore the very weak coupling J' between chains, turning this into a 1D problem. Of course, long-range magnetic order in 1D is destroyed by either thermal fluctuations or impurities [53]. However, even a very small interchain coupling J' will stabilize a magnetic state at $T = 0$. A quasi-1D system with both interchain coupling J' and easy-axis anisotropy K will exhibit long-range magnetic order at nonzero temperature.

The 1D Hamiltonian of this system is

$$\mathcal{H} = -J\sum_i \mathbf{S}_i \cdot \mathbf{S}_{i+1} - K\sum_i S_{iz}{}^2 - D\sum_i(-1)^i\mathbf{x} \cdot (\mathbf{S}_i \times \mathbf{S}_{i+1}). \tag{6.1}$$

In $MnCl_3$(bpy), the direction of the DM vector $\mathbf{D}$ along $\mathbf{x}$ conforms with the symmetry rules derived by Moriya [54] for materials with broken inversion symmetry. Reflecting the alternating positions of the bpy radical from one side of the chain to the other, the DM interaction alternates sign along the chain from one bond to the next.

To minimize the energy $E = \langle H \rangle$, we parameterize the spins in the magnetic unit cell as $\mathbf{S}_n = S(0, \sin\theta_n, \cos\theta_n)$ with $n = 1$ and 2:

$$\begin{aligned}\frac{E}{N} = &- JS^2\cos(\theta_2 - \theta_1) - \frac{1}{2}KS^2\{\cos^2\theta_1 + \cos^2\theta_2\}\\ &- DS^2\sin(\theta_2 - \theta_1).\end{aligned} \tag{6.2}$$

By symmetry, $\theta_2 = \pi - \theta_1$ and the energy is minimized when

$$\tan 2\theta_1 = \frac{2D}{K + 2|J|}. \tag{6.3}$$

As expected, $\theta_1 = 0$ when $D = 0$. This expression indicates that the canting of the spins by the DM interaction is opposed by both the easy-axis anisotropy and the exchange coupling.

The spin dynamics is then obtained by diagonalizing a 4×4 matrix. To transform into the local reference frame of each spin, we use the unitary rotation matrix $\underline{U}^r$ with $\phi_r = \pi/2$ to write

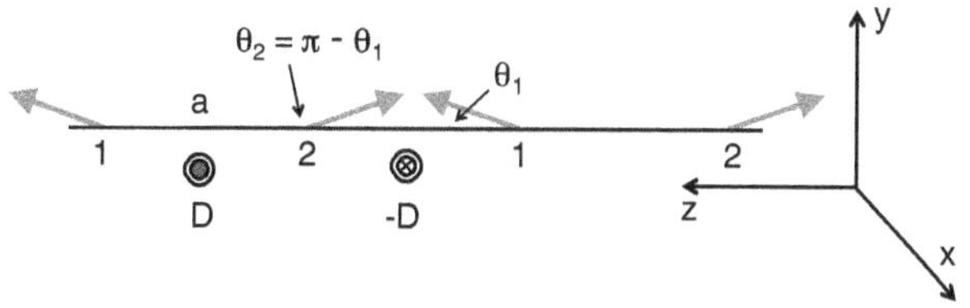

Figure 6.1. The ground state of a chain with AF interactions and an alternating DM interaction that cants the spins towards $\mathbf{y}$.

$$S_{rx} = -\bar{S}_{ry}, \tag{6.4}$$

$$S_{ry} = \cos\theta_r\, \bar{S}_{rx} + \sin\theta_r\, \bar{S}_{rz}, \tag{6.5}$$

$$S_{rz} = -\sin\theta_r\, \bar{S}_{rx} + \cos\theta_r\, \bar{S}_{rz}. \tag{6.6}$$

With the help of the shortcuts in equations (4.148)–(4.153) for the DM interaction, you can show that the nonzero matrix elements of $\underline{L}(\mathbf{q})$ are (exercise 6.3)

$$L_{11} = L_{22} = L_{33} = L_{44} = |J|S\cos(2\theta_1) - \frac{KS}{2}\{1 - 3\cos^2\theta_1\} + DS\sin(2\theta_1), \tag{6.7}$$

$$L_{12} = L_{21} = L_{34} = L_{43} = \frac{|J|S}{2}\{1 - \cos(2\theta_1)\}\cos(qa) - \frac{DS}{2}\sin(2\theta_1)\cos(qa), \tag{6.8}$$

$$L_{14} = L_{41} = L_{23} = L_{32} = -\frac{|J|S}{2}\{1 + \cos(2\theta_1)\}\cos(qa) - \frac{DS}{2}\sin(2\theta_1)\cos(qa), \tag{6.9}$$

$$L_{13} = L_{31} = L_{24} = L_{42} = -\frac{KS}{2}\sin^2\theta_1. \tag{6.10}$$

As required, $\underline{L}(\mathbf{q})$ is Hermitian.

We now solve for the eigenvalues and eigenvectors numerically. Using a rather large value of $D = 0.4$ meV, corresponding to a tilt of $\theta_1 = 0.058\,\pi$, the inelastic spectrum is plotted in figures 6.2(a) and (b) for $K = 0$ or 0.1 meV. With such a large D, you can clearly see the mode splitting near $L = 0.5$. The splitting is largest for $K = 0$, when one Goldstone mode survives due to rotational invariance about $\mathbf{x}$. That mode becomes gapped when $K > 0$.

The inelastic matrix elements for $J = -1$ meV, $K = 0.1$ meV, and $D = 0$ through 0.4 meV are given in table 6.2 for $\mathbf{q} = \mathbf{Q}$. When $D = 0$, the spins are aligned along $\pm\mathbf{z}$. The xx and yy matrix elements of $S^{(n)}_{\alpha\beta}(\mathbf{Q})$ for the degenerate modes are then identical. Since there are no spin fluctuations along $\mathbf{z}$, $S^{(n)}_{zz}(\mathbf{q}) = 0$ for all $\mathbf{q}$. When $D > 0$, modes 1 and 2 are dominated by spin fluctuations along $\mathbf{x}$ and $\mathbf{y}$, respectively. As D increases, spin fluctuations along $\mathbf{y}$ for mode 2 are suppressed compared to spin fluctuations along $\mathbf{x}$ for mode 1 so that $S^{(2)}_{yy}(\mathbf{Q}) < S^{(1)}_{xx}(\mathbf{Q})$. Correspondingly, $\omega_2(\mathbf{Q})$ increases while $\omega_1(\mathbf{Q})$ remains relatively unchanged. The small matrix element $S^{(2)}_{zz}(\mathbf{Q})$ for $D > 0$ indicates that mode 2 also contains some spin fluctuations along $\mathbf{z}$, which are allowed by the spin canting away from the z-axis.

What would THz spectroscopy reveal about this material? At $\mathbf{q} = 0$, the optical absorption is proportional to $\omega_n|\boldsymbol{\mu}_n \cdot \mathbf{h}|^2/N$ where $\boldsymbol{\mu}_n = \langle n|\mathbf{M}|0\rangle/\mu_B$ is the matrix

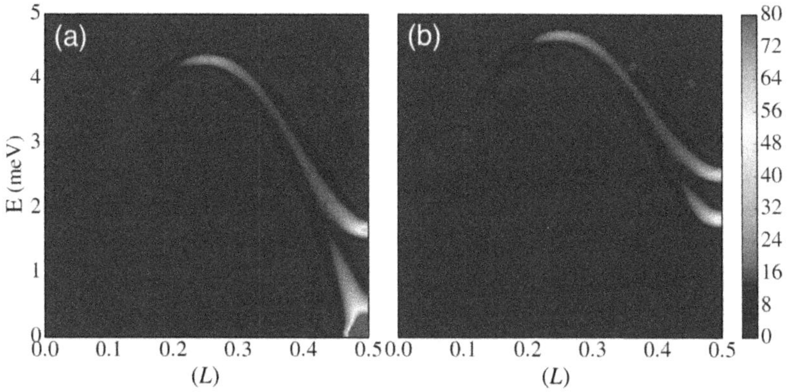

Figure 6.2. The inelastic spectrum of an AF-coupled, canted chain with $S = 2$, $J = -1$ meV, $D = 0.4$ meV, and (a) $K = 0$ or (b) 0.1 meV.

Table 6.2. Inelastic matrix elements and squared optical matrix elements for an AF chain with $S = 2$, $J = -1$ meV, and $K = 0.1$ meV. D and $\hbar\omega_n(\mathbf{q})$ (at $\mathbf{q} = 0$ or $\mathbf{Q}$) are in meV. Modes $n = 1$ and 2 are given on alternate lines. A small positive $D = 0^+$ is used to separate modes 1 and 2 in the first two lines.

D	θ_1/π	$\hbar\omega_n(\mathbf{q})$	$S_{xx}^{(n)}(\mathbf{Q})$	$S_{yy}^{(n)}(\mathbf{Q})$	$S_{zz}^{(n)}(\mathbf{Q})$	$\lvert\mu_{nx}\rvert^2/N$	$\lvert\mu_{ny}\rvert^2/N$	$\lvert\mu_{nz}\rvert^2/N$
0^+	0	1.83	1.14	0	0	0	1.75	0
		1.83	0	1.14	0	1.75	0	0
0.1	0.015	1.83	1.15	0	0	0	1.74	0.08
		1.88	0	1.12	0.0001	1.78	0	0
0.2	0.030	1.82	1.15	0	0	0	1.72	0.33
		2.02	0	1.06	0.0005	1.87	0	0
0.3	0.044	1.82	1.16	0	0	0	1.69	0.73
		2.23	0	0.98	0.001	2.00	0	0
0.4	0.058	1.80	1.17	0	0	0	1.66	1.27
		2.50	0	0.90	0.002	2.15	0	0

element of the magnetization operator $\mathbf{M} = 2\mu_B \sum_i \mathbf{S}_i$ and $\mathbf{h}$ is the orientation of THz magnetic field $\mathbf{H}^\omega$ (see section 4.4). Of course, this assumes that the electric polarization does not contribute to the optical absorption.

Table 6.2 contains the squared optical matrix elements for $D = 0$ through 0.4 meV. When $D = 0$, the x and y components of $|\mu_{n\alpha}|^2/N$ are identical and the z component vanishes, as expected from the vanishing SW amplitude along $\mathbf{z}$. For $D > 0$, the optical absorption with $\mathbf{h} = \mathbf{y}$ is dominated by mode 1 while the absorption with $\mathbf{h} = \mathbf{x}$ is dominated by mode 2. The difference between the optical absorption for $\mathbf{h} = \mathbf{x}$ and $\mathbf{y}$ is caused by the canting of the spins towards $\mathbf{y}$. Canting also produces the small optical weight $|\mu_{1z}|^2/N$. Like $S_{zz}^{(2)}(\mathbf{Q})$, $|\mu_{1z}|^2/N$ increases as D^2.

The optical weights $|\mu_{n\alpha}|^2/N$ at $\mathbf{q} = 0$ and the inelastic intensities $S_{\alpha\alpha}^{(n)}(\mathbf{Q})$ are complementary. When $D > 0$, large $S_{xx}^{(1)}(\mathbf{Q})$ and $S_{yy}^{(1)}(\mathbf{Q}) = 0$ are associated with large $|\mu_{1y}|^2/N$ and $|\mu_{1x}|^2/N = 0$. Similarly, large $S_{yy}^{(2)}(\mathbf{Q})$ and $S_{xx}^{(2)}(\mathbf{Q}) = 0$ are associated with large $|\mu_{2x}|^2/N$ and $|\mu_{2y}|^2/N = 0$. This complementarity arises because different components of the local spin disturbance $\delta\mathbf{S}_r(n,\mathbf{q})$ contribute to the optical absorption at $\mathbf{q} = 0$ and to the inelastic intensity at $\mathbf{q} = \mathbf{Q}$ [58]. As expected, the nonzero optical weights $\boldsymbol{\mu}_n$ are parallel to the average spin disturbance $(1/M)\sum_r \delta\,\mathbf{S}_r(n,\mathbf{q} = 0)$ from equation (4.93). Recall that the optical absorption $|\mu_{n\alpha}|^2/N$ is proportional to the inelastic intensity $S_{\alpha\alpha}^{(n)}(\mathbf{q})$ *only* when the latter is also evaluated at $\mathbf{q} = 0$.

6.3 A helix or cycloid produced by CE

Helices and cycloids are produced by the competition between nearest-neighbor exchange and some other energy (either AF next-neighbor exchange or DM interactions). For FM interactions between nearest neighbors, a helix in the xy plane and a cycloid in the xz plane with periods of $8a$ with and clockwise turn angles of 45° are sketched in figures 6.3(a) and (b), respectively. The cycloid in figure 6.3(c) is drawn for AF interactions between nearest neighbors. Compared to cycloid (b), every other spin is reversed in cycloid (c). This section treats helices or cycloids produced by CE; the next treats cycloids or helices produced by the DM interaction.

First, consider a cycloid produced by CE on a 1D chain along $\mathbf{x}$ with lattice constant a. Then

$$\mathcal{H} = -J_1 \sum_i \mathbf{S}_i \cdot \mathbf{S}_{i+1} - J_2 \sum_i \mathbf{S}_i \cdot \mathbf{S}_{i+2}. \tag{6.11}$$

Regardless of whether the nearest-neighbor exchange J_1 is FM or AF, the next-nearest neighbors will be aligned parallel to each other if the next-nearest-neighbor exchange J_2 is FM. Provided that $|J_2| > |J_1|/4$, an AF interaction $J_2 < 0$ generates the cycloid

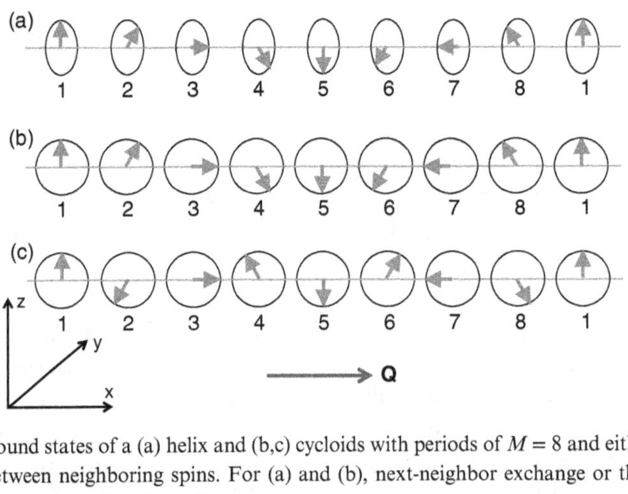

Figure 6.3. The ground states of a (a) helix and (b,c) cycloids with periods of $M = 8$ and either (a,b) FM or (c) AF interactions between neighboring spins. For (a) and (b), next-neighbor exchange or the DM interaction produces a clockwise turn angle $\Delta\theta = \pi/4$.

$$\mathbf{S}_r = S(\sin(Qra), 0, \cos(Qra)). \tag{6.12}$$

Although the plane of the spins is arbitrary, you can imagine that weak easy-plane anisotropy stabilizes the spins in the xz plane. For either FM or AF exchange J_1, the wavevector of the cycloid is given by

$$Q = \frac{1}{a}\cos^{-1}\left(\frac{J_1}{4|J_2|}\right). \tag{6.13}$$

If $J_1 > 0$, $Q \to 0$ as $|J_2| \to J_1/4$; if $J_1 < 0$, $Q \to \pi/a$ as $|J_2| \to -J_1/4$.

For $J_1 > 0$, the turn angle between neighboring spins is

$$\Delta\theta = Qa = \frac{2\pi p}{M}, \tag{6.14}$$

where p is an integer. Since $M\Delta\theta = 2\pi p$, the cycloid completes p full 2π rotations in the period Ma. As $|J_2| \to J_1/4$, $M \to \infty$ and $Q \to 0$. Rearranging equation (6.13), we obtain

$$J_2 = -\frac{J_1}{4}\sec\left(\frac{2\pi p}{M}\right). \tag{6.15}$$

The turn number p is restricted so that $\cos(\Delta\theta) > 0$ or $|p| < M/4$.

For $J_1 < 0$, we must treat the cases of even and odd M separately. For any M,

$$\Delta\theta = Qa = \pi + \frac{\pi p}{M}, \tag{6.16}$$

where the integer p is the number of full 2π rotations completed by the cycloid in distance $2Ma$ not including the AF oscillations. If M is odd, then p must also be odd so that $M\Delta\,\theta = (M{+}p)\pi$ is an even multiple of π. If M is even, then p must also be even so that $M\Delta\theta = (M + p)\pi$ is again an even multiple of π. With $M = 5$ and $p = 1$, $\Delta\theta = 6\pi/5$ or $-4\pi/5$. With $M = 8$ and $p = 2$, $\Delta\theta = 5\pi/4$ or $-3\pi/4$ as in figure 6.3(c). So we find

$$J_2 = \frac{J_1}{4}\sec\left(\frac{\pi p}{M}\right), \tag{6.17}$$

with even p for even M and odd p for odd M. Of course, physical quantities like $S(q,\omega)$ can only depend on the ordering wavevector Q and not on whether M is odd or even! The condition $\cos(\Delta\theta - \pi) > 0$ requires that $|p| < M/2$ for either even or odd M.

If there are M sites in the magnetic unit cell, then $\underline{L}$ is a $2M \times 2M$ matrix with

$$\begin{aligned} L_{rr} = {} & \frac{J_1S}{2}\left\{\cos(\theta_r - \theta_{[r+1]}) + \cos(\theta_r - \theta_{[r-1]})\right\} \\ & + \frac{J_2S}{2}\left\{\cos(\theta_r - \theta_{[r+2]}) + \cos(\theta_r - \theta_{[r-2]})\right\}, \end{aligned} \tag{6.18}$$

$$L_{r,[r\pm1]} = -\frac{J_1S}{4}\,G_1(r, [r \pm 1])\,e^{\pm iqa}, \tag{6.19}$$

$$L_{r,\,[r\pm 2]} = -\frac{J_2 S}{4}\, G_1(r,\,[r \pm 2])\, e^{\pm 2iqa}, \tag{6.20}$$

$$L_{r,\,[r\pm 1]+M} = -\frac{J_1 S}{4}\, G_2(r,\,[r \pm 1])\, e^{\pm iqa}, \tag{6.21}$$

$$L_{r,\,[r\pm 2]+M} = -\frac{J_2 S}{4}\, G_2(r,\,[r \pm 2])\, e^{\pm 2iqa}, \tag{6.22}$$

where r runs from 1 to M. All other matrix elements can be obtained from the symmetry relation $L_{rs}(\mathbf{q}) = L_{r+M,\,s+M}(\mathbf{q})$ and the fact that $\underline{L}(\mathbf{q})$ is Hermitian. To implement periodic boundary conditions, we define $[r] = \mathrm{mod}(r,\, M)$ so that $[M{+}s] = s$. The matrix elements above are given in terms of the functions

$$G_1(r,\, s) = \cos(\theta_r - \theta_s) + 1, \tag{6.23}$$

$$G_2(r,\, s) = \cos(\theta_r - \theta_s) - 1, \tag{6.24}$$

which were derived in section 4.9 (appendix 4.C) and simplified for $\phi_r = 0$.

It is now straightforward to numerically evaluate the inelastic spectrum for any cycloidal period. Figures 6.4(a) and (b) plot the spectra for cycloids with $M = 10$ ($J_2 = -0.31$ meV) and 20 ($J_2 = -0.26$ meV), respectively, both with FM nearest-neighbor exchange $J_1 = 1$ meV and $p = 1$. In the former case, strong inelastic peaks appear at $H = 0$, 0.1, and 0.2 (as well as at $1-H$). For $M = 20$, strong inelastic peaks appear at $H = 0$, 0.05, and 0.1 (and the corresponding $1-H$). Notice that the three SW branches are clones separated by $\Delta H = 1/M$ but with different intensities. For both $M = 10$ and 20, the side branches have much lower intensities than the SW branch centered at $H = 0.5$.

When the nearest-neighbor exchange $J_1 = -1$ meV is AF, the inelastic spectra are plotted in figures 6.4(c) and (d) for the same periods as in figures 6.4(a) and (b), now with $p = 2$. In order to clearly separate the peaks at $H = 0.5 \pm 1/M$, we used 'super-fine' wavevector and energy resolutions of $\Delta q = 0.001$ rlu and $\hbar\Delta\omega = 0.05$ meV. If we had used the same wavevector and energy resolutions $\Delta q = 0.02$ rlu and $\hbar\Delta\omega = 0.1$ meV as in figures 6.4(a) and (b), then the peaks at $H = 0.5 \pm 1/M$ would be hard to separate, especially for $M = 20$. The resulting spectra are very different than for FM nearest-neighbor interactions. As the period Ma increases, the peaks at $H = 0.5 \pm 1/M$ coalesce and the spectrum approaches that for a typical AF with a Goldstone mode at $H = 0.5$.

6.4 A cycloid produced by DM interactions

Now consider a cycloid propagating along $\mathbf{x}$ generated by a DM interaction D along $\mathbf{y}$. To make things more realistic, we also include easy-axis anisotropy K along $\mathbf{z}$.

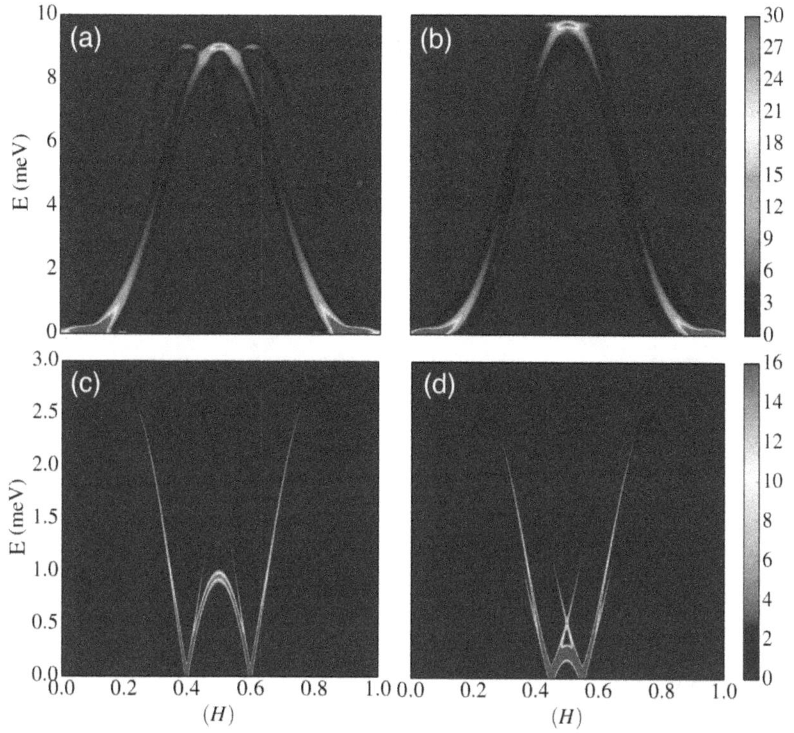

Figure 6.4. The inelastic spectrum of a cycloid produced by CE interactions with $S = 5/2$, (a,b) $J_1 = 1$ meV or (c,d) $J_1 = -1$ meV and periods of either (a,c) $M = 10$ ($J_2 = -0.31$ meV) or (b,d) $M = 20$ ($J_2 = -0.26$ meV). For (a) and (b), $p = 1$; for (c) and (d), $p = 2$.

This model roughly describes the cycloid observed in the room-temperature multiferroic $BiFeO_3$ [5–7] (see section 8.2 for more details). The nearest-neighbor exchange interaction J between sites i and $i + 1$ can be either FM ($J > 0$) or AF ($J < 0$), as in figures 6.3(b) and (c).

The Hamiltonian for this problem is

$$\mathcal{H} = -J\sum_i \mathbf{S}_i \cdot \mathbf{S}_{i+1} - K\sum_i S_{iz}{}^2 - D\sum_i \mathbf{y} \cdot (\mathbf{S}_i \times \mathbf{S}_{i+1}). \tag{6.25}$$

Once again, the cycloid has a period of Ma. The rth spin in the unit cell at $x = ra$ is given by

$$\mathbf{S}_r = S(\sin\theta_r, 0, \cos\theta_r) \tag{6.26}$$

with $\phi_r = 0$. Then the $2M \times 2M$ matrix $\underline{L}(\mathbf{q})$ has nonzero components

$$L_{rr} = L_{r+M,\, r+M} = \frac{JS}{2}\left\{\cos(\theta_r - \theta_{[r+1]}) + \cos(\theta_r - \theta_{[r-1]})\right\}$$
$$- \frac{KS}{2}\left\{1 - 3\cos^2\theta_r\right\} + \frac{DS}{2}\left\{\sin(\theta_{[r+1]} - \theta_r) + \sin(\theta_r - \theta_{[r-1]})\right\}, \tag{6.27}$$

$$L_{r,\,[r\pm1]} = -\frac{JS}{4}\, G_1(r, [r \pm 1])\, e^{\pm iqa} \mp \frac{DS}{4} \sin(\theta_{[r\pm1]} - \theta_r)\, e^{\pm iqa}, \tag{6.28}$$

$$L_{r,\, r+M} = L_{r+M,\, r} = -\frac{KS}{2} \sin^2\theta_r, \tag{6.29}$$

$$L_{r,\,[r\pm1]+M} = L_{r+M,\,[r\pm1]} = -\frac{JS}{4}\, G_2(r, [r \pm 1])\, e^{\pm iqa} \mp \frac{DS}{4} \sin(\theta_{[r\pm1]} - \theta_r)\, e^{\pm iqa}, \tag{6.30}$$

where r again runs from 1 to M.

6.4.1 With $K = 0$

In the absence of anisotropy, the ground state of the cycloid is given by equation (6.12) with classical energy

$$\frac{E}{N} = -JS\cos(Qa) - DS\sin(Qa). \tag{6.31}$$

Minimizing with respect to Q gives

$$Q = \frac{1}{a}\tan^{-1}\frac{D}{J}. \tag{6.32}$$

If the nearest-neighbor exchange $J > 0$ is FM, then $Q = 0$ when $D = 0$; if $J < 0$ is AF, then $Q = \pi/a$ when $D = 0$. As above, $\Delta\theta = Qa$ is the turn angle from site i to site $i + 1$.

The same considerations discussed in section 6.3 imply that for $J > 0$,

$$D = J\tan\left(\frac{2\pi p}{M}\right) \tag{6.33}$$

with integer p for any M. For $J < 0$,

$$D = J\tan\left(\frac{\pi p}{M}\right), \tag{6.34}$$

with even p for even M and odd p for odd M. Since $\cos(\Delta\theta) > 0$ for $J > 0$ and $\cos(\Delta\theta - \pi) > 0$ for $J < 0$, we require that $|p| < M/4$ for $J > 0$ and $|p| < M/2$ for $J < 0$.

With $M = 20$, the resulting inelastic spectra are plotted in figure 6.5(a) with $J = 1$ meV and $p = 1$ and in figure 6.5(b) with $J = -1$ meV and $p = 2$. For comparison, the corresponding spectra of a FM and an AF with $S = 5/2$ and $J = \pm 1$ meV (exchange only) are plotted in figures 6.5(c) and (d).

First consider the cycloid with FM interactions. The magnetic Bragg vectors are $Q = \pm \pi/10a$ or $H = \pm 1/20$. Figure 6.5(a) reveals that the intensity peaks at $H = 1/20$ and $1 - H = 19/20$. For $M = 20$, it is already difficult to distinguish between figures 6.5(a) and (c) for the cycloidal and FM spectra. Because D is usually much smaller then J, $M \gg 1$ and the dynamical spectrum of the cycloid would look very similar to the dynamical spectrum of a simple FM. The long-wavelength features produced by the cycloidal structure could then only be seen by magnifying the region near $H = 0$, a task quite suitable for optical spectroscopy.

For a cycloid with AF interactions, the inelastic intensity blows up near the ordering wavevectors $Q = \pi/a \pm \pi/10a$ or $H = 0.5 \pm 1/20$. The most noticeable distinction with the FM case is that the spectra near $H = 0$ and 0.5 are very different. Remember that for an AF with inversion symmetry, the SW frequencies themselves are symmetric about $H = 0.25$ and that the observed differences on either side of $H = 0.25$ come from the asymmetric intensities of the SW modes. For $M = 20$, the

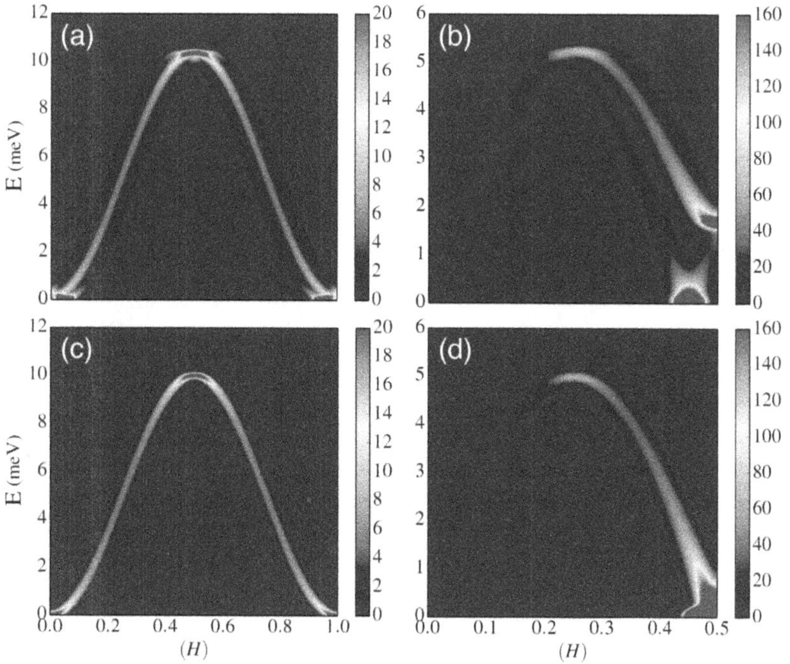

Figure 6.5. The inelastic spectrum of a cycloid generated by DM interactions with $S = 5/2$, a period of $M = 20$ and either (a) $J = 1$ meV (FM) or (b) $J = -1$ meV (AF). For comparison, we also plot the spectra for a corresponding (c) FM ($J = 1$ meV and $M = 1$) or (d) AF ($J = -1$ meV and $M = 2$) with only exchange. In all cases, $K = 0$. For (a), $p = 1$; for (b), $p = 2$.

spectrum in figure 6.5(b) exhibits a gap of 1.6 meV at $H = 0.5$. Since this gap decreases as M increases, it is very hard to distinguish between an AF cycloid with a very long period and a simple AF.

6.4.2 With $K > 0$

Because easy-axis anisotropy K bends the spins towards the z-axis, it squares off the $\cos(Qra)$ term in equation (6.12). In other words, anisotropy generates odd harmonics of Q. Although those higher harmonics may be rather small, they dramatically affect the SW frequencies near the ordering wavevector.

There are two ways to evaluate the ground state in the presence of anisotropy. An atomistic approach would numerically evaluate the phase θ_r of every spin by setting the transverse force to zero. This technique works well for up to a hundred spins or so. But for systems with several hundred spins or when both phases θ_r and ϕ_r are nonzero (doubling the number of variables), this approach quickly becomes intractable. Another disadvantage of the atomistic approach is that the size M of the unit cell must be fixed prior to minimizing the energy with periodic boundary conditions. However, the energy may be lower for a larger or smaller unit cell than the one assumed. The only way to determine the true energy minimum is to consider unit cells over a wide range of sizes.

We prefer a variational approach that parameterizes the spin state in terms of a few variables ($\ll 2M$). The variational parameters of the spin state are determined by minimizing the energy for a fixed set of exchange interactions and anisotropies. For a cycloid in the xz plane with higher harmonics produced by anisotropy along $\mathbf{z}$ [55], the spin state can be parameterized as

$$S_{rz} = S \sum_{l=0} C_{2l+1} \cos((2l + 1)Qra), \tag{6.35}$$

$$S_{rx} = \sqrt{S^2 - S_{rz}{}^2}\, \mathrm{sgn}\,(\sin(Qra)), \tag{6.36}$$

where C_{2l+1} are harmonic coefficients that satisfy $\sum_{l=0} C_{2l+1} = 1$. When $C_1 = 1$ and the higher-order coefficients $C_{2l+1\geqslant 3}$ vanish, this spin state reduces to the purely harmonic form of equation (6.12). Since the coefficients C_{2l+1} fall off rapidly with l, we can usually stop the summation at C_5 or C_7.

What about the even harmonics? In the absence of a magnetic field along $\mathbf{z}$ or $\mathbf{x}$, the spins S_{rz} and S_{rx} must be symmetric about zero. Therefore, only odd harmonics are possible. Another way to think about this is that anisotropy introduces the potential

$$-KS_{rz}{}^2 = -KS^2 \cos^2(Qra) = -\frac{KS^2}{2}\{1 + \cos(2Qra)\} \tag{6.37}$$

at each site. By perturbation theory, this potential generates additional terms with harmonics $Q + 2Q = 3Q$, $Q + 4Q = 5Q$, and so on. In the presence of a magnetic field, however, the perturbing potential

$$-2\mu_B \mathbf{B} \cdot \mathbf{S}_r = -2\mu_B S\{B_z \cos(Qra) + B_x \sin(Qra)\} \tag{6.38}$$

generates even harmonics starting with the constant term C_0, corresponding to the net FM moment.

If the sum over harmonics extends to C_5, then the spin state introduces the three variational parameters C_3, C_5, and Q since $C_1 = 1 - C_3 - C_5$. Three variables are much easier to handle than several hundred! The stability of the variational spin state can be checked by evaluating the net force transverse to the spin at each site (see appendix B of chapter 4). If the state minimizes the energy, then those forces must vanish or at least be very small.

To obtain the variational parameters, we minimize the energy constructed with the trial spin state in a large unit cell with up to 10^6 sites along the x axis. Since this is a 1D problem, the energy minimization is very efficient. If the calculated wavevector Q of the cycloid is different than the one desired, then the DM interaction is raised or lowered until the target wavevector is reached.

The disadvantage of the variational approach is that it breaks down for small M, when higher harmonics can dominate the spin state. Recall that the Fourier transform of a step function contains an infinite number of terms that fall off very slowly with their order. So the variational approach only works in the limit of large M. On the other hand, the atomistic approach must be applied over a wide range of M to obtain the spin state with the lowest energy per site. So neither the atomistic nor the variational approach is well suited to determine a commensurate spin state with small M. We shall discuss this further in section 6.8 on the 'inverse problem'.

Using the variational approach, we have evaluated the spin state for $|J| = 1$ meV and $K = 0.05$ meV in both the FM and AF cases. For the even value $M = 20$, $D = 0.334|J|$ and for either FM or AF exchange, $C_1 = 1.057$, $C_3 = -0.061$ and $C_5 = 0.004$. These coefficients imply that

$$\frac{1}{M}\sum_{r=1}^{M} S_{rz}{}^2 = \frac{S^2}{2}\{C_1^2 + C_3^2 + C_5^2\} = 0.56\, S^2 > 0.5\, S^2, \tag{6.39}$$

confirming that the spins favor the z-axis.

The resulting inelastic spectra are plotted in figure 6.6. For a cycloid with FM interactions, the changes from the case $K = 0$ are relatively minor. A SW gap of $2SK = 0.25$ meV opens up for the FM of figure 6.6(c). A much smaller gap appears in the cycloidal spectrum of figure 6.6(a). The suppression of the SW gap for a cycloid compared to that for a FM is easy to understand. In the absence of higher harmonics when $K = 0$, the cycloid can freely rotate about $\mathbf{y}$, producing a Goldstone mode that vanishes at $H = 0$ and 1. Even when $K \neq 0$, a non-distorted cycloid without higher harmonics is still rotationally invariant because the anisotropy energy $-KN\langle S_{rz}{}^2\rangle$ does not change as the cycloid rotates about $\mathbf{y}$. But with the higher harmonics *produced* by anisotropy, the cycloid is no longer rotationally

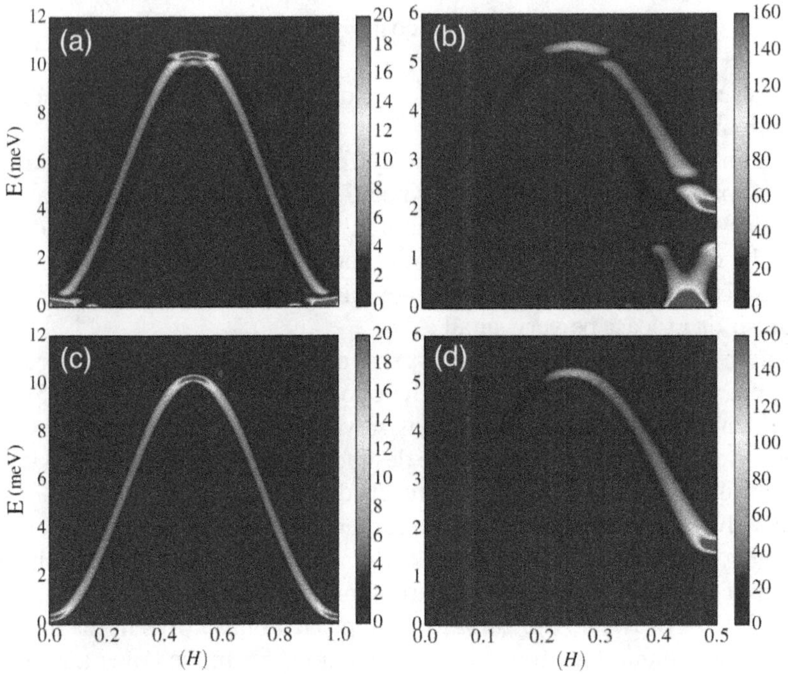

Figure 6.6. The inelastic spectrum of a cycloid generated by DM interactions with $S = 5/2$, a period of $M = 20$ and either (a) $J = 1$ meV (FM) or (b) $J = -1$ meV (AF). For comparison, we also plot the spectrum for a corresponding (c) FM ($J = 1$ meV and $M = 1$) or (d) AF ($J = -1$ meV and $M = 2$) with only exchange. All parameters are the same as in figure 6.5 except $K = 0.05$ meV.

invariant. Consequently, the SW gap for the distorted cycloid is quadratic in K and of order $SK^2/J \sim 6 \times 10^{-3}$ meV while the SW gap for a FM is linear in K and given by the much larger value $2SK = 0.25$ meV.

For a cycloid with AF interactions, the changes created by $K = 0.05$ meV are more dramatic. At $H = 0.5$, the upper SW branch still appears at about 2 meV. But a gap of about 0.75 meV opens between the upper and lower SW branches. In addition to the magnetic Bragg peak at $H = 0.45$, we also find a much less intense peak at $H = 0.35$. Since $3H - 1 = 0.35$, this peak is produced by the third harmonic of Q. The fifth harmonic at $5H - 2 = 0.25$ is too weak to be seen in figure 6.6(b). Because the inelastic intensity of an AF blows up at the ordering wavevectors $(2l + 1)Q$, higher harmonics are much easier to see in a cycloid with AF interactions than in one with FM interactions. Bear in mind that the calculated *dynamical* intensities do not include the *elastic* contributions at wavevectors $H = (2l + 1)0.45$ proportional to ${C_{2l+1}}^2S^2$.

Once again and for the same reasons given above, the SW gap for a cycloid with AF interactions is much smaller than for a simple AF.

6.4.3 Optical spectroscopy and cycloidal modes

Rather than expand the regions around Q to see the fine structure in the INS spectra produced by the cycloidal structure, we evaluate the optical mode frequencies and absorption at $\mathbf{q} = 0$. For a cycloid with AF interactions, the SW modes at $Q = (2\pi/a)(0.5 + \delta)$ can be roughly classified (see below) as either in-cycloidal plane or out-of-cycloidal plane modes, denoted as Φ_n and Ψ_n, respectively. Recall that the mode frequencies are the same at any multiple m of Q: $\omega_n(0) = \omega_n(mQ)$. So the THz mode frequencies at $q = 0$ are identical to the INS mode frequencies at Q.

Figure 6.7 plots the spectroscopic mode frequencies versus K for three different even values of M and $p = 2$. Professional or aspiring spectroscopists must forgive us for using an energy scale of meV rather than cm^{-1} or THz (1 meV = 8.064 cm^{-1} = 0.2418 THz) but this allows us to more easily compare the spectroscopic mode frequencies with previous INS results. For example, panel (c) of figure 6.7 for M = 20 corresponds to panels (b) of figures 6.5 and 6.6.

When $K = 0$, equation (4.93) implies that the spin disturbances associated with modes $\Phi_{\pm n}$ or $\Psi_{\pm n}$ at wavevector $q = 0$ or Q are (ignoring some constant prefactors)

$$\delta\mathbf{S}_r(\Phi_{\pm n}) = \sqrt{S}\,\{\xi_1^{(n)}(-1)^r\mathbf{t}(ra) + \xi_2^{(n)}\mathbf{y}\}\mathrm{e}^{\pm 2\pi i n\delta r}, \tag{6.40}$$

$$\delta\mathbf{S}_r(\Psi_{\pm n}) = \sqrt{S}\,\{\rho_1^{(n)}(-1)^r\mathbf{y} + \rho_2^{(n)}\mathbf{t}(ra)\}\mathrm{e}^{\pm 2\pi i n\delta r}, \tag{6.41}$$

where

$$\mathbf{t}(ra) = \cos(2\pi\delta r)\,\mathbf{x} - \sin(2\pi\delta r)\,\mathbf{z} \tag{6.42}$$

is a unit vector tangential to the cycloid at site r and all four constants $\xi_{1,2}^{(n)}$ and $\rho_{1,2}^{(n)}$ may be complex [58]. Doubly-degenerate modes $\Phi_{\pm n}$ and $\Psi_{\pm n}$ with $n > 0$ correspond to modulations of the cycloid with wavevectors $\pm 2\pi n\delta/a$. For finite M, each mode except Φ_0 contains a mixture of in-cycloidal plane and out-of-cycloidal plane components. With $\xi_2^{(0)} = 0$ for all M, Φ_0 is a Goldstone mode corresponding to a uniform in-plane rotation about $\mathbf{y}$. However, Ψ_0 is not a Goldstone mode because its out-of-plane and in-plane components cost DM and exchange energies, respectively. As M increases, $\xi_2^{(n)}$ and $\rho_2^{(n)}$ both decrease but they remain largest for the modes with higher n [58].

Due to the perturbing potential $-K\sum_i S_{iz}^{\;2}$, each pair $\Phi_{\pm 1}$ or $\Psi_{\pm 1}$ is split by the energy (exercise 6.8)

$$\frac{K}{M}\sum_{r=1}^{M}\left\langle \Phi_{-1}\left| S_{rz}^{\;2}\right| \Phi_{+1}\right\rangle \propto KS, \tag{6.43}$$

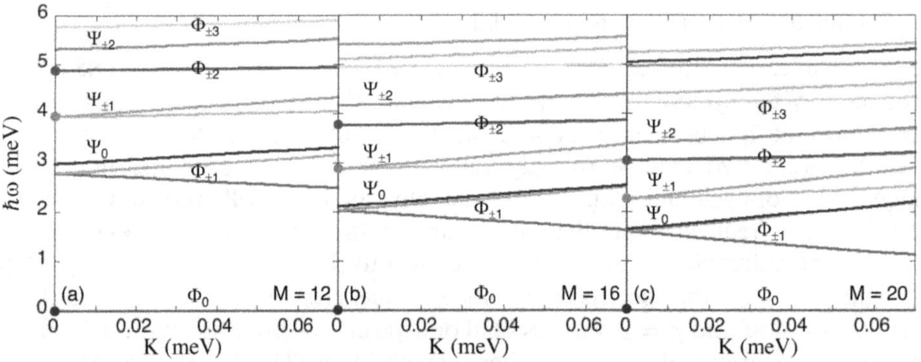

Figure 6.7. The evolution of the spectroscopic modes for (a) $M = 12$, (b) 16, and (c) 20. The cycloid has AF nearest-neighbor exchange $J = -1$ meV with $S = 5/2$. Φ_n are in-plane modes and Ψ_n are out-of-plane modes. INS active modes at $K = 0$ are denoted by solid points (see the text). All take $p = 2$.

$$\frac{K}{M}\sum_{r=1}^{M}\left\langle \Psi_{-1}\left|S_{rz}^{\;2}\right|\Psi_{+1}\right\rangle \propto KS. \tag{6.44}$$

This linear mode splitting is evident in figure 6.7.

In the continuum limit $M \to \infty$, $\xi_2^{(n)} = \rho_2^{(n)} = 0$ so that $\Phi_{\pm n}$ are purely in-cycloidal plane and $\Psi_{\pm n}$ are purely out-of-cycloidal plane [50]. De Sousa and Moore [56] predicted that Ψ_0 and $\Phi_{\pm 1}$ are degenerate in the continuum limit when $K = 0$. We find that Ψ_0 and $\Phi_{\pm 1}$ are nearly degenerate when $K = 0$ and $M = 20$. As seen in figure 6.7, the small splitting of those modes at $K = 0$ grows as M decreases and is quite visible when $M = 12$ with $\omega(\Psi_0) > \omega(\Phi_{\pm 1})$. When $K > 0$, the $\Phi_{\pm 1}$ modes break up into $\Phi_1^{(\mathrm{u})} = (\Phi_{+1} + \Phi_{-1})/\sqrt{2}$ and $\Phi_1^{(\mathrm{l})} = (\Phi_{+1} - \Phi_{-1})/\sqrt{2}$ where u = upper and l = lower. With increasing K, $\omega(\Psi_0)$ eventually dips below $\omega(\Phi_1^{(\mathrm{u})})$. For $M = 20$, $\omega(\Phi_1^{(\mathrm{u})})$ and $\omega(\Psi_0)$ cross at about $K = 0.06$ meV, just as they do for $BiFeO_3$ [56].

What about the INS intensities and optical absorption? For $K = 0$, equation (4.94) implies that modes with nonzero INS weight $S_{\alpha\alpha}^{(n)}(Q)$ are Φ_0 ($\alpha = x$ or z), Ψ_1 ($\alpha = y$), and Φ_2 ($\alpha = x$ or z). These three modes are denoted by solid points in figure 6.7. Using equation (4.95), the only modes with any optical weight at $q = 0$ are $\Psi_{\pm 1}$ when $\mathbf{h} = \mathbf{x}$ or $\mathbf{z}$. For $K > 0$, Φ_0 becomes optically active when $\mathbf{h} = \mathbf{y}$ and hybridizes with $\Phi_{\pm 2}$ [58]. As discussed in section 8.2, similar results were found for $BiFeO_3$ [55].

6.5 Comparison of CE and DM cycloids

This brief section answers the age-old question, 'Is it possible to distinguish between cycloids generated by CE and DM interactions?' Admittedly, this is mostly of academic interest. As indicated by table 6.1, cycloids produced by DM interactions alone are rather rare. In fact, $BiFeO_3$ [5–7] and MnSi [38, 39] are among the few known cases where weak DM interactions produce a long-wavelength cycloid or

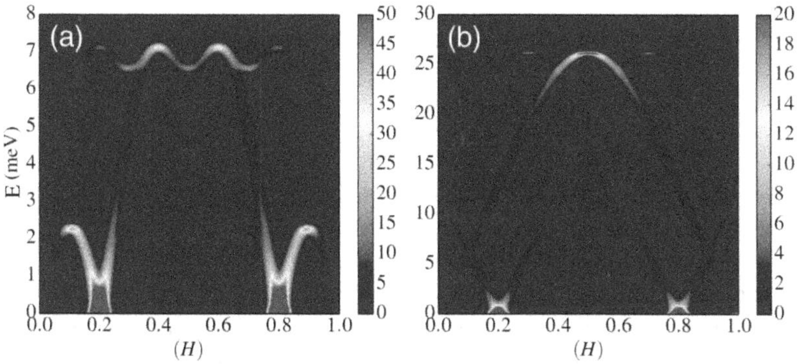

Figure 6.8. The inelastic spectrum of a cycloid generated by (a) CE or (b) DM interactions, both with $J_1 = 1$ meV, $K = 0$, $S = 5/2$, $M = 5$, and $p = 1$.

helix. However, it is natural to wonder if the mechanism that creates a cycloid or helix leaves a definitive signature.

As usual, distinctions are easiest to make when the cycloid has a short period. Figure 6.8 plots the inelastic spectra of two cycloids propagating along **x** with periods of $5a$ and $p = 1$. With FM nearest-neighbor exchange $J_1 = 1$ meV, the cycloid is produced by either (a) CE ($J_2 = -0.81$ meV) or (b) DM interactions ($D = 3.08$ meV). Such a large, unrealistic DM interaction is used to prove a point: the cycloid produced by CE has all SW branches contained within the first BZ between $H = 0$ and 1 but the cycloid produced by DM interactions has SW branches that originate from beyond the first BZ at $H = -1/M = -0.2$ and $1+1/M = 1.2$.

It is much more difficult to distinguish between long-period cycloids ($M \gg 1$) produced by CE and DM interactions without magnifying the spectrum around $H = 1/M$ or $1-1/M$. However, the distinction established for $M = 5$ still holds: a cycloid produced by DM interactions has SW branches that originate from beyond the first BZ while a cycloid produced by CE does not.

6.6 Incommensurate cycloids in 2D or 3D

We now treat a cycloid that is incommensurate in 2D or 3D, which means that the wavelengths $\lambda_\alpha = 2\pi/Q_\alpha$ are incommensurate with the lattice in two or three directions. Such a cycloid can be produced by CE. At first sight, this problem appears intractable. Since a 1D unit cell of length $\lambda'_\alpha \geqslant \lambda_\alpha$ is required to impose periodic boundary conditions in the α direction, this problem appears to require a 3D unit cell of dimensions $\lambda'_x\,\lambda'_y\,\lambda'_z$, corresponding to a $2M \times 2M$ $\underline{L}$ matrix with $M = \lambda'_x\,\lambda'_y\,\lambda'_z/abc$, which could be huge.

Surprisingly, 2D or 3D incommensurate order can be treated as if it were 1D. The key is to rewrite $\mathbf{Q} = (H, K, L)$ in the form [57]

$$\begin{aligned}\mathbf{Q} &\approx \mathbf{Q}_0 + \frac{p}{M_{\mathrm{inc}}}\{n_1\mathbf{b}_1 + n_2\mathbf{b}_2 + n_3\mathbf{b}_3\} \\ &= (H_0, K_0, L_0) + \frac{p}{M_{\mathrm{inc}}}(n_1, n_2, n_3),\end{aligned} \tag{6.45}$$

where $\mathbf{b}_i$ are the reciprocal-lattice vectors and n_i are integers. The prefactor p/M_{inc} involves integers p and M_{inc} such that M_{inc} is not divisible by p if $p > 1$. The term $\mathbf{Q}_0 = (H_0, K_0, L_0)$ includes any commensurate part of the wavevector $\mathbf{Q}$ with M_0 spins. For example, imagine that a DM interaction acting on a G-type AF produces a cycloid propagating along (1,0,−1) as in $BiFeO_3$ [5–7]. Then its ordering wavevector is $\mathbf{Q} = (0.5 + \delta, 0.5, 0.5 - \delta) = \mathbf{Q}_0 + \delta\,(1, 0, -1)$, where $\mathbf{Q}_0 = (0.5, 0.5, 0.5)$ with $M_0 = 2$. When $\mathbf{Q} \neq \mathbf{Q}_0 \neq 0$, the total number of spins in the unit cell is $M = M_0\, M_{\mathrm{inc}}$.

Our goal is to find p, M_{inc}, and n_i such that $(p/M_{\mathrm{inc}})n_1 \approx H - H_0$ and so on. Since

$$\mathbf{R} = m_1\mathbf{a}_1 + m_2\mathbf{a}_2 + m_3\mathbf{a}_3 \tag{6.46}$$

and $\mathbf{a}_i \cdot \mathbf{b}_j = 2\pi\delta_{ij}$,

$$(\mathbf{Q} - \mathbf{Q}_0)\cdot\mathbf{R} \approx \frac{2\pi p}{M_{\mathrm{inc}}}\{m_1n_1 + m_2n_2 + m_3n_3\} = \frac{2\pi p}{M_{\mathrm{inc}}}\, r, \tag{6.47}$$

$$r = m_1n_1 + m_2n_2 + m_3n_3. \tag{6.48}$$

Keep in mind that r is only defined modula M_{inc} so that a site with $r = 0$ is equivalent to one with $r = M_{\mathrm{inc}}$. Following the notation introduced in section 6.3, p is the number of complete 2π rotations of the cycloid in period $M_{\mathrm{inc}}a$.

Approximating $\mathbf{Q} - \mathbf{Q}_0 = (H - H_0, K - K_0, L - L_0)$ by $p\,(n_1, n_2, n_3)/M_{\mathrm{inc}}$ incurs the error

$$\epsilon = \frac{|(H - H_0, K - K_0, L - L_0) - p(n_1, n_2, n_3)/M_{\mathrm{inc}}|}{|(H - H_0, K - K_0, L - L_0)|}. \tag{6.49}$$

We also introduce the efficiency $\eta = M_{\mathrm{inc}}\,\epsilon$ to reflect the difficulty of diagonalizing a large matrix. This method reduces a 3D problem to a 1D problem. Like visiting Paris, that is always a good idea[2].

Even when $\mathbf{Q}$ is commensurate, this method can save time by decreasing the size of $\underline{L}$. For example, the apparent unit cell for $\mathbf{Q} = (0.75, 0.5, 0.25)$ has dimensions $\lambda_x' = 4a$, $\lambda_y' = 2b$, and $\lambda_z' = 4c$ with $M = 16$ sites. But rewriting $\mathbf{Q}$ in the form of equation (6.45) with $\mathbf{n} = (3, 2, 1)$, we find $p = 1$ and $M = 4$. This reduces the size of $\underline{L}$ from 32×32 to 8×8, which is 16 times smaller!

A smaller M can sometimes be obtained by setting $\mathbf{Q}_0$ to zero rather than subtracting it from $\mathbf{Q}$. For example, consider $\mathbf{Q} = (0.18, 0.48, 0.24)$. You could take $\mathbf{Q}_0 = (0, 0.5, 0)$ so that $\mathbf{Q} - \mathbf{Q}_0 = (0.18, -0.02, 0.24)$ with $\mathbf{n} = (9, -1, 12)$, $p = 1$,

[2] Starring in the title role of the 1954 movie 'Sabrina,' Audrey Hepburn said 'Paris is always a good idea'.

Table 6.3. Approximations for $\mathbf{Q} = (0.229, 0.5, -0.458)$ with $\mathbf{Q}_0 = (0, 0.5, 0)$.

$(\mathbf{Q} - \mathbf{Q}_0)_{\text{app}}$	$\mathbf{n}$	p	M_{inc}	ϵ	η
(0.2, 0, −0.5)	(2, 0, −5)	1	10	0.1	1
(0.25, 0, −0.45)	(5, 0, −9)	1	20	0.04	0.8
(0.22, 0, −0.46)	(11, 0, −23)	1	50	0.02	1
(0.2308, 0, −0.4615)	(1, 0, −2)	3	13	0.008	0.1
(0.23, 0, −0.46)	(1, 0, −2)	23	100	0.005	0.5
(0.229, 0, −0.458)	(1, 0, −2)	229	1000	$\sim 10^{-3}$	1

$M_0 = 2$, $M_{\text{inc}} = 50$, and $M = 100$. But it is more sensible to take $\mathbf{Q}_0 = 0$ with $\mathbf{n} = (3, 8, 4)$, $p = 3$, and $M = 50$. For the AF cycloids in sections 6.3 and 6.4, we took $Q_0 = 0$ and $Qa = \pi(M + p)/M$ with even $M + p$.

As listed in table 6.1, the AF5 phase of Co-doped $MnWO_4$ [15–17] supports a cycloid with wavevector $\mathbf{Q} = (0.229, 0.5, -0.458)$. Taking $\mathbf{Q}_0 = (0, 0.5, 0)$ with $M_0 = 2$, some approximations for $\mathbf{Q} - \mathbf{Q}_0 = (0.229, 0, -0.458)$ and their errors are listed in table 6.3. The most cost-effective approximation is to replace $\mathbf{Q} - \mathbf{Q}_0$ by (0.231, 0,−0.462) with $M_{\text{inc}} = 13$, $M = 26$, $p = 3$, $\epsilon = 0.008$, and $\eta = 0.1$.

To demonstrate this technique with a simple model, consider the coupled chains repeating forever in the $\pm\mathbf{x}$ directions sketched in figure 6.9. A single AF exchange J_1 couples spins within each chain. Another AF interaction J_2 couples neighboring spins on adjacent chains. We also include the zig-zag interactions J_3 and $J_4 \neq J_3$. Due to CE, we expect $\mathbf{Q}$ to be incommensurate in 2D.

With easy-plane anisotropy $K < 0$ along $\mathbf{y}$, the spins lie within the xz plane:

$$\mathbf{S}(\mathbf{R}) = S(\sin(\mathbf{Q} \cdot \mathbf{R}), 0, \cos(\mathbf{Q} \cdot \mathbf{R})). \tag{6.50}$$

Taking $\mathbf{R} = (m_1 a, 0, m_3 c)$, we obtain $\mathbf{Q} \cdot \mathbf{R} = m_1 a Q_x + m_3 c Q_z$ and

$$\begin{aligned} \frac{E}{NS^2} = &- J_1 \cos(Q_z c) - J_2 \cos(Q_x a) - J_3 \cos(Q_x a - Q_z c) \\ &- J_4 \cos(Q_x a + Q_z c), \end{aligned} \tag{6.51}$$

which must be minimized as a function of Q_x and Q_z. For example, exchange constants $J_n = \{-4,-2, 1,-3\}$ meV give $\mathbf{Q} = (0.284, 0, 0.367)$[3].

Imagine that the ordering wavevector is *known* to be $\mathbf{Q}_{\text{ex}} = (0.284, 0, 0.367)$ based on elastic measurements. The rough approximation $\mathbf{Q} = (0.28, 0, 0.36)$ implies

$$\mathbf{Q} \cdot \mathbf{R} \approx \frac{2\pi}{25}(7m_1 + 9m_3) \tag{6.52}$$

[3] Actually, we could set either J_3 or J_4 to zero and still get an incommensurate cycloid in two-dimensions: for $J_n = \{-4,-2,0,-3\}$ meV, $\mathbf{Q} = (0.174, 0, 0.427)$.

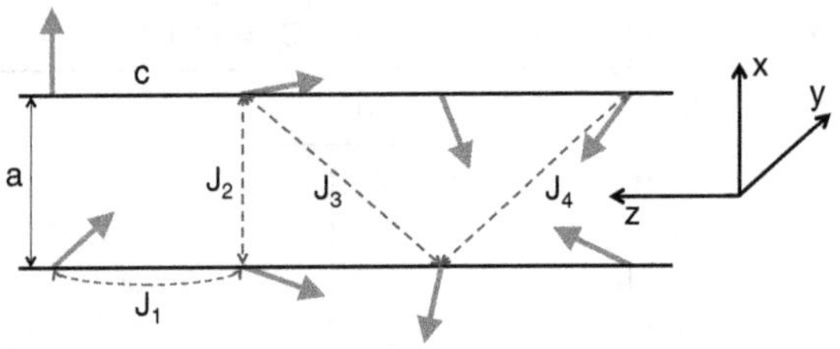

Figure 6.9. An incommensurate cycloid produced by competing interactions in 2D.

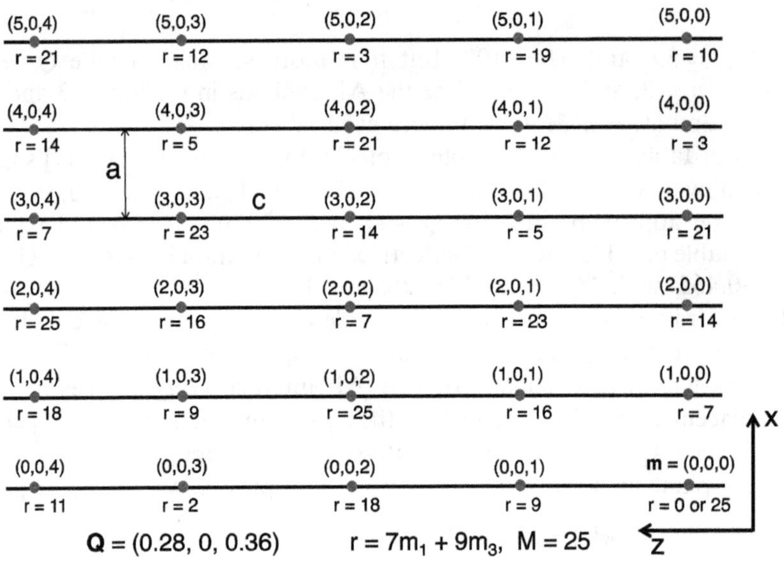

Figure 6.10. Points on a 2D lattice indexed by r.

with $p = 1$, $\mathbf{n} = (7, 0, 3)$, $M = 25$, and an error of $\epsilon = 0.017$. The approximate spin state is then

$$\mathbf{S}(\mathbf{R}) = S(\sin(2\pi r/25),\, 0,\, \cos(2\pi r/25)) \tag{6.53}$$

with integer $r = M\mathbf{Q}{\cdot}\mathbf{R}/2\pi = 7m_1 + 9m_3$ running from 1 to 25. For this approximation, the lattice sites on a 2D lattice are indexed by $\{m_1, 0, m_3\}$ and r in figure 6.10 (exercise 6.11).

When the exchange couplings and ordering wavevector are not matched closely enough, the SW spectrum may exhibit imaginary frequencies over a wide range of wavevectors, indicating that the spin state is unstable. That is the case if we take $J_n = \{-4, -2, 1, -3\}$ meV while approximating $\mathbf{Q}_{\text{ex}} = (0.284, 0, 0.367)$ by $\mathbf{Q} = (0.28, 0, 0.36)$. A better approximation for $\mathbf{Q}_{\text{ex}}$ is $\mathbf{Q} = (0.288, 0, 0.368)$ with $p = 2$, $\mathbf{n} = (18, 0, 23)$, $M_{\text{inc}} = 125$, and $\epsilon = 0.009$. Using that improved approximation, those same exchange constants do not produce any significant singularities.

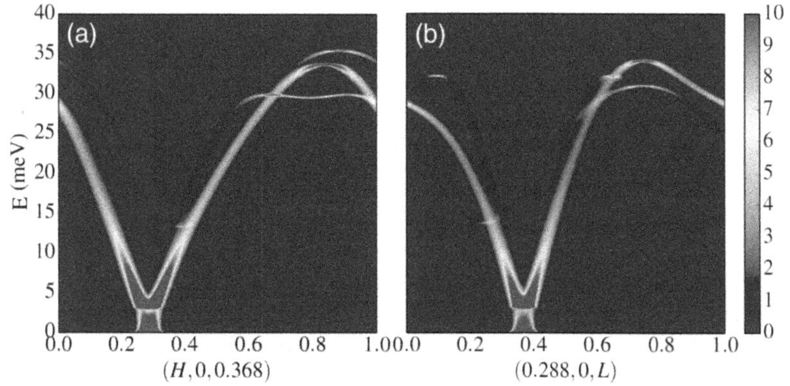

Figure 6.11. The inelastic spectrum of the model described in figure 6.9 with $S = 5/2$, $J_k = \{-4,-2, 1,-3\}$ meV, and $K = -0.05$ meV along (a) $(H, 0, 0.368)$ or (b) $(0.288, 0, L)$.

Together with easy-plane anisotropy $K = -0.05$ meV perpendicular to $\mathbf{y}$, those exchange parameters produce the SW spectra plotted in figures 6.11(a) and (b) for $\mathbf{q} = (H, 0, 0.368)$ or $(0.288, 0, L)$. Easy-plane anisotropy lifts one of the two Goldstone modes with a gap of about 3.1 meV. The other Goldstone mode survives with $\omega(\mathbf{Q}) = 0$ due to the invariance of the spin state when it is rotated about $\mathbf{y}$.

In practice, elastic scattering only determines $\mathbf{Q}$ within a few percent. So it is permissible to shift $\mathbf{Q}$ from its observed value by $\pm 2\%$ to achieve a better error ϵ or efficiency η. While the assumed $\mathbf{Q}$ may not be consistent with a group theory symmetry analysis of the crystal, that discrepancy will not affect the accuracy of the calculated mode frequencies. Fits to the measured SW spectra for different approximate values for $\mathbf{Q}$ should produce values for J_n and K that are close to one another. Of course, wavevectors $\mathbf{Q}$ with smaller errors ϵ will produce more accurate exchange and anisotropy parameters.

6.7 A helix produced by GF on a TLA

Due to GF, a CL state cannot minimize all AF nearest-neighbor exchange interactions $J_1 < 0$ on the triangular lattice sketched in figure 6.12. So the spins of a triangular-lattice AF (TLA) find a NC state that minimizes the energy.

If $J_1 < 0$ is the only exchange interaction, then neighboring spins on each triangle will rotate by 120° from each other. Using the hexagonal wavevector notation $(H, K, L)_h$ introduced in appendix D of chapter 4, $(H, H, 0)_h = (4\pi/a)H\mathbf{x}$ lies along a hexagonal axis and $(H, -H, 0)_h = (4\pi/\sqrt{3}a)H\mathbf{y}$ lies perpendicular to one. As given in table 6.1 for $YMnO_3$ [9, 10] and $CuCrO_2$ [22, 23], the ordering wavevector of the 120° phase is $(0.33, 0.33, 0)_h = (4\pi/3a)\mathbf{x}$. Its spectrum is evaluated in exercise 6.15.

Now consider the slightly more complicated case when J_2 and J_3 are also nonzero. With stacked hexagonal planes, this model roughly describes $CuFeO_2$ [13, 14], which is further discussed in section 7.5. For $CuFeO_2$, $\mathbf{Q} \approx (0.2, 0.2)_h$, which can be rewritten as $(0.8\pi/a)\mathbf{x}$. So the spin state has a period of about $2.5a$. The index L can be ignored when the hexagonal planes are decoupled. In the absence of any other interactions, the ordering wavevector $\mathbf{Q} = (0.2, 0.2)_h$ is obtained by taking $J_1 = -0.1855$ meV, $J_2 = -0.10$ meV, and $J_3 = -0.15$ meV, which are close to the known exchange interactions in $CuFeO_2$ [14]. Using the language of section 6.6, wavevector $\mathbf{Q} = (0.2, 0.2)_h$ corresponds to $p = 1$, $M = 5$, and $\mathbf{n} = (1, 1)$. As in $CuFeO_2$, we assume that $S = 5/2$ spins form a helix in the yz plane normal to the hexagonal axis $\mathbf{x}$. The resulting inelastic spectra are plotted in figures 6.13(a) and (b) for $\mathbf{q} = (H, H)_h$ and $(H,-H)_h$, respectively. Not surprisingly, the inelastic intensity diverges at $\mathbf{Q}$.

How close are these predicted spectra to those measured for $CuFeO_2$? Comparing figure 6.13(a) with the spectrum plotted figure 1.5, we see that the observed spectrum has a second, less intense plume of intensity originating at $\mathbf{q} = (0.3, 0.3)_h$. This secondary plume arises from a weak lattice distortion produced by the alternating displacements of oxygen atoms away from and towards the neighboring Fe^{3+} ions [59] with a periodicity of $a\mathbf{x}$ and wavevector $\mathbf{q}_{lat} = (0.5, 0.5)_h$. That distortion

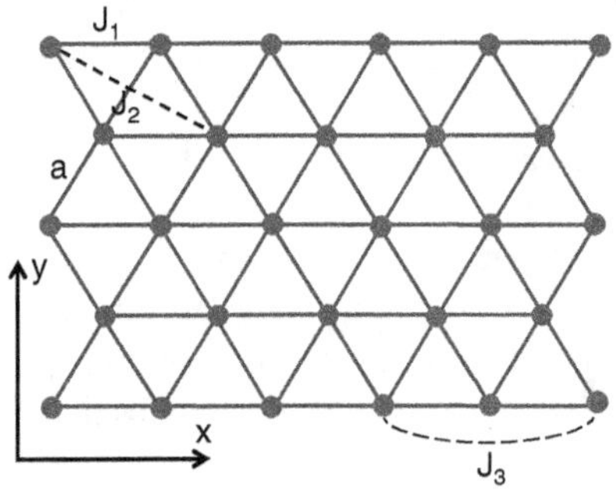

Figure 6.12. A triangular or hexagonal lattice with exchange interactions J_1, J_2, and J_3 and lattice constant a.

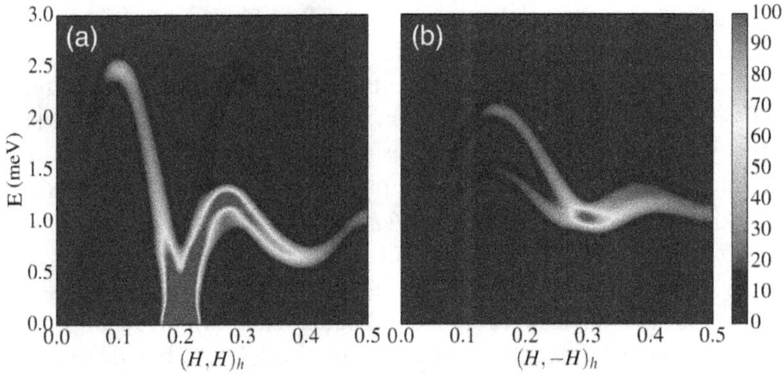

Figure 6.13. Inelastic spectra of a spin helix on a TLA with $S = 5/2$, $J_1 = -0.1855$ meV, $J_2 = -0.10$ meV, and $J_3 = -0.15$ meV.

generates another harmonic term in the spin state [14] with wavevector $\mathbf{q}_{\text{lat}} - \mathbf{Q} = (0.3, 0.3)_h$. As discussed further in section 7.5, other features of the observed spectrum are created by the weak easy-axis anisotropy along $\mathbf{z}$ and the AF coupling between neighboring hexagonal planes.

6.8 The inverse problem

In section 6.6, we used experimental estimates for $\mathbf{Q}$ to evaluate the microscopic interactions (J, D, K, etc). To achieve that goal, $\mathbf{Q}$ was shifted within experimental error to a more tractable value with a smaller unit cell.

The inverse problem is to evaluate the spin state for fixed set of microscopic interactions. This is much more complicated than it sounds. A small change in the exchange constants can produce a huge change in the period of the cycloid or helix. For example, imagine that a given set of exchange parameters produces a cycloid with wavevector $\mathbf{Q} = (0.2, 0, 0)$, corresponding to a period of $\lambda = 5a$ ($M = 5$) or $p = 1$ cycloidal rotation in a distance of $5a$. A very small change in the exchange constants might shift the wavevector to $\mathbf{Q} = (0.201, 0, 0)$, corresponding to a period of $\lambda = 1000\text{a}$ ($M = 1000$) or $p = 201$ cycloidal rotations in a distance of $1000a$!

Because infinitesimal changes in the starting assumptions produce huge changes in the outcome, the inverse problem is not well defined. Nevertheless, the SW spectrum evaluated with $\mathbf{Q} = (0.2, 0, 0)$ will be almost identical to the one evaluated with $\mathbf{Q} = (0.201, 0, 0)$. So fits to the observed SW spectra using these two wavevectors will yield exchange and anisotropy parameters that are very close to each other, even if one wavevector is consistent with the crystal symmetry and the other is not. Therefore, it makes sense to replace $\mathbf{Q}$ (within experimental error) by an approximate value with the smallest possible unit cell. The uncertainty in $\mathbf{Q}$ then translates into uncertainties in the estimated microscopic interactions.

6.9 Exercises

1. Would the SW spectrum of an AF chain with alternating DM interactions change if D switched sign?
2. Add easy-plane anisotropy E to the AF chain with alternating DM interactions and re-evaluate the SW intensity.
3. Show that equations (4.148)–(4.153) reduce to equations (6.7)–(6.10) for the canted chain.
4. Revisit the AF on a square lattice in the xy plane with field applied along $\mathbf{z}$, parallel to the easy-axis anisotropy K. Are the SW frequencies $\omega(\mathbf{Q})$ continuous at B_{SF}? What is the effect of easy-plane anisotropy E perpendicular to $\mathbf{y}$?
5. For a cycloid generated by CE in 1D with FM nearest-neighbor exchange $J_1 > 0$, is there a sudden change in the inelastic spectrum as $|J_2|$ falls below $J_1/4$?

6. How would easy-plane anisotropy affect the inelastic spectra for a cycloid produced by CE? Would a Goldstone mode still survive? Try it for $p = 1$ and $M = 5$ with $J_1 = 1$ meV.
7. Compare the inelastic spectra of cycloids produced by the DM interaction with either FM or AF nearest-neighbor interactions but with the same turn angle. Hint: this is a trick question!
8. Show that the degenerate Φ_1 modes and the degenerate Ψ_1 modes are linearly split by the potential $V = K\sum_i S_{iz}^2$.
9. Consider the sequence of Φ_n and Ψ_n cycloidal modes in the limit where the period of the cycloid diverges or $Q \to Q_0 = \pi/a$. How many distinct modes survive? What happens to the coefficients $\xi_2^{(n)}$ and $\rho_2^{(n)}$ in this limit?
10. How might you approximate $\mathbf{Q} = (-0.214, 0.5, 0.457)$, the ordering wavevector of the AF2 phase of $MnWO_4$? What are the errors for each approximation?
11. Using the definitions of $S_x(r)$ and $S_z(r)$ for $\mathbf{Q} = (0.28, 0, 0.36)$, draw the spins for each point in figure 6.10. This may remind you of those connect-the-dots drawings you did as a kid!
*12. Write a code that optimizes $\eta = M_{\text{inc}}\, \epsilon$ for any $\mathbf{Q} - \mathbf{Q}_0$.
13. Calculate the energy for the spin helix on a TLA with three exchange constants. Show that $\mathbf{Q} = (0.2, 0.2)_h$ minimizes this energy when $J_1 = -0.1855$ meV, $J_2 = -0.10$ meV, and $J_3 = -0.15$ meV.
14. Evaluate the inelastic spectra for the helix on a TLA with the exchange constants given above. How would easy-plane anisotropy affect the spectra?
15. Evaluate the inelastic spectra for the 120° phase with $\mathbf{Q} = (0.33, 0.33)_h$ when $J_1 = -1$ meV and all other exchanges are neglected but with easy-plane anisotropy E.
*16. Remember those phase factors $\exp\{i(\alpha_s - \alpha_r)\}$ and $\exp\{i(\alpha_s + \alpha_r)\}$ that multiplied $G_1(r, s)$ and $G_2(r, s)$ when the unitary matrix $\underline{U}^i$ was generalized in exercise 4.25? Do you think those phase factors can change the predictions of SW theory? Check this for a cycloid produced by CE.

References

[1] Cheong S-W and Mostovoy M 2007 Multiferroics: a magnetic twist for ferroelectricity *Nat. Mater.* **6** 13–20

[2] Tokura Y, Seki S and Nagaosa N 2014 Multiferroics of spin origin *Rep. Prog. Phys.* **77** 076501

[3] Mirebeau I and Petit S 2014 Magnetic frustration probed by inelastic neutron scattering: Recent examples *J. Magn. Magn. Mater.* **350** 209–16

[4] Benedek N A and Fennie C J 2011 Hybrid improper ferroelectricity: a mechanism for controllable polarization-magnetization coupling *Phys. Rev. Lett.* **106** 107204

[5] Sosnowska I, Neumaier T P and Steichele E 1982 Spiral magnetic ordering in bismuth ferrite *J. Phys. C: Solid State Phys.* **15** 4835

[6] Popov Y F, Zvezdin A K, Vorobev G P, Kadomtseva A M, Murashev V A and Rakov D N 1993 Linear magnetoelectric effect and phase transitions in bismuth ferrite *JETP Lett.* **57** 69
[7] Sosnowska I and Zvezdin A K 1995 Origin of the long period magnetic ordering in $BiFeO_3$ *J. Magn. Magn. Mater.* **140-4** 167–8
[8] Park J-G, Le M D, Jeong J and Lee S 2014 Structure and spin dynamics of multiferroic $BiFeO_3$ *J. Phys.: Condens. Matter* **26** 433202
[9] Sato T J, Lee S-H, Katsufuji T, Masaki M, Park S, Copley J R D and Takagi H 2003 Unconventional spin fluctuations in the hexagonal antiferromagnet $YMnO_3$ *Phys. Rev.* B **68** 014432
[10] Singh A K, Patnaik S, Kaushik S D and Siruguri V 2010 Dominance of magnetoelastic coupling in multiferroic hexagonal $YMnO_3$ *Phys. Rev.* B **81** 184406
[11] Ackermann M, Brüning D, Lorenz T, Becker P and Bohatý L 2013 Thermodynamic properties of the new multiferroic material $(NH_4)_2[FeCl_5(H_2O)]$ *New J. Phys.* **15** 123001
[12] Rodríguez-Velamazán J A, Fabelo O, Millán Á, Campo J, Johnson R D and Chapon L 2016 Magnetically-induced ferroelectricity in the $(ND_4)_2[FeCl_5(D_2O)]$ molecular compound *Sci. Rep.* **5** 14475
[13] Seki S, Yamasaki Y, Shiomi Y, Iguchi S, Onose Y and Tokura Y 2007 Impurity-doping-induced ferroelectricity in the frustrated antiferromagnet $CuFeO_2$ *Phys. Rev.* B **75** 100403
[14] Haraldsen J T, Ye F, Fishman R S, Fernandez-Baca J A, Yamaguchi Y, Kimura K and Kimura T 2010 Multiferroic phase of doped delafossite $CuFeO_2$ identified using inelastic neutron scattering *Phys. Rev.* B **82** 020404
[15] Ye F, Chi S, Fernandez-Baca J A, Cao H, Liang K-C, Wang Y, Lorenz B and Chu C W 2012 Magnetic order and spin-flop transitions in the cobalt-doped multiferroic $Mn_{1-x}Co_xWO_4$ *Phys. Rev.* B **86** 094429
[16] Kumar C M N, Xiao Y, Lunkenheimer P, Loidl A and Ohl M 2015 Crystal structure, incommensurate magnetic order, and ferroelectricity in $Mn_{1-x}Cu_xWO_4$ $(0 \leqslant x \leqslant 0.19)$ *Phys. Rev.* B **91** 235149
[17] Takahashi Y, Kibayashi S, Kaneko Y and Tokura Y 2016 Versatile optical magnetoelectric effects by electromagnons in $MnWO_4$ with canted spin-spiral plane *Phys. Rev.* B **93** 180404
[18] Kimura T, Goto T, Shintani H, Ishizaka K, Arima T and Tokura Y 2003 Magnetic control of ferroelectric polarization *Nature* **426** 55–8
[19] Kenzelmann M, Harris A B, Jonas S, Broholm C, Schefer J, Kim S B, Zhang C L, Cheong S-W, Vajk O P and Lynn J W 2005 Magnetic inversion symmetry breaking and ferroelectricity in $TbMnO_3$ *Phys. Rev. Lett.* **95** 087206
[20] Fabrizi F, Walker H C, Paolasini L, de Bergevin F, Boothroyd A T, Prabhakaran D and McMorrow D F 2009 Circularly polarized x rays as a probe of noncollinear magnetic order in multiferroic $TbMnO_3$ *Phys. Rev. Lett.* **102** 237205
[21] Johnson R D, Chapon L C, Khalyavin D D, Manuel P, Radaelli P G and Martin C 2012 Giant improper ferroelectricity in the ferroaxial magnet $CaMn_7O_{12}$ *Phys. Rev. Lett.* **108** 067201
[22] Poienar M, Damay F, Martin C, Hardy V, Maignan A and André G 2009 Structural and magnetic properties of $CuCr_{1-x}M_xO_2$ by neutron powder diffraction *Phys. Rev.* B **79** 014412
[23] Frontzek M *et al* 2011 Magnetic excitations in the geometric frustrated multiferroic $CuCrO_2$ *Phys. Rev.* B **84** 094448

[24] Tomiyasu K, Fukunaga J and Suzuki H 2004 Magnetic short-range order and reentrant-spin-glass-like behavior in $CoCr_2O_4$ and $MnCr_2O_4$ by means of neutron scattering and magnetization measurements *Phys. Rev.* B **70** 214434

[25] Yamasaki Y, Miyasaka S, Kaneko Y, He J-P, Arima T and Tokura Y 2006 Magnetic reversal of the ferroelectric polarization in a multiferroic spinel oxide *Phys. Rev. Lett.* **96** 207204

[26] Lawes G *et al* 2005 Magnetically driven ferroelectric order in $Ni_3V_2O_8$ *Phys. Rev. Lett.* **95** 087205

[27] Ehlers G, Podlesnyak A A, Hahn S E, Fishman R S, Zaharko O, Frontzek M, Kenzelmann M, Pushkarev A V, Shiryaev S V and Barilo S 2013 Incommensurability and spin dynamics in the low-temperature phases of $Ni_3V_2O_8$ *Phys. Rev.* B **87** 214418

[28] Masuda T, Zheludev A, Bush A, Markina M and Vasiliev A 2004 Competition between helimagnetism and commensurate quantum spin correlations in $LiCu_2O_2$ *Phys. Rev. Lett.* **92** 177201

[29] Seki S, Yamasaki Y, Soda M, Matsuura M, Hirota K and Tokura Y 2008 Correlation between spin helicity and an electric polarization vector in quantum-spin chain magnet $LiCu_2O_2$ *Phys. Rev. Lett.* **100** 127201

[30] Schrettle F, Krohns S, Lunkenheimer P, Hemberger J, Büttgen N, Krug von Nidda H-A, Prokofiev A V and Loidl A 2008 Switching the ferroelectric polarization in the $S = 1/2$ chain cuprate $LiCuVO_4$ by external magnetic fields *Phys. Rev.* B **77** 144101

[31] Sanjeewa L D, Garlea V O, McGuire M A, McMillen C D, Cao H and Kolis J W 2016 Structural and magnetic characterization of the one-dimensional $S = 5/2$ antiferromagnetic chain system $SrMnVO_4(OH)$ *Phys. Rev.* B **93** 224407

[32] Shinozaki S-i, Okutani A, Yoshizawa D, Kida T, Takeuchi T, Yamamoto S, Risset O N, Talham D R, Meisel M W and Hagiwara M 2016 Antiferromagnetic order in single crystals of the $S = 2$ quasi-one-dimensional chain $MnCl_3$(bpy) *Phys. Rev.* B **93** 014407

[33] Fishman R S, Shinozaki S-i, Okutani A, Yoshizawa D, Kida T, Hagiwara M and Meisel M W 2016 Long-range magnetic order and interchain interactions in the $S = 2$ chain system $MnCl_3$(bpy) *Phys. Rev.* B **94** 104435

[34] Zheludev A, Shirane G, Sasago Y, Kiode N and Uchinokura K 1996 Spiral phase and spin waves in the quasi-two-dimensional antiferromagnet $Ba_2CuGe_2O_7$ *Phys. Rev.* B **54** 15163–70

[35] Katsura H, Nagaosa N and Balatsky A V 2005 Spin current and magnetoelectric effect in noncollinear magnets *Phys. Rev. Lett.* **95** 057205

[36] Sergienko I A and Dagotto E 2006 Role of the Dzyaloshinskii–Moriya interaction in multiferroic perovskites *Phys. Rev.* B **73** 094434

[37] Arima T-h 2007 Ferroelectricity induced by proper-screw type magnetic order *J. Phys. Soc. Jpn* **76** 073702

[38] Ishikawa Y, Shirane G, Tarvin J A and Kohgi M 1977 Magnetic excitations in the weak itinerant ferromagnet MnSi *Phys. Rev.* B **16** 4956–70

[39] Kugler M *et al* 2015 Band structure of helimagnons in MnSi resolved by inelastic neutron scattering *Phys. Rev. Lett.* **115** 097203

[40] Solovyev I V 2013 Origin of multiferroicity in $MnWO_4$ *Phys. Rev.* B **87** 144403

[41] Kézsmárki I, Kida N, Murakawa H, Bordács S, Onose Y and Tokura Y 2011 Enhanced directional dichroism of terahertz light in resonance with magnetic excitations of the multiferroic $Ba_2CoGe_2O_7$ oxide compound *Phys. Rev. Lett.* **106** 057403

[42] Bordács S *et al* 2012 Chirality of matter shows up via spin excitations *Nat. Phys.* **8** 734–8

[43] Kézsmárki I *et al* 2014 One-way transparency of four-coloured spin-wave excitations in multiferroic materials *Nat. Commun.* **5** 3203
[44] Takahashi Y, Shimano R, Kaneko Y, Murakawa H and Tokura Y 2011 Magnetoelectric resonance with electromagnons in a perovskite helimagnet *Nat. Phys.* **8** 121
[45] Kézsmárki I, Nagel U, Bordács S, Fishman R S, Lee J H, Yi H T, Cheong S-W and Rõõm T 2015 Optical diode effect at spin-wave excitations of the room-temperature multiferroic $BiFeO_3$ *Phys. Rev. Lett.* **115** 127203
[46] Kuzmenko A M, Dziom V, Shuvaev A, Pimenov A, Schiebl M, Mukhin A A, Ivanov V Y, Gudim I A, Bezmaternykh L N and Pimenov A 2015 Large directional optical anisotropy in multiferroic ferroborate *Phys. Rev.* B **92** 184409
[47] Bordács S, Kocsis V, Tokunaga Y, Nagel U, Rõõm T, Takahashi Y, Taguchi Y and Tokura Y 2015 Unidirectional terahertz light absorption in the pyroelectric ferrimagnet $CaBaCo_4O_7$ *Phys. Rev.* B **92** 214441
[48] Takahashi Y, Kibayashi S, Kaneko Y and Tokura Y 2016 Versatile optical magnetoelectric effects by electromagnons in $MnWO_4$ with canted spin-spiral plane *Phys. Rev. B* **93** 180404
[49] Nii Y, Sasaki R, Iguchi Y and Onose Y 2017 Microwave magnetochiral effect in the non-centrosymmetric magnet CuB_2O_4 *J. Phys. Soc. Jpn* **86** 024707
[50] Rovillain P, de Sousa R, Gallais Y, Sacuto A, Méasson M A, Colson D, Forget A, Bibes M, Barthélémy A and Cazayous M 2010 Electric-field control of spin waves at room temperature in multiferroic $BiFeO_3$ *Nat. Mater.* **9** 975
[51] Kuzmenko A M *et al* 2018 Switching of magnons by electric and magnetic fields in multiferroic borates *Phys. Rev. Lett.* **120** 027203
[52] Shuvaev A, Dziom V, Pimenov A, Schiebl M, Mukhin A A, Komarek A C, Finger T, Braden M and Pimenov A 2013 Electric field control of terahertz polarization in a multiferroic manganite with electromagnons *Phys. Rev. Lett.* **111** 227201
[53] Yosida K 1996 *The Theory of Magnetism* (Berlin: Springer)
[54] Moriya T 1960 Anisotropic superexchange interaction and weak ferromagnetism *Phys. Rev.* **120** 91–8
[55] Fishman R S, Furukawa N, Haraldsen J T, Matsuda M and Miyahara S 2012 Identifying the spectroscopic modes of multiferroic $BiFeO_3$ *Phys. Rev.* B **86** 220402
[56] de Sousa R and Moore J E 2008 Optical coupling to spin waves in the cycloidal multiferroic $BiFeO_3$ *Phys. Rev.* B **77** 012406
[57] Fishman R S 2018 Pinning, rotation, and metastability of $BiFeO_3$ cycloidal domains in a magnetic field *Phys. Rev.* B **97** 014405
[58] Fishman R S, Rõõm T and deSousa R arXiv:1809.09680
[59] Terada N, Mitsuda S, Ohsumi H and Tajima K 2006 Spin-driven crystal lattice distortion in frustrated magnet $CuFeO_2$: synchrotron x-ray diffraction study *J. Phys. Soc. Jpn* **75** 023602

Randy S Fishman, Jaime A Fernandez-Baca and Toomas Rõõm

Chapter 7

Inelastic neutron-scattering case studies

Neutrons: the kinder, gentler probe of condensed matter.

—Dr John Axe [1]

7.1 Introduction

INS has been widely used to study the spin dynamics of magnetic materials since Brockhouse performed the first experiments on magnetite in 1957 [2]. As discussed in chapter 2, neutron scattering directly measures the function $S_{\alpha\beta}(\boldsymbol{\kappa},\omega)$, which is proportional to the imaginary part of the general susceptibility. This chapter presents four case studies of neutron-scattering experiments that explore the SWs of various materials. These experiments were performed on triple-axis and time-of-flight spectrometers at different neutron facilities using a variety of samples ranging from amorphous materials to powders and single crystals. The magnetic systems range from isotropic FMs and AFs to incommensurate helices. While the power of INS is fully realized in measurements on single crystals, polycrystalline specimens must be used when single crystals cannot be grown in sufficient quantity (usually a few grams). But even INS measurements on polycrystalline samples can provide important information not available through other techniques.

Section 7.2 describes the measurements by Fernandez-Baca *et al* [3] on amorphous FeB alloys, which are among the best known examples of isotropic FMs. Despite the absence of a periodic lattice, these amorphous alloys exhibit the basic features of a FM material. Section 7.3 describes the work of Campo *et al* [4, 5] on a single-crystal sample of the easy-axis molecular AF $A_2[FeX_5(H_2O)]$ (A = K or Rb, X = Cl or Br). This study illustrates how rhombic anisotropy splits the otherwise degenerate SW branches of an AF, as previously discussed in section 5.5.

doi:10.1088/978-1-64327-114-9ch7

Because magnetic and ferroelectric degrees of freedom are coupled [6], multiferroics have attracted considerable attention due to the potential magnetic control of ferroelectricity (and vice versa) in electronic devices. The work of Walker *et al* [7] on a polycrystalline sample of the multiferroic metal–organic framework $(CH_3)_2NH_2[Mn(HCO_2)_3]$ is discussed in section 7.4. Metal–organic frameworks (MOFs) provide enormous flexibility to tune the exchange interactions between nearest-neighbor moments [8]. Finally, section 7.5 describes the studies of Ye, Haraldsen, and collaborators [9–11] on the frustrated triangular AF $CuFeO_2$. In its pure form, this material is a CL AF. But it can be driven into a NC multiferroic state by a magnetic field or by doping.

7.2 Amorphous FMs

Due to their chemical, bond, and topological disorder, amorphous metals (sometimes referred to as metallic glasses) exhibit the same short-range structural order as a random dense packing of hard spheres. Amorphous metals can be fabricated by rapid quenching from the melt or by deposition techniques without the metallurgic complications, such as composition gradients, that commonly affect single crystals. Since the first reported metallic glass in the 1960s, there has been considerable interest in understanding how structural disorder and the consequent lack of a reciprocal lattice modifies phenomena that were traditionally attributed to periodic systems [12].

Because ferromagnetism is produced by short-range exchange interactions, Gubanov [13] proposed that a periodic arrangement of atoms is not strictly necessary for FM order. Although chemical and bond disorder produce a distribution of the magnitudes and possibly the signs of the exchange interactions, they should not destroy ferromagnetism provided that the average exchange interaction remains positive (FM) and that there are no significant *local* magnetic anisotropies.

In amorphous rare-earth compounds, the single-ion crystal-field anisotropy is large compared with the exchange energy and varies randomly in magnitude and direction from site to site, severely disrupting the exchange energies and reducing the effective interactions. If the system falls below the lower marginal dimensionality, long-range magnetic order is destroyed and magnetic properties are typical of spin glasses and random-field systems [14].

We focus on cases where the *local* anisotropy is weak compared to the exchange interaction, which is typically the case for most metallic glasses. Indeed, metallic glasses provide some of the best known examples of isotropic FMs. Due to their liquid-like short-range structural order, Bloch's theorem cannot be applied to amorphous materials. So the only reciprocal-lattice vector is $\boldsymbol{\tau} = (0, 0, 0)$ and the wavevectors probed by a scattering experiment are $\boldsymbol{\kappa} = \mathbf{q}$. In other words, the wavevectors for all excitations must be measured from the origin, which is the only proper reciprocal-lattice point. Because these materials are isotropic, the energies of excitations (lattice, electronic, or magnetic) depend only on the magnitude q.

For any isotropic FM, SW theory can be applied by defining an average magnetization density. In the hydrodynamic [15] or continuum [16] limit of small wavevectors, the excitations are the familiar SWs with the dispersion [15, 17]

$$E_{SW} = \Delta + D(T)q^2 + F(T)q^4 + \cdots, \tag{7.1}$$

where E_{SW} is the SW energy ($\hbar\omega$) which we will refer to simply as E, and Δ is a small energy gap due to possible pseudo-dipolar interactions. The SW stiffness $D(T)$ depends on the short-range exchange interactions, regardless of whether the electron spins are localized or itinerant, and on whether the structure is amorphous or crystalline. There is no definitive evidence for the higher-order $F(T)$ term in equation (7.1) in amorphous systems. At larger wavevectors the local randomness of the system dominates the excitation spectrum, which then consists of a general density of excitations rather than of well-defined SWs.

Figure 7.1 shows INS measurements on a 20 g sample of melt-spun amorphous ribbons of the alloy $Fe_{86}B_{14}$ at $T = 495$ K [3], below the Curie temperature $T_C = 556$ K. Due to the high neutron-absorption cross section of ^{10}B, the ribbons were made with 98.5% of low-absorbing ^{11}B. The data shown in this figure were collected at the

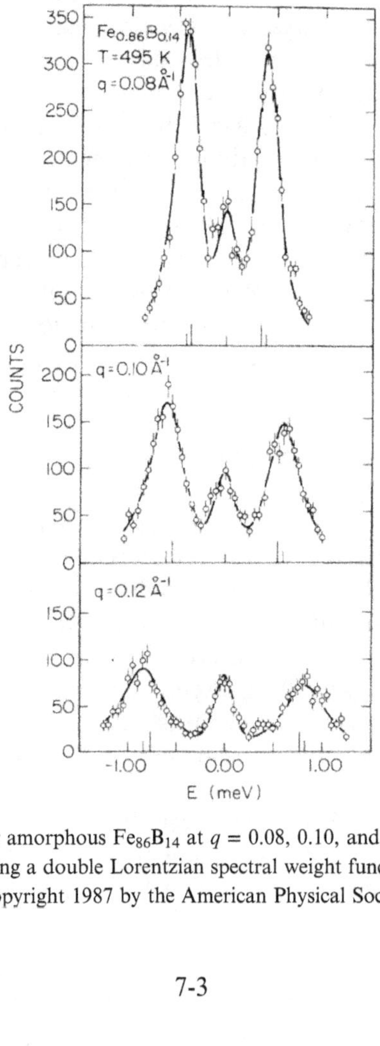

Figure 7.1. Constant-q scans for amorphous $Fe_{86}B_{14}$ at q = 0.08, 0.10, and 0.12 $Å^{-1}$, and T = 495 K. Solid curves are a least-squares fits using a double Lorentzian spectral weight function [3]. Reprinted from [3] with permission from the authors. Copyright 1987 by the American Physical Society.

HB3 triple-axis spectrometer at the HFIR of ORNL[1]. Other measurements were performed at the old BT9 triple-axis instrument of the NCNR at NIST[2].

The scans in figure 7.1 are for wavevectors $q = 0.08, 0.10$, and 0.12 Å^{-1}. The peak at the elastic-scattering position $E = 0$ is produced by nuclear incoherent and static magnetic scattering. Notice that SWs are found on both the energy gain ($E < 0$) side, where the neutron gains energy as it destroys a SW excitation, and the energy loss ($E > 0$) side, where the neutron loses energy as it creates a SW. As discussed in section 4.3, the probabilities for the creation or annihilation of a SW are given by $n_B(\omega)$ for energy gain and $1 + n_B(\omega)$ for energy loss. At $T = 0$, there would be no peak on the left and a much weaker peak on the right. At high temperatures $k_B T \gg \hbar\omega$, the peaks are essentially equal. Due to kinematic restrictions, the scans do not extend far enough in energy to reach the background [18].

As expected from equation (7.1), the peaks shift to higher energy with increasing q. They also broaden in energy and lose intensity due to the reduction in $n_B(\omega)$. An attempt to fit the spectra shown in this figure with a double Dirac delta function in the imaginary susceptibility of equation (2.8) did not yield satisfactory results, suggesting that the SW excitations exhibit some intrinsic broadening. This is not surprising considering the high temperatures of these measurements and the intrinsic disorder of amorphous alloy. To account for intrinsic SW broadening, the imaginary susceptibility $\mathrm{Im}\,\chi^{\mathrm{ss}}_{\alpha\beta}(\mathbf{q},\omega)$ was approximated by a double Lorentzian:

$$\mathrm{Im}\,\chi^{\mathrm{ss}}_{\alpha\alpha}(\mathbf{q},\omega) = \frac{\hbar}{\pi}\left\{S_{\alpha\alpha}(\mathbf{q})\frac{2\Gamma}{4(\hbar\omega - \hbar\omega_{\mathbf{q}})^2 + \Gamma^2} - S_{\alpha\alpha}(-\mathbf{q})\frac{2\Gamma}{4(\hbar\omega + \hbar\omega_{\mathbf{q}})^2 + \Gamma^2}\right\}, \quad (7.2)$$

where one Lorentzian represents the creation of a SW at energy $\hbar\omega_{\mathbf{q}}$ and the other represents the annihilation of a SW at energy $-\hbar\omega_{\mathbf{q}}$ as in equations (2.8) and (4.45). When $\Gamma \ll \hbar\omega_{\mathbf{q}}$, SW excitations are well defined and Γ is the full width at half maximum (FWHM) for the intrinsic SW linewidth, which is inversely related to the SW lifetime $\tau \sim 1/\Gamma$.

The solid curves in figure 7.1 are fits to the double Lorentzian of equation (7.2) convoluted with the instrumental resolution. These fits allow the SW energies and intrinsic linewidths to be extracted at several temperatures. The SW energies E are plotted against q^2 in figure 7.2. The straight lines indicate that the quadratic term of equation (7.1) satisfactorily describes the SW dispersion at all temperatures. These fits yield the SW stiffness $D(T)$ for each temperature and a small SW gap $\Delta \approx 0.07$ meV that may arise from pseudo-dipolar interactions or, perhaps, from systematic errors in these difficult measurements. As expected from SW theory [19, 20], $D(T)$ decreases as $T \to T_C$.

Finally, we turn to the intrinsic linewidths of the SW excitations in the long-wavelength limit. The theoretical wavevector and temperature dependences of these linewidths depend on the type of damping. For interactions between SWs at nonzero temperatures, the linewidths should follow the theoretical relation [21, 22]

[1] https://neutrons.ornl.gov

[2] www.ncnr.nist.gov

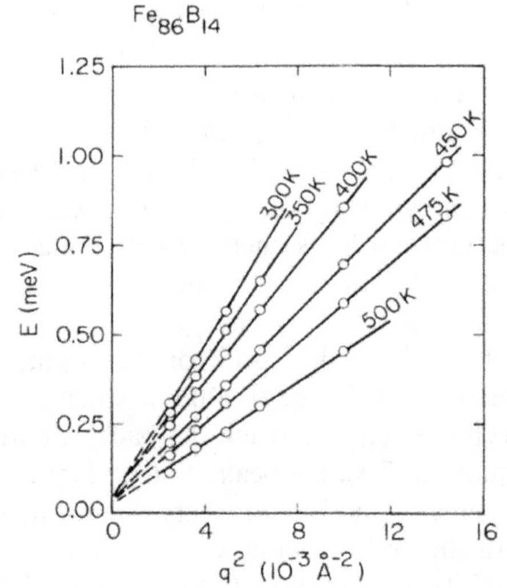

Figure 7.2. SW energies versus q^2 for amorphous $Fe_{86}B_{14}$ at selected temperatures, illustrating the quadratic dispersion for small q [3]. The slope of each line is the SW stiffness $D(T)$, which decreases as $T \to T_C$. Reprinted from [3] with permission from the authors. Copyright 1987 by the American Physical Society.

$$\Gamma(q, T) \propto q^4 \left[T \ln \left(\frac{k_B T}{E_{SW}} \right) \right]^2 . \tag{7.3}$$

Figure 7.3 plots the SW linewidths obtained from fits to the double Lorentzian function of equation (7.2) for $Fe_{86}B_{14}$ [3]. For a variety of wavevectors and temperatures, the intrinsic SW linewidths collapse onto a single universal curve when plotted against equation (7.3). Agreement is also observed in other amorphous magnetic systems, suggesting that conventional SW theory predicts the dispersion relations and linewidths.

7.3 An easy-axis AF

This section discusses the SWs of the easy-axis molecular AF $A_2[FeX_5(H_2O)]$ (A = K, Rb; X = Cl, Br), which has been studied extensively by Campo *et al* [4, 5]. Members of the *erythrosiderite* family, these materials are orthorhombic with space group Pnma. The structural unit cell consists of eight symmetrically equivalent A^+ ions and four isolated $[FeX_5(H_2O)]^{2-}$ octahedra arranged in two planes perpendicular to the *b*-axis (with coordinates $y = 1/4$ and $3/4$ respectively). Each octahedron has a six-fold coordinated Fe^{3+}: five of the six bonds are with X atoms and one is with the O atom of an H_2O molecule.

Besides ionic bonds between the structural building blocks, pronounced H-bonding along the *b*-axis via O–H–X between neighboring $[FeX_5(H_2O)]^{2-}$ octahedra stabilizes the crystal structure. H-bonded octahedra form zig-zag chains along **b**, with antiparallel Fe–O bonds on adjacent octahedra [23]. Despite the geometric isolation between $[FeX_5(H_2O)]^{2-}$ octahedra, long superexchange pathways of the

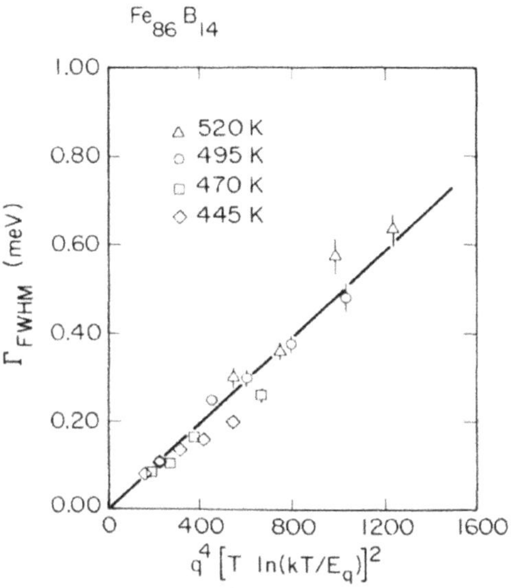

Figure 7.3. Intrinsic SW linewidth Γ versus $q^4[T\ \ln(k_BT\ /\ E_{SW})]^2$ for amorphous $Fe_{86}B_{14}$ at q = 0.08, 0.10, and 0.12 $Å^{-1}$, and T = 445, 470, 495, and 520 K [3]. The observed SW linewidths are consistent with the predicted broadening from equation (7.3). Reprinted from [3] with permission from the authors. Copyright 1987 by the American Physical Society.

type Fe–O–X–Fe and Fe–X–X–Fe are remarkably effective at transmitting magnetic interactions and create AF order with relatively high transition-temperatures from 6 to 23 K [4, 24]. Figure 7.4(a) displays the unit cell of $K_2[FeCl_5H_2O]$ and shows the four $[FeCl_5(H_2O)]^{2-}$ octahedra. This figure also indicates the superexchange pathways Fe–O–Cl–Fe mediated by the H-bonds (i.e. Fe–O–H–Cl–Fe pathways labeled as J_1) that are most effective at transmitting the magnetic interactions [4].

At low temperatures, $A_2[FeX_5(H_2O)]$ (A = K or Rb, X = Cl or Br) are AFs. Figure 7.4(b) projects the Fe^{3+} ions onto the ac plane. In this structure, the AF unit cell is the same as the nuclear cell with Fe^{3+} spins at $y = 1/4$ aligned parallel to each other and antiparallel to the Fe^{3+} spins at $y = 3/4$ [24]. These spins are aligned parallel or antiparallel to the a-axis. The relatively high AF ordering temperatures of these materials have been attributed to the delocalization of the Fe^{3+} spins toward the ligands [4, 5].

Figure 7.4(b) illustrates the five magnetic interaction pathways between the Fe^{3+} ions and the first shell of neighboring ions [4]. As explained above, J_1 propagates along the b-axis through double-exchange pathways Fe–O–H–Cl–Fe mediated by H-bonds. Interactions labeled as J_2 and J_3 connect Fe atoms in different b planes through Fe–Cl–Cl–Fe pathways (with no O atoms involved), while the interactions labeled as J_4 and J_5 connect Fe atoms in the same b plane through Fe–Cl–O–Fe pathways without H-bonds. Interestingly, the Fe^{3+} spins order in an incommensurate cycloidal structure [25, 26] and this system becomes multiferroic [23] when the alkali A ions in this structure are replaced by ammonium and when X = Cl.

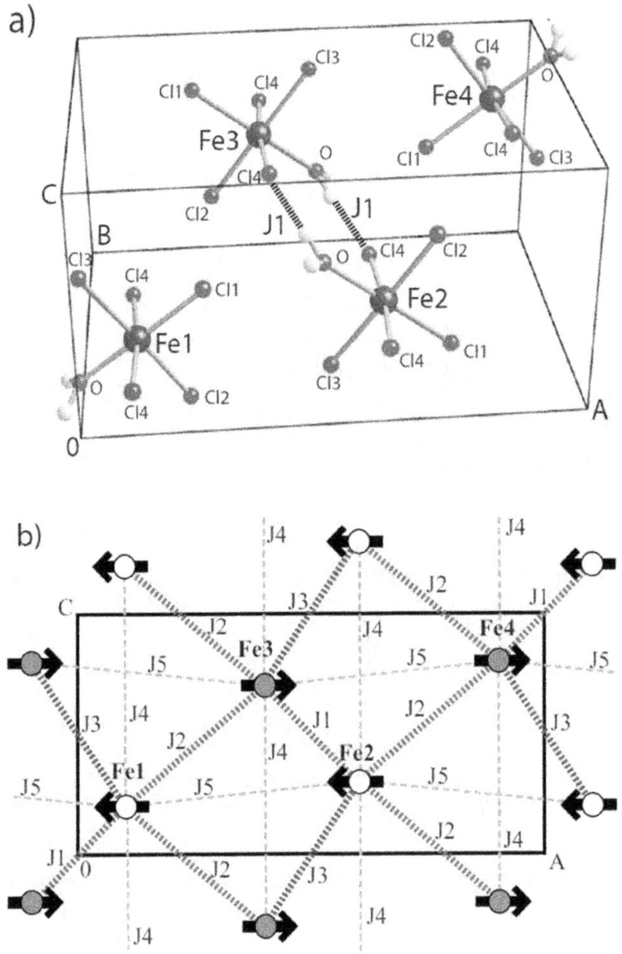

Figure 7.4. (a) The structure of $K_2[FeCl_5H_2O]$. K atoms are omitted for clarity. Fe atoms labeled as Fe1 and Fe2 are in the $y = 1/4$ plane while Fe atoms labeled as Fe3 and Fe4 are in the $y = 3/4$ plane. J_1 is mediated by the Fe–O–H–X–Fe superexchange paths between Fe2 and Fe3 atoms. (b) The five magnetic interactions discussed in the text projected onto the ac plane along with the magnetic structure. Empty circles correspond to Fe ions in the $y = 1/4$ plane whereas filled circles correspond to the Fe ions in the $y = 3/4$ plane. Reprinted from [4] with permission from the authors. Copyright 2008 by the American Physical Society.

To estimate the strengths of the superexchange interactions, Campo *et al* performed INS measurements on the *IN*12 cold-neutron triple-axis spectrometer at the Institute Laue Langevin[3] in Grenoble. Because the well-known incoherent scattering of ^1H produces a high neutron background and large absorption, deuterium (D or ^2H) was substituted for ^1H in the sample preparation (a common practice in neutron-scattering experiments). This experiment used a 1 cm^3 single

[3] www.ill.eu

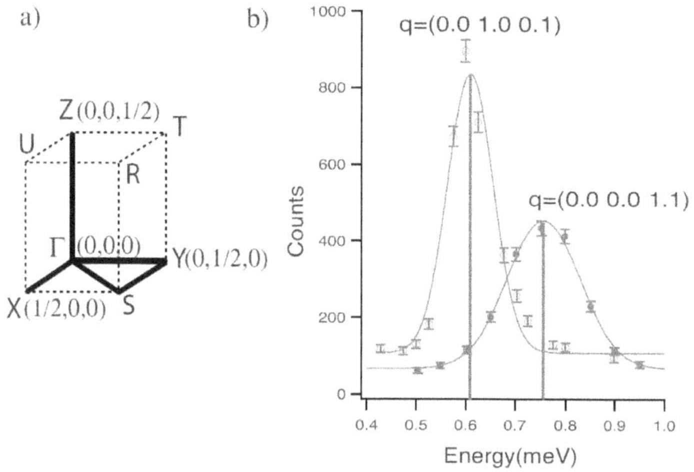

Figure 7.5. (a) Scheme of a BZ in $K_2[FeCl_5(D_2O)]$. Γ is the zone center. *X, Y, Z, U, T, S,* and *R* are zone boundaries along high-symmetry directions. The SW excitation wavevector **q** is measured from Γ, which may be any reciprocal-lattice point $\boldsymbol{\tau}$. The scattering wavevector is $\boldsymbol{\kappa} = \boldsymbol{\tau} + \mathbf{q}$. (b) Two energy scans for $\boldsymbol{\kappa} = (0, 0, 1.1)$ (filled circles) and $\boldsymbol{\kappa} = (0, 1, 0.1)$ (empty circles), corresponding to $\mathbf{q} = (0, 0, 0.1)$ measured from the reciprocal-lattice points $\boldsymbol{\tau} = (0, 0, 1)$ and $\boldsymbol{\tau} = (0, 1, 0)$. This experiment was performed at the *IN*12 triple-axis spectrometer at the Institut Laue Langevin[3]. Reprinted from [4] with permission from the authors. Copyright 2008 by the American Physical Society.

crystal of $K_2[FeCl_5(D_2O)]$ to measure the SWs along $(0, 1, \zeta)$, $(0, 1 + \zeta, 0)$, $(0, 1.5, \zeta)$, $(0, 1 + \zeta, \zeta)$, $(0, 0, 1 + \zeta)$, and $(\zeta, 0, 1)$ at $T = 1.6$ K, which is much lower than the Néel temperature of 14.1 K. Since the AF unit cell of $K_2[FeCl_5(D_2O)]$ is the same as the nuclear cell, the magnetic ordering wavevector of the AF structure is $\mathbf{Q} = (0, 0, 0)$. Thus, the magnetic and lattice periodicities are the same and the excitation wavevectors **q** can be measured from any reciprocal-lattice wavevector $\boldsymbol{\tau}$. Figure 7.5(a) shows an arbitrary BZ, where Γ is the zone center and *X, Y, Z, U, T, S* and *R* are the zone boundaries along high-symmetry directions.

Well-defined SWs were observed at all wavevectors within selected BZs with an energy gap at the zone center, as expected for an AF with uniaxial anisotropy. Surprisingly, the anisotropy gaps measured at $\boldsymbol{\tau} = (0, 0, 1)$ and $\boldsymbol{\tau} = (0, 1, 0)$ are not equal. Figure 7.5(b) plots energy scans for $\mathbf{q} = (0, 0, 0.1)$ measured from two equivalent reciprocal-lattice points with $\boldsymbol{\kappa} = (0, 0, 1.1)$ and $\boldsymbol{\kappa} = (0, 1, 0.1)$, respectively. Campo *et al* [4] attributed the different energy gaps in these scans to a rhombic lattice distortion. To understand this effect, Campo *et al* introduced a Hamiltonian that includes both uniaxial and rhombic anisotropy:

$$\mathcal{H} = -\frac{1}{2}\sum_{i,j,\alpha,\beta} J_{i,j;\alpha,\beta}\mathbf{S}_{i,\alpha}\cdot\mathbf{S}_{j,\beta} - \sum_{i,\alpha}\sigma_\alpha A_\alpha S^z_{i,\alpha} - \sum_{i,\alpha} E_\alpha\{(S^x_{i,\alpha})^2 - (S^y_{i,\alpha})^2\}, \quad (7.4)$$

where the indexes i, j refer to the magnetic unit cells and α, β refer to the magnetic ions inside each magnetic unit cell: $\mu_\alpha = \pm 1$ represents the orientation of the spin along the z-axis for atom α. This expression also defines

$$A_\alpha = \sigma_\alpha \, g_\alpha \mu_B B + 2D_\alpha S_\alpha, \tag{7.5}$$

where g_α is the gyromagnetic ratio (no longer isotropic), B is the magnetic field along $\mathbf{z}$, and D_α is the uniaxial anisotropy. The last term in equation (7.4) is the rhombic anisotropy E. Like the easy-plane anisotropy in section 5.5, this term lifts the two-fold degeneracy of the AF SWs. After solving this Hamiltonian using linear SW theory, Campo *et al* performed a least-squares fit to their experimental data with the five exchange parameters described above. The model predictions in figure 7.6 show SW branches that are split by the rhombic anisotropy. The splitting is most pronounced near the BZ center Γ. Hence, the discrepancy between the anisotropy gaps measured at $\boldsymbol{\tau} = (0, 0, 1)$ and $\boldsymbol{\tau} = (0, 1, 0)$ can be explained by the lifting of the two-fold degeneracy of the lower SW branch [4].

Because INS only measures spin fluctuations perpendicular to the scattering wavevector $\boldsymbol{\kappa}$ (see equation (4.50)), the measured SW energies fall onto one of the two split branches. Near the zone center $\boldsymbol{\tau} = (0, 0, 1)$ (branches labeled as $(0, 0, 1 + \zeta)$ and $(\zeta, 0, 1)$ in the lower panel of figure 7.6), the experiment measures spin fluctuations perpendicular to **c**; near $\boldsymbol{\tau} = (0, 1, 0)$ (branch labeled as $(0, 1, \zeta)$), the experiment measures spin fluctuations perpendicular to **b**. Spin fluctuations perpendicular to **c** lie in a higher SW branch than spin fluctuations perpendicular to **b**. The energies of these two branches at Γ (see figure 7.6) correspond to the two different

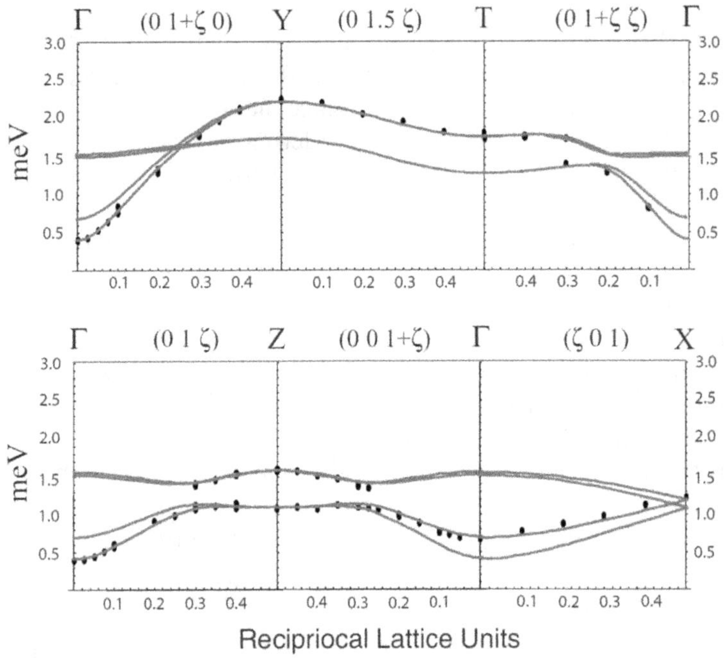

Figure 7.6. SW dispersion curves for $K_2[FeCl_5(D_2O)]$. Black dots are measured values, as described in the text. Solid red lines are the best fit using a model with rhombic anisotropy. Reprinted from [4] with permission from the authors. Copyright 2008 by the American Physical Society.

anisotropy gaps. These measurements are consistent with the reported spin-flop transition that reorients the spins along the *c*-axis [27].

7.4 A multiferroic metal–organic framework

There has been a great deal of effort in the last few years to find new magnetic materials with emergent properties that can be used in electronic devices or data storage. Due to their discrete molecular building blocks, the exchange interactions in MOFs can be tuned in the design of complex magnets [8]. This section examines the multiferroic MOF $(CH_3)_2NH_2[Mn(HCO_2)_3]$ recently studied by Walker *et al* [7].

The first reported metal–organic multiferroic [28], $(CH_3)_2NH_2[Mn(HCO_2)_3]$, belongs to the dimethylammonium metal formate family of MOFs. It crystallizes in the three-dimensional perovskite-like structure shown in figure 7.7 with trigonal symmetry (space group $R\bar{3}c$) at room temperature. The structural unit cell consists of one Mn^{2+} cation at the origin, three formate HCO_2^- groups bonding to the metal cation, and one disordered dimethylammonium (DMA) cation $(CH_3)_2NH_2^+$ [29]. Each Mn^{2+} ion is connected to its six nearest metal-ion neighbors through formate bridges, which can generally mediate either FM or AF coupling [30]. The coordination polyhedra around each Mn^{2+} ion can be described as distorted octahedra MnO_6 that form a perovskite-like framework. DMA cations reside in the cages of this framework (analogous to the A sites of a perovskite structure) and are linked to the oxygen atoms by strong H-bonds. Like the atoms on the A sites of a perovskite, the DMA cations also help stabilize the three-dimensional framework.

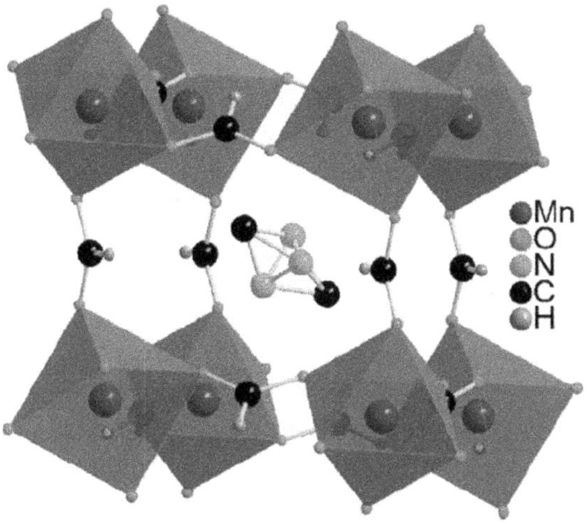

Figure 7.7. Building block of $(CH_3)_2NH_2[Mn(HCO_2)_3]$, showing the perovskite-like structure ABX_3, where A is DMA, B is Mn^{2+}, and X is the formate HCO_2^-. DMA lies at the center of the BX_3-type cage formed by Mn and the formate ions. The N atoms of DMA are disordered over three positions at room temperature. Reprinted from [28] with permission from the authors. Copyright 2009 American Chemical Society.

At room temperature, the DMA cations are three-fold disordered with two carbon atoms lying along the three-fold axis of the cages [29].

Since its ferroelectricity and magnetism arise at different temperatures [31], $(CH_3)_2$ $NH_2[Mn(HCO_2)_3]$ is a proper or type-I multiferroic [28]. Driven by the ordering of the DMA cations, its ferroelectricity is accompanied by a first-order structural transition from the trigonal R$\bar{3}$c to the monoclinic Cc space group at $T_t \approx$ 187 K [32]. As discussed above, its magnetism is driven by the interactions between neighboring Mn^{2+} ion spins mediated by the formate bridges. Although this system is an AF, a weak FM component was reported at T_N = 8.5 K [29].

Because suitable single crystals were not available, Walker *et al* [7] studied a powder sample. As in the previous case study, deuterium (D or ^{2}H) was substituted for ^{1}H in the sample preparation. Measurements on a 1.3 g $(CD_3)_2ND_2[Mn(DCO_2)_3]$ sample were performed at the LET time-of-flight (TOF) direct geometry chopper spectrometer at the ISIS facility of Rutherford Appleton Laboratory[4].

As discussed in chapter 2, a neutron time-of-flight chopper spectrometer is well suited to measure $S_{\alpha\beta}(\boldsymbol{\kappa},\omega)$ in extended regions of $\{\boldsymbol{\kappa},\omega\}$ space. Due to the polycrystalline sample, however, the INS spectra do not exhibit sharp SW peaks but rather broad features due to the scattering from small crystallites at random orientations. Since there are no well-defined crystal directions, all measurements were performed as a function of the magnitude of the scattering wavevector. So instead of measuring $S_{\alpha\beta}(\boldsymbol{\kappa},\omega)$, the experiments measured $S_{\alpha\beta}(|\boldsymbol{\kappa}|,\omega)$, as described in section 5.6.

Data from this type of experiment is often represented as a color intensity map. While these maps provide a good visualization tool, quantitative analysis is often performed by extracting constant-E or constant-$|\boldsymbol{\kappa}|$ cuts from the color intensity map. A constant-E cut is obtained by integrating the intensity in a range $-\Delta E < E < \Delta E$ along the κ-axis. A constant-$|\boldsymbol{\kappa}|$ cut is obtained by integrating the intensity in a range $-\Delta\kappa < \kappa < \Delta\kappa$ along the E-axis. Figures 7.8(b)–(d) show color intensity maps of the experiment. While this book refers to the neutron-scattering wavevector as $\boldsymbol{\kappa}$, Walker *et al* refer to it as **Q** and we will use that convention for the rest of this section.

Figure 7.8(a) plots an elastic scan ($-0.25 < E < 0.25$ meV) of the intensity map along the Q-axis for various temperatures, revealing the AF Bragg peaks below $T_N = 8.5$ K. At the lowest temperature of $T = 5$ K, figure 7.8(b) shows SWs emerging from the BZ centers. While this intensity map does not provide direct detailed information about the SW dispersion, SWs appear up to a maximum energy transfer of about 0.8 meV and there is no indication of a SW energy gap. Comparison with figures 5.10(a) and (c) confirms that this compound is an AF with the greatest intensity at the top of the spectrum. As the temperature is raised towards T_N, the SW excitations broaden and soften in energy, as seen in figure 7.8(c). As T increases Above T_N, the SW excitations coalesce into a quasi-elastic peak that becomes more symmetric about the elastic line in figure 7.8(e). Of course, the scattering at negative

[4] www.isis.stfc.ac.uk

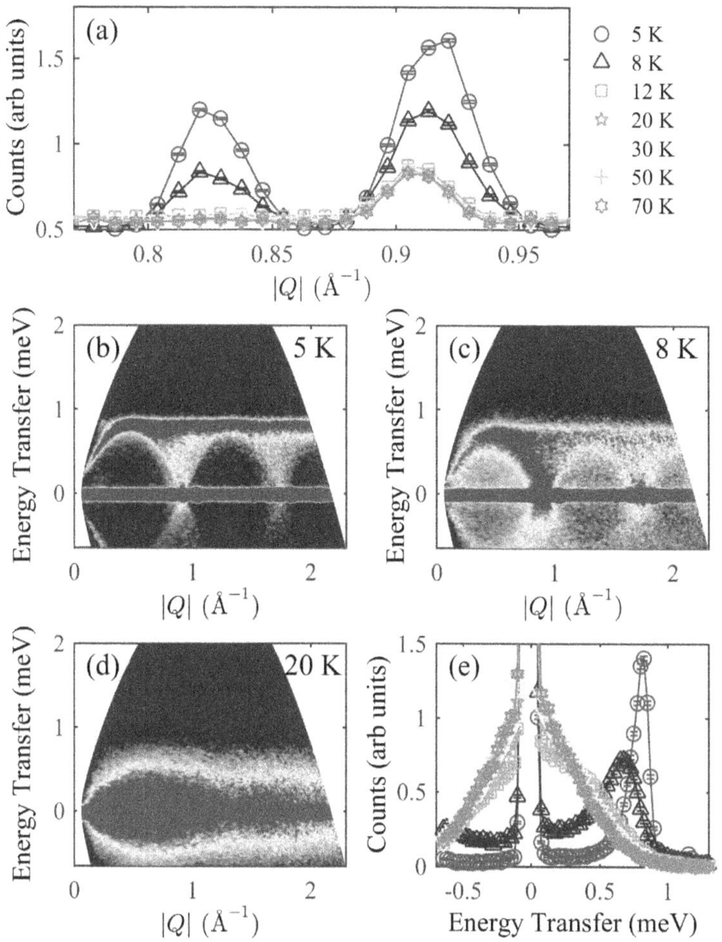

Figure 7.8. Summary of INS measurements for a powder sample of the MOF $(CD_3)_2ND_2[Mn(DCO_2)_3]$. (a) Cut along the elastic line for $-0.25 < E < 0.25$ meV as a function of temperature shows the magnetic Bragg peaks at (0 0 1), (1 1 0), and (2 0 $\bar{1}$) below $T_N = 8.5$ K. (b)–(d) Color intensity maps of energy transfer against momentum transfer at T = 5, 8, and 20 K. (e) A constant-Q cut through the magnetic scattering for $0.3 < |Q| < 0.7$ Å^{-1} as a function of temperature using the same color scheme as in (a). Reprinted from [7] with permission from the authors. Copyright 2017 by the American Physical Society.

energies in figures 7.8(b)–(d) must satisfy the detailed balance condition of equation (4.47), integrated over the orientations of $\boldsymbol{\kappa}$:

$$S(\kappa, -\omega) = e^{-\omega/k_B T} S(\kappa,\omega) \tag{7.6}$$

for $\omega > 0$.

To model the observed magnetic spectrum, Walker *et al* performed systematic simulations of the SW dispersions, the spin–spin correlation function, and the

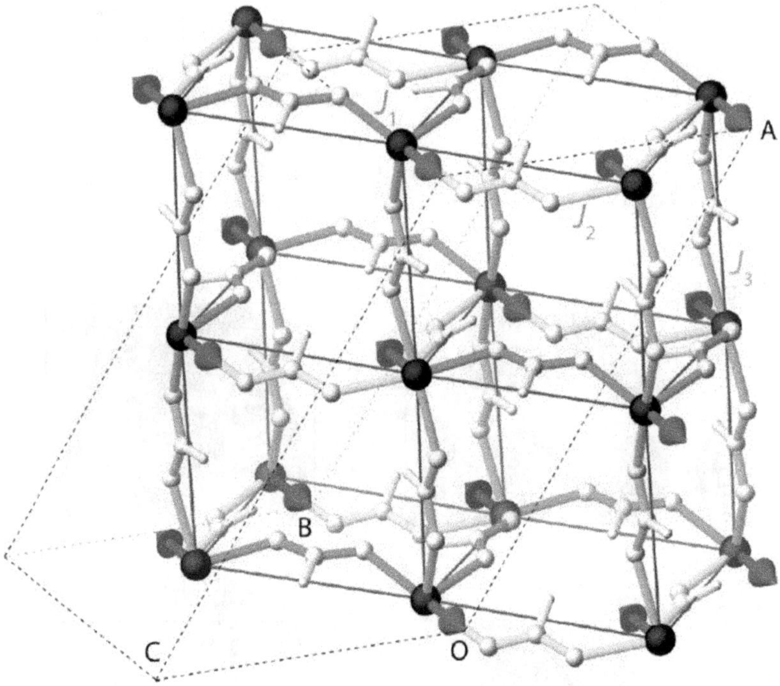

Figure 7.9. Simplified low-temperature *Cc* structure of $(CD_3)_2ND_2[Mn(DCO_2)_3]$, depicting the Mn ions linked by formate bridges, which are colored red, yellow, and cyan for the three nearest-neighbor magnetic interactions. The network is composed of quasi-cubes but the DMA ions sitting within these quasi-cubes are omitted for clarity. The Mn moments order as a G-type AF. Reprinted from [7] with permission from the authors. Copyright 2017 by the American Physical Society.

neutron-scattering cross section using the SpinW code[5] [33] based on the standard Heisenberg Hamiltonian

$$\mathcal{H} = -\frac{1}{2}\sum_{i,j} J_{i,j}\,\mathbf{S}_i \cdot \mathbf{S}_j, \tag{7.7}$$

with three nearest-neighbor exchange interaction constants J_1, J_2, and J_3 between the divalent $S = 5/2$ Mn ions shown in figure 7.9. Using three different exchange constants to describe the interactions between nearest-neighbor Mn ions is reasonable given the monoclinic symmetry of this structure. Due to the absence a SW gap in the measured spectra, this Hamiltonian does not include an easy-axis anisotropy term. Nevertheless, the measurements do not rule out easy-plane anisotropy.

This analysis basically consisted in evaluating the scattering function $S(\mathbf{Q},\omega)$ for many combinations of J_1, J_2, and J_3 and comparing it with the experimental color intensity plot of figure 7.8(b). The broadening and smearing of the SW branches were calculated by using an energy-dependent instrumental resolution to powder average along different directions in reciprocal space. Maps of χ^2 were constructed

[5] www.psi.ch/spinw/

by comparing the calculated and observed spectra to identify regions with the lowest values of χ^2. This tedious process does not guarantee an absolute minimum or even a unique solution.

To illustrate the difficulty of this analysis, figure 7.10 compares the data with the calculations for four different sets of J_1, J_2, and J_3. Figure 7.10(a) shows plots of constant-Q cuts along E near a BZ boundary, where the data are represented by the blue circles and the calculations by various line schemes. Some data features, such as the energy cutoff at the high-energy side of the peak, are not well represented by the calculations. This may be due to an oversimplified resolution function, an over-simplified powder averaging, or perhaps just an inadequate model. Note that the calculations did not account for any intrinsic SW broadening. Figure 7.10(b) shows four different constant-E cuts along Q, which also illustrate that it is very difficult to account for all features of the data.

While it may be necessary to perform new INS measurements once a single crystal becomes available, measurements on a polycrystal already provide important information about this material. First, there is no appreciable SW energy gap. Second, the SW bandwidth is around 0.8 meV. Third, since a calculation with $J_1 = J_2 = J_3$ yielded very poor agreement with the data, J_1, J_2, and J_3 must be different. The best fit was obtained for $J_1 \approx 3\,J_2 \approx 3\,J_3$, which suggests that the structure consists of weakly interacting spin zig-zag chains.

7.5 Spin states of a TLA

This section describes work on *delafossite* $CuFeO_2$ by Ye *et al* [9] and Haraldsen *et al* [10, 11]. Although $CuFeO_2$ starts out as a model TLA with a CL magnetic structure, it becomes NC and multiferroic by applying an external magnetic field [34] or by doping [35, 36]. The multiferroic phase is best described as a distorted helix [10] with alternating large and small turn angles. Section 7.5.1 will discuss the AF phase of pure $CuFeO_2$ [9]. Section 7.5.2 will then discuss the NC incommensurate phase of doped $CuFe_{1-x}Ga_xO_2$ ($x = 0.035$) [10, 11]. Experiments were performed on single crystals at the NIST Center for Neutron Research (NCNR)[6] and at the HFIR and SNS of ORNL[7].

7.5.1 An ↑ ↑ ↓ ↓ AF

At room temperature, $CuFeO_2$ is hexagonal (space group R$\bar{3}$m) with Fe layers stacked in an *ABC* sequence along the *c*-axis. The Fe^{3+} ions form a layered TLA, with magnetic layers separated by nonmagnetic layers of Cu and O. The *ABC* stacking of the hexagonal layers is illustrated in figure 7.11(a), where the Cu and O atoms have been omitted for simplicity. Although a weak low-temperature lattice distortion changes the lattice symmetry from hexagonal R$\bar{3}$m to monoclinic C2/m, we use hexagonal notation (see appendix D of chapter 4) for consistency with the current literature.

[6] www.ncnr.nist.gov
[7] https://neutrons.ornl.gov

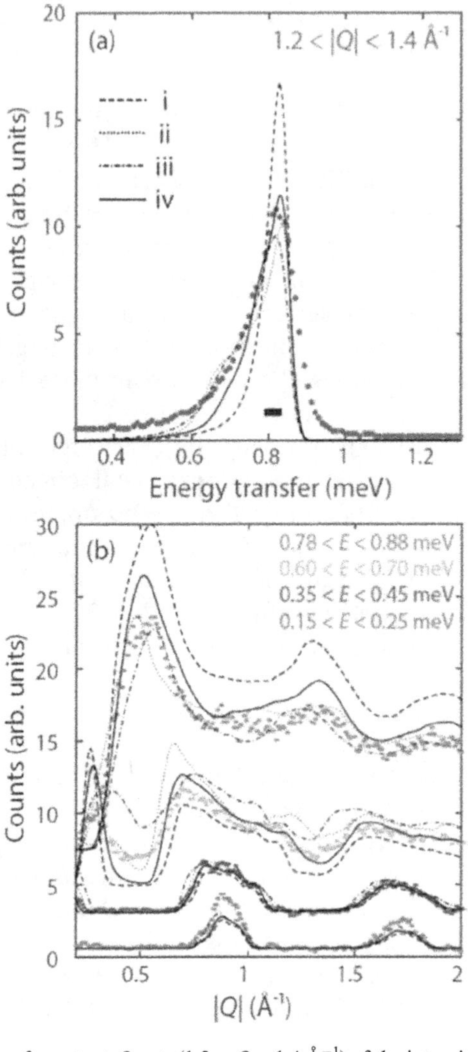

Figure 7.10. (a) Comparison of constant-Q cuts ($1.2 < Q < 1.4$ Å^{-1}) of the intensity maps of figure 7.8 along E (blue circles), and simulations using different exchange parameters. (b) Comparison of four constant-E cuts of the intensity maps of figure 7.8 along Q and simulations using different exchange parameters. These cuts have been offset vertically for clarity. The horizontal bar in (a) represents the instrumental energy resolution. Reprinted from [7] with permission from the authors. Copyright 2017 by the American Physical Society.

Despite the orbital singlet of Fe^{3+} ($L = 0$, $S = 5/2$), the Fe spins form the Ising-like 4-sublattice ↑ ↑ ↓ ↓ state with ordering wavevector $\mathbf{Q} = (1/4, 1/4, 3/2)$ [37] shown in figure 7.11(b). As illustrated in figure 7.11(c), the Fe^{3+} spins lie along **c** and the spins of adjacent layers are AF coupled.

To determine the magnetic interactions, Ye *et al* [9] performed high-resolution INS measurements with cold neutrons. Most measurements were made on a single

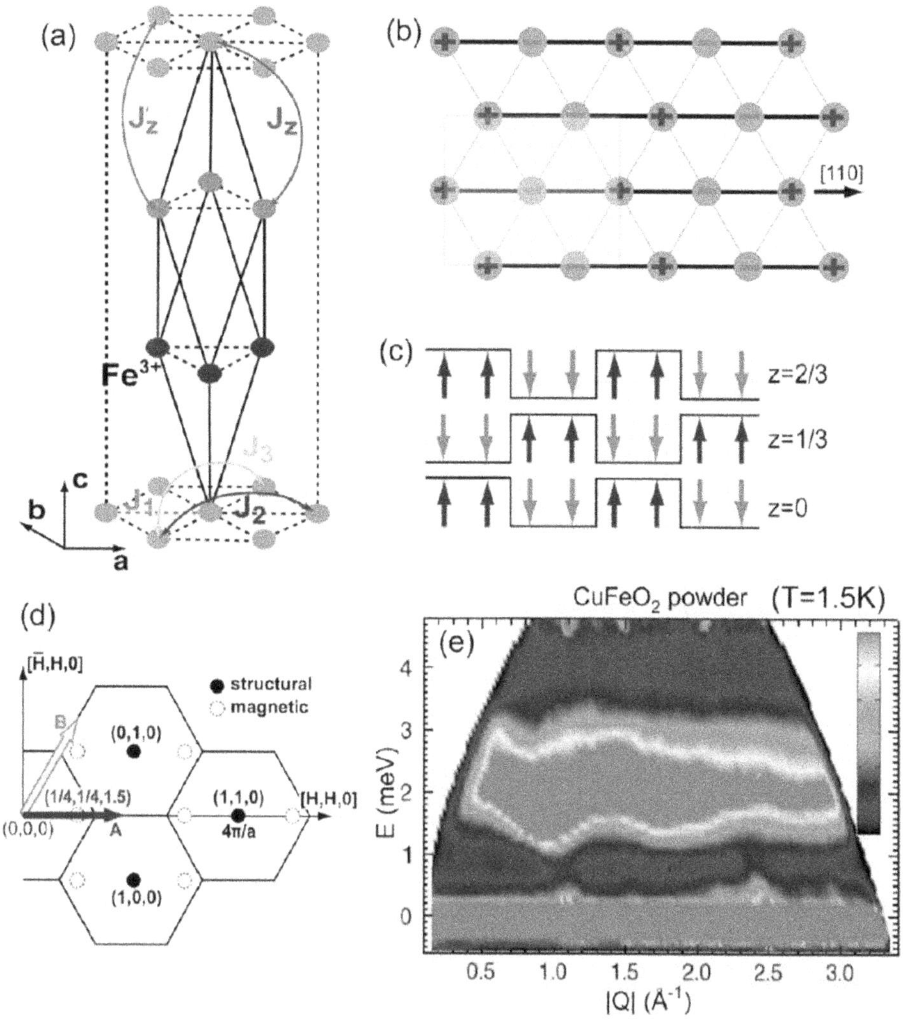

Figure 7.11. (a) Schematics of the Fe^{3+} lattice in $CuFeO_2$. The *ABC* stacking sequence of hexagonal layers is illustrated with three different colors representing Fe atoms on different layers. The relevant exchange interactions between neighboring Fe^{3+} ions are also shown. (b) The low-temperature ↑ ↑ ↓ ↓ spin configuration in the hexagonal plane. The magnetic unit cell is shown as a yellow shaded area. (c) The ↑ ↑ ↓ ↓ layers are AF coupled along the *c*-axis. (d) Schematic diagram of the (*H*, *K*, 0) scattering plane in reciprocal space. Structural Bragg peaks are represented as filled circles; empty circles are the projections of the magnetic Bragg peaks onto the (*H*, *K*, 0) scattering plane. (e) Color intensity map of INS measurements made using the time-of-flight Disk Chopper Spectrometer at NCNR[8] with a polycrystalline sample show the SW energy bandwidth. Reprinted from [9] with permission from the authors. Copyright 2007 by the American Physical Society.

crystal of $CuFeO_2$ using the cold-neutron triple-axis spectrometer SPINS at the NIST Center for Neutron Research (NCNR). But to estimate the SW bandwidth, some preliminary measurements were made on a polycrystalline sample at the Disk

[8] www.ncnr.nist.gov

Chopper Spectrometer, also at NCNR. As seen in figure 7.11(e), these preliminary measurements revealed that the SW bandwidth lies between 0.9 and 3.2 meV. Triple-axis measurements were then constrained to energy transfers $E < 3.9$ meV.

Magnetic Bragg peaks were observed at $\mathbf{Q} = (1/4, 1/4, 3/2)$ and equivalent positions displaced by reciprocal-lattice wavevectors $\boldsymbol{\tau}$. Figure 7.11(d) shows the $(H, K, 0)$ plane, where filled circles represent the structural (nuclear) Bragg peaks while open circles project the magnetic Bragg peaks onto the $(H, K, 0)$ plane.

Ye *et al* first studied the magnetic excitations along $(H, H, 3/2)$ (the solid arrow in figure 7.11(d)). However, the color intensity map of figure 7.12(a) revealed multiple energy branches that could not be explained by a simple 4-sublattice model. They realized that the single crystal of $CuFeO_2$ has three equally populated magnetic domains or twins with ordering wavevectors rotated by 60° with respect to each other in the basal plane. INS probes the excitations of all three domains. The lower-lying excitations were identified as the SW branch along $(H, H, 3/2)$. Higher-energy branches were produced by unintended scans of the twins along the open arrow in figure 7.11(d). Because two of the three twinned domains contribute, the higher-energy branches are more intense than the low-energy branch. This was subsequently verified by Nakajima *et al* [39, 40], who were able to eliminate two of the hexagonal domains by applying strain prior to the INS measurements shown in figure 1.4.

As explained in chapter 2, the scattering wavevector is $\boldsymbol{\kappa} = \boldsymbol{\tau} + \mathbf{Q} + \mathbf{q}'$. So the SW wavevector $\mathbf{q}'$ is measured from the magnetic zone center $\boldsymbol{\tau} + \mathbf{Q}$. In an intensity color map of energy versus momentum transfer, SWs are expected to emerge from $\mathbf{Q}$ and equivalent positions $\boldsymbol{\tau} + \mathbf{Q}$, as illustrated in figure 7.8(b) in the previous section. As first noticed by Terada *et al* [41, 42], the SW branch does not have a minimum at $\mathbf{Q} = (1/4, 1/4, 3/2)$. Instead, two minima appear on either side of $\mathbf{Q}$ at $\mathbf{Q}_{m1} \approx (0.21, 0.21, 3/2)$ and $\mathbf{Q}_{m2} \approx (0.29, 0.29, 3/2)$ with an energy gap of about 0.9 meV.

The scattering profiles along $(0, 0, L)$ show much cleaner features due to the equivalent contributions of all three domains [9]. Magnetic excitations along $(0,0, L)$ are dispersive, ranging in energy from 1.3 meV at the zone center to 2.5 meV at the zone boundary. This indicates that hexagonal planes are magnetically coupled and that treating $CuFeO_2$ as a 2D TLA is an oversimplification.

In order to model the SWs in $CuFeO_2$, Ye *et al* started with the generic Heisenberg Hamiltonian:

$$\mathcal{H} = -\frac{1}{2}\sum_{i,j} J^{\parallel}_{i,j}\, \mathbf{S}_i \cdot \mathbf{S}_j - \frac{1}{2}\sum_{i,j} J^{\perp}_{i,j}\, \mathbf{S}_i \cdot \mathbf{S}_j - D\sum_i {S_{iz}}^2, \tag{7.8}$$

where $J^{\parallel}_{i,j}$ are the in-plane exchange interactions (J_1, J_2, J_3 defined in figure 7.11(a)), $J^{\perp}_{i,j}$ is the interaction between adjacent planes (a single simplified out-of-plane interaction J_z), and D is the single-ion anisotropy. Using an HP transformation and a $1/S$ expansion, they derived an analytic form for the SW dispersion. While the SW

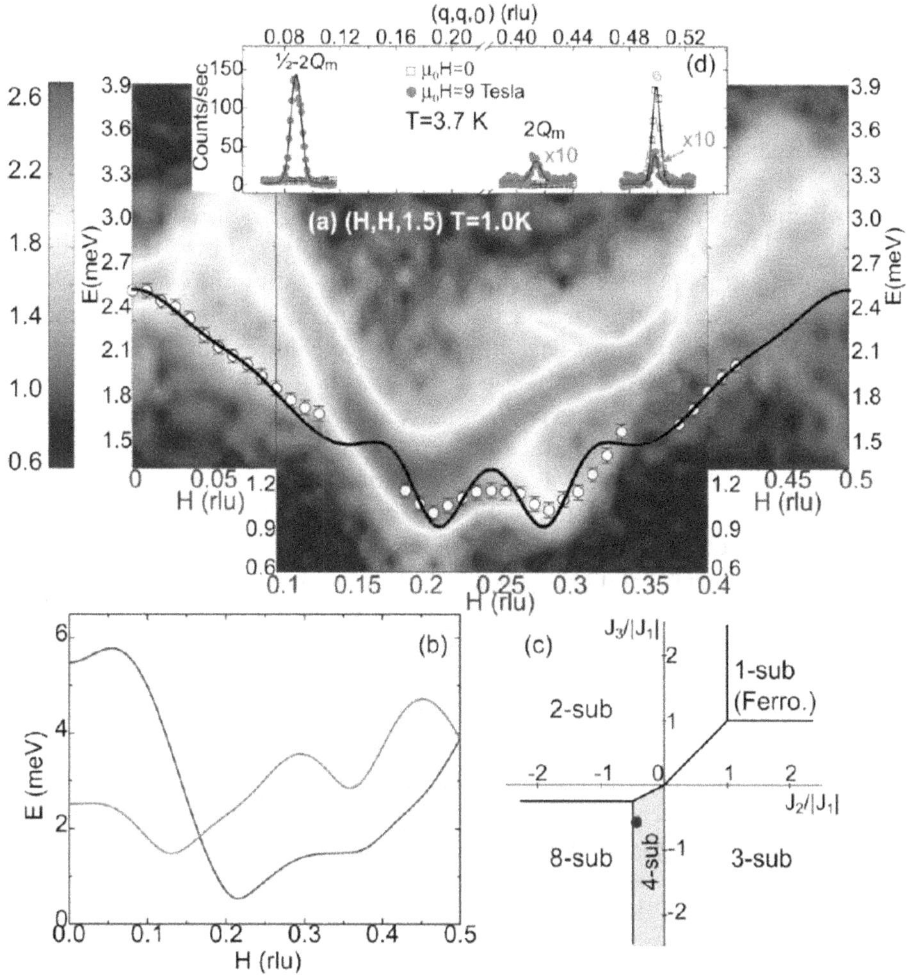

Figure 7.12. (a) SW dispersion along (H, H, 3/2). Points are experimental data and the solid line denotes a fit to the model described in the text. Only the data from the lowest branch are included in the analysis. (b) Predicted contribution from the twins along the path denoted by the open arrow in figure 7.11(d). (c) Phase diagram from [38]. Copyright 1995 by the Physical Society of Japan. The solid symbol is the value obtained from our measurements. Reprinted from [9] with permission from the authors. Copyright 2007 by the American Physical Society.

energy bandwidth is mainly controlled by the out-of-plane interactions $J_{i,j}^{\perp}$, the spectrum along (0, 0, L) strongly depends on the anisotropy D.

By comparing the observed and predicted SW spectra, Ye *et al* obtained $DS = 0.17$ meV, which is larger than the anisotropy estimated from magnetization measurements [43]. They then performed a global least-square fit to the observed SW energies along (H, H, 3/2), (0, 0, L), (0.21, 0.21, L), and (0.25, 0.25, L). The solid line in figure 7.12(a) gives the result of this global fit with in-plane AF interactions $J_1S = -1.14$ meV, $J_2S = -0.50$ meV, and $J_3S = -0.65$ meV. Higher-order interactions within each plane did not improve the fits. Using the simplified out-of-plane interaction scheme described above, only one out-of-plane exchange interaction $J_zS = -0.3$ meV is needed to characterize the SW dispersion along $(0, 0, L)$.

As first suggested by Petrenko *et al* [43], the SWs in $CuFeO_2$ can be well described by a 3D Heisenberg Hamiltonian with single-ion anisotropy. Interestingly, the in-plane exchange constants are consistent with the Monte-Carlo predictions of Takagi and Mekata [38] for a 2D Ising TLA. Figure 7.12(c) shows the stable spin configurations as functions of $J_2/|J_1|$ and $J_3/|J_1|$. Representing the predicted in-plane exchange constants, the filled circle falls in the region of stability for the 4-sublatice $\uparrow\uparrow\downarrow\downarrow$ phase. Hence, a mean-field approach effectively explains the long-range magnetic order in $CuFeO_2$.

Perhaps the most important experimental results are the observed SW frequency dips at $\mathbf{Q}_{m1} \approx (0.21, 0.21, 3/2)$ and $\mathbf{Q}_{m2} \approx (0.29, 0.29, 3/2)$, symmetric about the magnetic BZ center $\mathbf{Q} = (1/4, 1/4, 3/2)$ (see figure 7.12(a)). Remarkably, these are the same wavevectors associated with the incommensurate magnetic order of multiferroic $CuFeO_2$ in an applied magnetic field [34, 44, 45]! Hence, the softening of the SW branch at $\mathbf{Q}_{m1}$ and $\mathbf{Q}_{m2}$ provides a dynamic precursor of the multiferroic phase. By reducing the single-ion anisotropy D without modifying the exchange constants, the spin gap closes while $\mathbf{Q}_{m1}$ and $\mathbf{Q}_{m2}$ remain relatively unchanged.

In any AF, the SW modes linearly split in a magnetic field and a critical value of $B_c \approx \Delta/2\mu_B$ destroys the local stability of the AF state (see exercise 6.4). For $\Delta = 0.9$ meV, this estimate gives $B_c \approx 7.7$ T, which is in reasonable agreement with the observation that incommensurate magnetic order appears when $B > 7.0$ T [34]. The softening of the magnetic excitations in an applied magnetic field reveals a close connection between the destabilization of the $\uparrow\uparrow\downarrow\downarrow$ spin structure and the formation of a new type of magnetic order. As discussed in the next section, the AF state can also be locally destabilized by chemical substitution.

7.5.2 A distorted helix

The SWs in lightly doped $CuFeO_2$ were first studied by Terada *et al* [41], who noticed that the anisotropy observed in undoped $CuFeO_2$ disappears with a doping of 2% Al. The reported SW spectrum of $CuFe_{0.98}Al_{0.02}O_2$ contains gapless modes emerging from wavevectors $\mathbf{Q}_1 = (q, q, 3/2)$ and $\mathbf{Q}_2 = (1/2 - q, 1/2 - q, 3/2)$ with $q \approx 0.2$. This observation was not fully appreciated until the discovery that $CuFeO_2$ becomes multiferroic either by applying a high magnetic field [34] or by doping [36]. In both cases, the multiferroic state is associated with a magnetic structure described by wavevectors $\mathbf{Q}_1$ and $\mathbf{Q}_2$.

It was quickly recognized that the multiferroic state of $CuFeO_2$ cannot be explained by the SC model [46] (also called the inverse DM mechanism [47]), which predicts that the electric polarization is perpendicular to both the spin rotation axis and the magnetic ordering wavevector $\mathbf{Q}$. In multiferroic $CuFeO_2$, the electric polarization is parallel to both [35, 36]. Instead, Arima [48] explained that the multiferroic behavior of $CuFeO_2$ was caused by *p–d* orbital hybridization, which is allowed by the lowered crystal symmetry due to a lattice distortion. Indeed, a lattice distortion produced by displacements of the oxygen atoms with wavevector $\mathbf{q}_{\text{lat}} = (1/2, 1/2, 0)$ (illustrated in figure 7.13) had been reported earlier [49, 50].

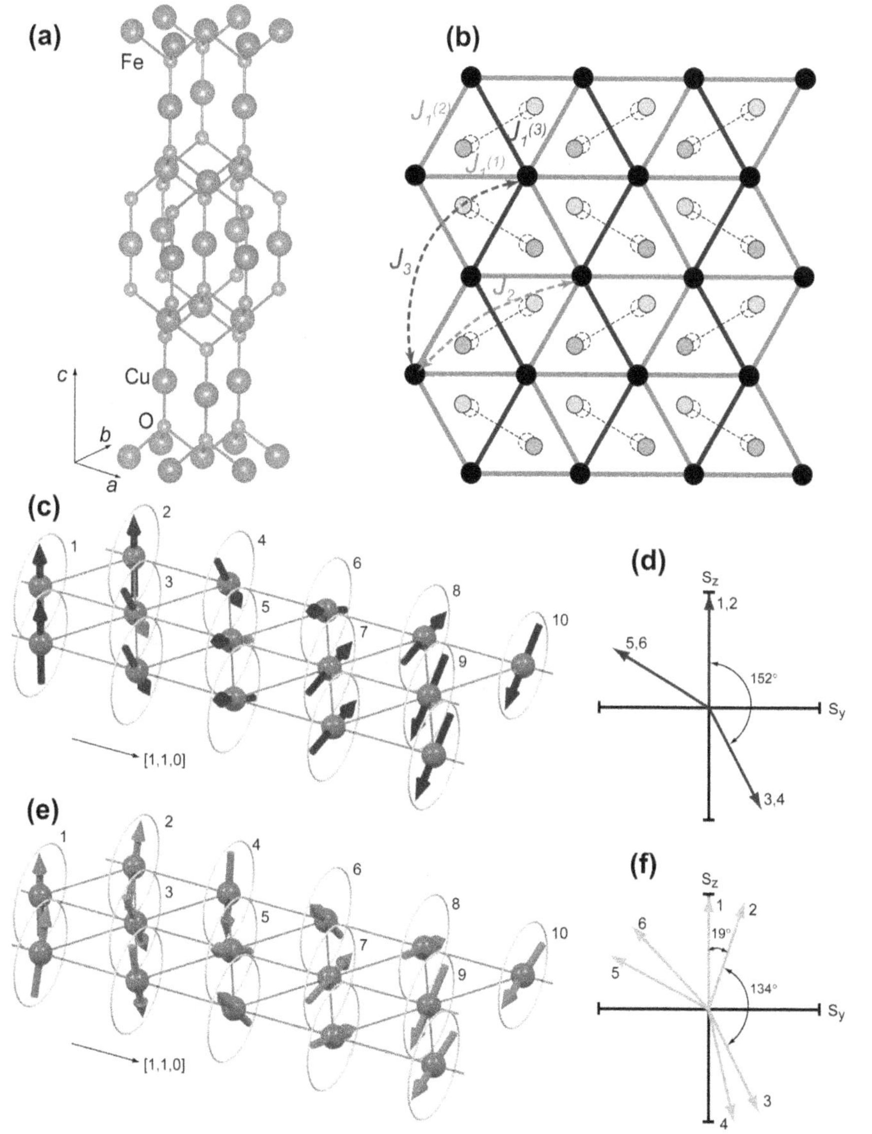

Figure 7.13. (a) The Fe^{3+} lattice in $CuFeO_2$. (b) Schematics of the Fe^{3+} hexagonal layers showing the in-plane exchange interactions. The orange and blue colored circles illustrate the displacements of oxygen atoms lying above and below the Fe^{3+} hexagonal plane with wavevector $\mathbf{q}_{\text{lat}} = (0.5, 0.5, 0)$. The nearest-neighbor interaction J_1 is separated into $J_1^{(1)}$, $J_1^{(2)}$, and $J_1^{(3)}$ to account for this lattice distortion. The magnetic structure and yz projections of the [(c) and (d)] proper spiral and [(e) and (f)] the distorted helix [51]. Reprinted from [10] with permission from the authors. Copyright 2010 by the American Physical Society.

Nakajima *et al* [35] proposed that the magnetic structure of multiferroic $CuFeO_2$ is the 'proper' spiral pictured in figure 7.13(c). In this state, pairs of spins on neighboring sites are aligned parallel to each other and rotate together as a helix propagating along (1, 1, 0). The two wavevectors $\mathbf{Q}_1$ and $\mathbf{Q}_2$ then correspond to the propagation wavevectors of the coupled helices. On the other hand, Fishman and Okamoto [51] predicted that the magnetic ground state is the distorted helix with alternating large and small turn angles, sketched in figure 7.13(e), which they claimed to have lower energy than the proper spiral. This spin configuration contains a main propagation wavevector $\mathbf{Q}_1 \approx (0.21, 0.21, 1.5)$ produced by competing exchange interactions and a secondary propagation wavevector $\mathbf{Q}_2 = \mathbf{q}_{\text{lat}} - \mathbf{Q}_1 \approx (0.29, 0.29, 1.5)$ produced by the lattice distortion with wavevector $\mathbf{q}_{\text{lat}} = (0.5, 0.5, 0)$.

Since they both predict the observed wavevectors $\mathbf{Q}_1$ and $\mathbf{Q}_2$, elastic neutron scattering alone is insufficient to distinguish between these two magnetic structures. On the other hand, INS is sensitive to the actual spin configuration and provides a 'dynamical fingerprint' of the multiferroic ground state. To extract information from this fingerprint, Haraldsen, Ye, and collaborators [10, 11] performed INS measurements on a single crystal of multiferroic, 3.5% Ga-doped $CuFeO_2$ using the Cold Neutron Chopper Spectrometer (CNCS) at the SNS and the HB1 triple-axis spectrometer at the HFIR, both at ORNL. Figures 7.14(a) and (b) show color intensity maps of energy versus momentum transfer along $(H, H, 3/2)$. Notable features are the 'gapless' excitations at $\mathbf{Q}_1$ and $\mathbf{Q}_2$, with the weaker branches emerging from $\mathbf{Q}_2$ in figure 7.14(b). Also notice the 'shoulder' at $H \approx 0.08$ and the intensity 'hole' around $H \approx 0.30$ and $E \approx 1.0$ meV in figure 7.14(a).

In order to evaluate the scattering function $S(\boldsymbol{\kappa},\omega)$ and identify the fingerprint, Haraldsen *et al* used the Hamiltonian

$$\mathcal{H} = -\frac{1}{2}\sum_{i,j} J_{i,j}\,\mathbf{S}_i \cdot \mathbf{S}_j - D\sum_i S_{iz}{}^2, \tag{7.9}$$

where $J_{i,j}$ is the exchange coupling and D is the easy-axis anisotropy. Spin harmonics were incorporated by defining $S_z(\mathbf{R})$ within any hexagonal plane as

$$S_z(\mathbf{R}) = \sum_{l=0} C_{2l+1}\cos\big(Q_1(2l+1)x\big) + \sum_{l=0} B_{2l+1}\sin\big(Q_2(2l+1)x\big), \tag{7.10}$$

where the higher $C_{2l+1>1}$ harmonics of Q_1 are produced by the anisotropy D, as described in section 6.4.2. The $B_{2l+1>0}$ harmonics of $Q_2 = q_{\text{lat}} - Q_1$ are produced by the lattice distortion with wavevector $\mathbf{q}_{\text{lat}} = (2\pi/a)\mathbf{x}$. The squares of these harmonics are proportional to the observed elastic intensities at odd multiples of $\mathbf{Q}_1$ and $\mathbf{Q}_2$. The constant C_1 is chosen so that the maximum value for $S_z(\mathbf{R})$ is $S = 5/2$. The perpendicular spin $S_y(\mathbf{R})$ is then given by

$$S_y(\mathbf{R}) = \sqrt{S^2 - S_z(\mathbf{R})^2}\ \text{sgn}\,[g(\mathbf{R})], \tag{7.11}$$

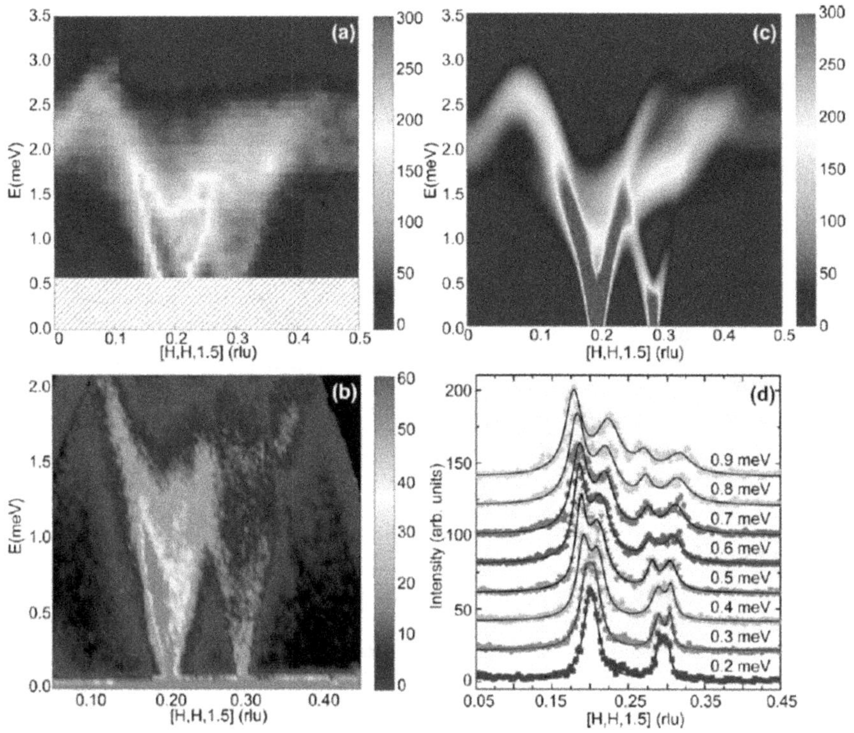

Figure 7.14. Color intensity maps of energy versus momentum for $CuFe_{1-x}Ga_xO_2$ ($x = 0.035$). Experiments performed at (a) the HB1 triple-axis spectrometer at the HFIR and (b) the time-of-flight Cold Neutron Chopper Spectrometer (CNCS) at the SNS, both at ORNL. (c) Simulation of the INS data using intralayer and interlayer exchange interactions, single-ion anisotropy, and a lattice distortion, as explained in the text. (d) Constant-energy cuts of the CNCS intensity color map along $(H, H, 1.5)$. Reprinted from [10] with permission from the authors. Copyright 2010 by the American Physical Society.

where

$$g(\mathbf{R}) = \sin(Q_1 x) + G_1 \cos(Q_2 x) \tag{7.12}$$

and G_1 is an additional variational parameter. Hexagonal planes were realistically stacked along the c-axis in an ABC pattern and neighboring planes were coupled by the three exchange interactions J_{zn}. For a given set of exchange interactions and anisotropy, the 3D spin configuration was obtained by minimizing the energy $E = \langle \mathcal{H} \rangle$ with respect to the variational parameters $C_{2l+1>0}$, $B_{2l+1>0}$, G_1, and Q_1 on a large unit cell with length $\sim 10^4 a$.

Spin fluctuations about this 3D spin configuration were evaluated [10, 11] using the linear SW theory described in chapter 4. In the language of section 6.6, the ordering wavevectors $\mathbf{Q}_1 = (0.21, 0.21, 1.5)$ and $\mathbf{Q}_2 = (0.29, 0.29, 1.5)$ were approximated by $\mathbf{Q}_1 \approx \mathbf{Q}_0 + (2/10)(1, 1, 0)$ and $\mathbf{Q}_2 \approx \mathbf{Q}_0 + (3/10)(1, 1, 0)$, respectively, both with $M_{\rm inc} = 10$ and $\mathbf{Q}_0 = (0, 0, 1.5)$. Since $M_0 = 2$, $M = M_0 M_{\rm inc} = 20$ and $\underline{L}(\mathbf{q})$ is a 40×40 matrix.

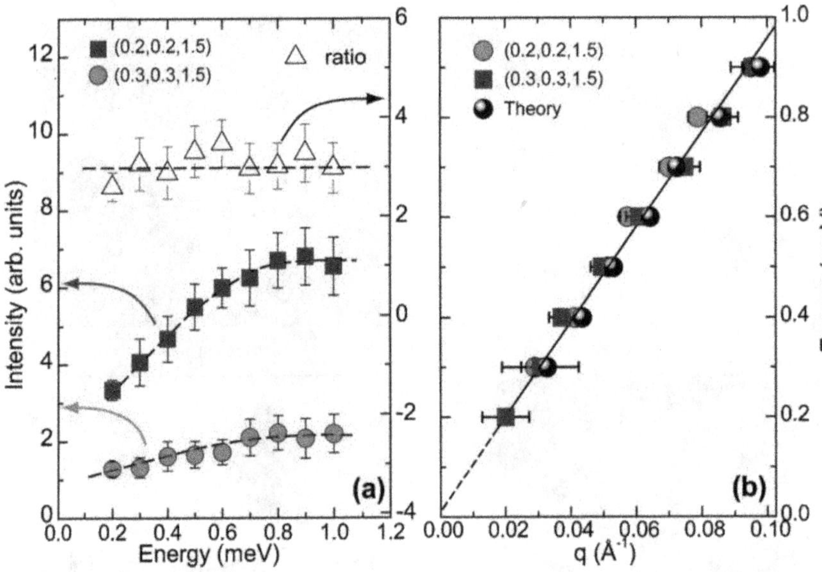

Figure 7.15. (a) Intensity versus energy for the two modes at $\mathbf{Q}_1 = (0.2, 0.2, 3/2)$ (blue filled squares) and $\mathbf{Q}_2 = (0.3, 0.3, 3/2)$ (red filled circles) along $(H, H, 3/2)$. The ratio of mode intensities (open triangles) shows that the primary mode is three times as intense as the secondary mode. The simulated spectrum is consistent with this intensity ratio. (b) SW velocities for the two modes along $\boldsymbol{\kappa} = (H, H, 3/2)$ together with the predicted values (black circles). Reprinted from [10] with permission from the authors. Copyright 2010 by the American Physical Society.

Figure 7.14(c) shows a color intensity map for $\boldsymbol{\kappa} = (H, H, 3/2)$, including the contributions of all three hexagonal domains based on an optimal set of exchange parameters. Strongly affected by the lattice distortion, the twin domains account for the spectral weight around $E \approx 1.5$ meV and $H \approx 0.30$. The model accurately produces the intensity 'hole' at $H \approx 0.30$ as well as the shoulder at $H \approx 0.08$, which is caused by the interlayer interactions [10, 11].

The main SW mode is associated with the wavevector $\mathbf{Q}_1$ while the secondary and weaker SW mode is associated with the wavevector $\mathbf{Q}_2 = \mathbf{q}_{\text{lat}} - \mathbf{Q}_1$. Their overall magnitudes are consistent with the calculated values. Although the model predicts four harmonics, only the B_1 component has enough intensity to be observed in this experiment. Figure 7.15 also compares the measured and predicted SW velocities.

These results demonstrate that lattice distortions play an important role in determining the magnetic ground state of 3.5% Ga-doped $CuFeO_2$. The calculated values of $S(\boldsymbol{\kappa},\omega)$ are in remarkable agreement with the INS data, which provides a 'dynamical fingerprint' of the multiferroic phase. As a bonus, comparison between the fitted exchange and anisotropy parameters in the CL and NC phases of $CuFeO_2$ reveal that the dominant effect of doping is to reduce the easy-axis anisotropy D, releasing the spins from the z-axis in the $\uparrow\uparrow\downarrow\downarrow$ state to form the distorted helical state.

7.6 Summary

In this chapter, we have presented four case studies of INS experiments in various materials. Fernandez-Baca *et al* [3] illustrated how amorphous FeB alloys provide some of the best examples of isotropic FMs. Campo *et al* [4, 5] illustrated how rhombic anisotropy splits the otherwise degenerate SW branches of the easy-axis AF $A_2[FeX_5(H_2O)]$ (A = K, Rb; X= Cl, Br). Based on measurements of a polycrystalline sample, Walker *et al* [7] provided important and useful information about the multiferroic magnetic MOF $(CH_3)_2NH_2[Mn(HCO_2)_3]$. Finally, Ye and Haraldsen *et al* [9–11] described the magnetic structure and spin dynamics of the frustrated TLA $CuFeO_2$ in both its AF and distorted helix phases. In all four case studies, the observed magnetic excitations could be described reasonably well as the SWs of a Heisenberg Hamiltonian with additional anisotropy terms when required.

References

[1] Axe J 1989 Neutrons: the kinder, gentler probe of condensed matter *Materials Research Society Meeting* **166** p 3

[2] Brockhouse B N 1957 Scattering of neutrons by spin waves in magnetite *Phys. Rev.* **106** 859–64

[3] Fernandez-Baca J A, Lynn J W, Rhyne J J and Fish G E 1987 Long-wavelength spin-wave energies and linewidths of the amorphous invar alloy $Fe_{100-x}B_x$ *Phys. Rev.* B **36** 8497–511

[4] Campo J, Luzón J, Palacio F, McIntyre G J, Millán A and Wildes A R 2008 Understanding magnetic interactions in the series A_2FeX_5 H_2O (A = K, Rb; X = Cl, Br). II. Inelastic neutron scattering and DFT studies *Phys. Rev.* B **78** 054415

[5] Luzón J, Campo J, Palacio F, McIntyre G J and Milláan A 2008 Understanding magnetic interactions in the series A_2FeX_5 H_2O (A = K, Rb; X = Cl, Br). I. Spin densities by polarized neutron diffraction and DFT calculations *Phys. Rev.* B **78** 054414

[6] Kimura T, Goto T, Shintani H, Ishizaka K, Arima T and Tokura Y 2003 Magnetic control of ferroelectric polarization *Nature* **426** 55

[7] Walker H C, Duncan H D, Le M D, Keen D A, Voneshen D J and Phillips A E 2017 Magnetic structure and spin-wave excitations in the multiferroic magnetic metal-organic framework $(CD_3)_2ND_2[Mn(DCO_2)_3]$ *Phys. Rev.* B **96** 094423

[8] Kurmoo M 2009 Magnetic metal-organic frameworks *Chem. Soc. Rev.* **38** 1353–79

[9] Ye F, Fernandez-Baca J A, Fishman R S, Ren Y, Kang H J, Qiu Y and Kimura T 2007 Magnetic interactions in the geometrically frustrated triangular lattice antiferromagnet $CuFeO_2$ *Phys. Rev. Lett.* **99** 157201

[10] Haraldsen J T, Ye F, Fishman R S, Fernandez-Baca J A, Yamaguchi Y, Kimura K and Kimura T 2010 Multiferroic phase of doped delafossite $CuFeO_2$ identified using inelastic neutron scattering *Phys. Rev.* B **82** 020404

[11] Haraldsen J T and Fishman R S 2010 Effect of interlayer interactions and lattice distortions on the magnetic ground state and spin dynamics of a geometrically frustrated triangular-lattice antiferromagnet *Phys. Rev.* B **82** 144441

[12] Moorjani K and Coey J M D 1984 *Magnetic Glasses, Methods and Phenomena 6* 1st edn (Amsterdam: Elsevier)

[13] Gubanov A I 1960 Quasi-classical theory of amorphous ferromagnetics *Sov. Phys. Solid State* **2** 468–71

[14] Lynn J W and Rhyne J J 1988 Spin dynamics of amorphous magnets *Spin Waves and Magnetic Excitations, part II* ed A S Borovik-Romanov and S K Sinha (Amsterdam: North Holland) ch 4

[15] Halperin B I and Hohenberg P C 1969 Hydrodynamic theory of spin waves *Phys. Rev.* **188** 898–918
[16] Herring C and Kittel C 1952 The theory of spin waves in ferromagnetic media *Phys. Rev.* **88** 1435
[17] Keffer F 1966 Spin waves *Handbuck der Physik, part 2* vol 18 ed S Flugge (Berlin: Springer)
[18] Lynn J W and Fernandez-Baca J A 1995 Neutron scattering studies of the spin dynamics of amorphous alloys *The Magnetism of Amorphous Metals and Alloys* ed J A Fernandez-Baca and W-Y Ching (Singapore: World Scientific) pp 221–60
[19] Dyson F J 1956 General theory of spin-wave interactions *Phys. Rev.* **102** 1217–30
[20] Dyson F J 1956 Thermodynamic behavior of an ideal ferromagnet *Phys. Rev.* **102** 1230–44
[21] Brooks Harris A 1968 Energy width of spin waves in the Heisenberg ferromagnet *Phys. Rev.* **175** 674–9
[22] Brooks Harris A 1969 Energy width of spin waves in the Heisenberg ferromagnet *Phys. Rev.* **184** 606
[23] Ackermann M, Bruning D, Lorenz T, Becker P and Bohaty L 2013 Thermodynamic properties of the new multiferroic material $(NH_4)_2[FeCl_5(H_2O)]$ *New J. Phys.* **15** 123001
[24] Gabas M, Palacio F, Rodriguez-Carvajal J and Visser D 1995 Magnetic structures of the three-dimensional Heisenberg antiferromagnets K_2 $[FeCl_5\ D_2O]$ and Rb_2 $[FeCl_5\ D_2O]$ *J. Phys.: Condens. Matter* **7** 4725
[25] Rodriguez-Velamazan J A, Fabelo O, Millan A, Campo J, Johnson R D and Chapon L 2015 Magnetically-induced ferroelectricity in the $(ND_4)_2$ $[FeCl_5(D_2O)]$ molecular compound *Sci. Rep.* **5** 14475
[26] Tian W *et al* 2016 Spin-lattice coupling mediated multiferroicity in $(ND_4)_2[FeCl_5D_2O]$ *Phys. Rev.* B **94** 214405
[27] Campo J, Palacio F, Paduan-Filho A, Becerra C C, Fernandez-Díaz M T and Rodríguez-Carvajal J 1997 Anomalous AF-SF line boundary in the phase diagram of Rb_2 $[(Fe_{1-x}In_x)$ $Cl_5H_2O]$ solid solutions *Phys. B: Condens. Matter* **234–236** 622–4
[28] Jain P, Ramachandran V, Clark R J, Zhou H D, Toby B H, Dalal N S, Kroto H W and Cheetham A K 2009 Multiferroic behavior associated with an order–disorder hydrogen bonding transition in metal–organic frameworks (MOFs) with the perovskite ABX_3 architecture *J. Am. Chem. Soc.* **131** 13625–7
[29] Wang X-Y, Gan L, Zhang S-W and Gao S 2004 Perovskite-like metal formates with weak ferromagnetism and as precursors to amorphous materials *Inorg. Chem.* **43** 4615–25
[30] Colacio E, Ghazi M, Kivekäs R and Moreno J M 2000 Helical-chain copper(II) complexes and a cyclic tetranuclear copper(II) complex with single syn-anti carboxylate bridges and ferromagnetic exchange interactions *Inorg. Chem.* **39** 2882–90
[31] Khomskii D March 2009 Classifying multiferroics: mechanisms and effects *Phys. Online* J **2** 20
[32] Sanchez-Andujar M *et al* 2014 First-order structural transition in the multiferroic perovskite-like formate $[(CH_3)_2NH_2][Mn(HCOO)_3]$ *Cryst. Eng. Commun.* **16** 3558–66
[33] Toth S and Lake B 2015 Linear spin wave theory for single-Q incommensurate magnetic structures *J. Phys.: Condens. Matter* **27** 166002
[34] Kimura T, Lashley J C and Ramirez A P 2006 Inversion-symmetry breaking in the noncollinear magnetic phase of the triangular-lattice antiferromagnet $CuFeO_2$ *Phys. Rev.* B **73** 220401

[35] Nakajima T, Mitsuda S, Kanetsuki S, Prokes K, Podlesnyak A, Kimura H and Noda Y 2007 Spin noncollinearlity in multiferroic phase of triangular lattice antiferromagnet $CuFe_{1-x}Al_xO_2$ *J. Phys. Soc. Jpn* **76** 043709
[36] Seki S, Yamasaki Y, Shiomi Y, Iguchi S, Onose Y and Tokura Y 2007 Impurity-doping-induced ferroelectricity in the frustrated antiferromagnet $CuFeO_2$ *Phys. Rev.* B **75** 100403
[37] Mitsuda S, Matsumoto Y, Wada T, Kurihara K, Urata Y, Yoshizawa H and Mekata M 1995 Magnetic ordering of $CuFe_{1-x}Al_xO_2$ *Phys. B: Condens. Matter* 213–4 194–6
[38] Takagi T and Mekata M 1995 New partially disordered phases with commensurate spin density wave in frustrated triangular lattice *J. Phys. Soc. Jpn* **64** 4609–27
[39] Nakajima T, Mitsuda S, Haku T, Shibata K, Yoshitomi K, Noda Y, Aso N, Uwatoko Y and Terada N 2011 Spin-wave spectrum in single-domain magnetic ground state of triangular lattice antiferromagnet $CuFeO_2$ *J. Phys. Soc. Jpn* **80** 014714
[40] Nakajima T, Mitsuda S, Haraldsen J T, Fishman R S, Hong T, Terada N and Uwatoko Y 2012 Magnetic interactions in the multiferroic phase of $CuFe_{1-x}Ga_xO_2$ ($x = 0.035$) refined by inelastic neutron scattering with uniaxial-pressure control of domain structure *Phys. Rev.* B **85** 144405
[41] Terada N, Mitsuda S, Oohara Y, Yoshizawa H and Takei H 2004 Anomalous magnetic excitation on triangular lattice antiferromagnet $CuFeO_2$ *J. Magn. Magn. Mater.* **272–276** E997–8
[42] Terada N, Mitsuda S, Fujii T and Petitgrand D 2007 Inelastic neutron scattering study of frustrated heisenberg triangular magnet $CuFeO_2$ *J. Phys.: Condens. Matter* **19** 145241
[43] Petrenko O A, Lees M R, Balakrishnan G, de Brion S and Chouteau G 2005 Revised magnetic properties of $CuFeO_2$—a case of mistaken identity *J. Phys.: Condens. Matter* **17** 2741
[44] Ajiro Y, Asano T, Takagi T, Mekata M, Katori H A and Goto T 1994 High-field magnetization process in the triangular lattice antiferromagnet $CuFeO_2$ up to 100 T *Phys. B: Condens. Matter* **201** 71–4
[45] Mitsuda S, Uno T, Mase M, Nojiri H, Takahashi K, Motokawa M and Arai M 1999 Neutron diffraction study of triangular lattice antiferromagnet $CuFeO_2$ under high magnetic field *J. Phys. Chem. Solids* **60** 1249–51
[46] Katsura H, Nagaosa N and Balatsky A V 2005 Spin current and magnetoelectric effect in noncollinear magnets *Phys. Rev. Lett.* **95** 057205
[47] Sergienko I A and Dagotto E 2006 Role of the Dzyaloshinskii–Moriya interaction in multiferroic perovskites *Phys. Rev.* B **73** 094434
[48] Arima T 2007 Ferroelectricity induced by proper-screw type magnetic order *J. Phys. Soc. Jpn* **76** 073702
[49] Ye F, Ren Y, Huang Q, Fernandez-Baca J A, Dai P, Lynn J W and Kimura T 2006 Spontaneous spin-lattice coupling in the geometrically frustrated triangular lattice antiferromagnet $CuFeO_2$ *Phys. Rev.* B **73** 220404
[50] Terada N, Mitsuda S, Ohsumi H and Tajima K 2006 Spin-driven crystal lattice distortion in frustrated magnet $CuFeO_2$: synchrotron x-ray diffraction study *J. Phys. Soc. Jpn* **75** 023602
[51] Fishman R S and Okamoto S 2010 Noncollinear magnetic phases of a triangular-lattice antiferromagnet and of doped $CuFeO_2$ *Phys. Rev.* B **81** 020402

Chapter 8

THz spectroscopy case studies

Here comes the Sun and I say, it's alright.

—The Beatles

8.1 Introduction

The usefulness of THz spectroscopy was dramatically demonstrated by the discovery of electromagnons in multiferroic manganites [1–3]. Electromagnons are SWs that can be excited by a THz electric field $\mathbf{E}^{\omega}$. Multiferroics with both FE and magnetic order may carry electromagnons that are simultaneously magnetic- and electric-dipole active. In other words, the matrix elements of both the magnetization $\mathbf{M}$ and the polarization $\mathbf{P}$ in equations (4.61–4.64) are nonzero. As described in chapter 3, the selection rules provided by polarization-dependent studies on a complete set of crystal orientations are required to obtain the optical strengths of each SW mode [3].

Multiferroics also provide a new way to manipulate THz radiation with an applied electric field [4], which can either shift the SW frequency [5] or change the magnetic domain structure [6, 7]. Gigantic non-reciprocal directional dichroism in $Ba_2CoGe_2O_7$ produces an optical diode that transmits THz light in one direction but not in the opposite direction [8, 9] at certain SW frequencies. Non-reciprocal directional dichroism in the THz range has also been observed in perovskites [10], $Ca_2CoSi_2O_7$ and $Sr_2CoSi_2O_7$ [11], $BiFeO_3$ [12, 13], ferroborate [14], $CaBaCo_4O_7$ [15], $MnWO_4$ [16] and CuB_2O_4 [17]. Hence, THz radiation probes several new effects in multiferroic materials.

This chapter examines two multiferroic materials, $BiFeO_3$ and $Ba_2CoGe_2O_7$. In $BiFeO_3$, the $S = 5/2$ Fe^{3+} spins form a long-wavelength spin cycloid below $T_N = 640K$ [18]. In contrast, the $S = 3/2$ Co^{2+} ions in $Ba_2CoGe_2O_7$ form a canted square-lattice AF below $T_N = 6.7$ K [19, 20]. While SW theory provides an almost

doi:10.1088/978-1-64327-114-9ch8

perfect description for the spin dynamics of $BiFeO_3$, it falls short for the spin dynamics in $Ba_2CoGe_2O_7$. A hallmark of the dynamic linear magnetoelectric effect, non-reciprocal directional dichroism is found in both materials.

8.2 A cycloid produced by DM interactions

This section extends the results of theory section 6.4 to the multiferroic $BiFeO_3$. Whereas the simpler theory in section 6.4 was applied to a one-dimensional lattice, we now examine a cubic lattice with hexagonal planes normal to the static polarization along one of the cubic diagonals. The single DM interaction responsible for the cycloid in section 6.4 is now joined by an additional DM interaction normal to the hexagonal planes.

8.2.1 Model for $BiFeO_3$

The cycloidal spin state of $BiFeO_3$ is primarily driven by the DM interaction D_1 on a hexagonal lattice. While it can lie along any of the cubic diagonals, the FE polarization $\mathbf{P}$ of $BiFeO_3$ is typically taken to lie along the unit vector $\mathbf{z}' = [1, 1, 1]$. In addition to D_1, a strong AF exchange J_1 couples spins on neighboring hexagonal layers normal to $\mathbf{z}'$, a weaker AF exchange J_2 couples neighboring spins on each hexagonal layer, a weak easy-axis anisotropy K acts along $\mathbf{z}'$, and a DM interaction D_2 acts between neighboring layers. This second DM interaction tilts the cycloid and modulates the magnetization in the plane perpendicular to $\mathbf{z}'$ with a net zero magnetic moment [21–24]. The Hamiltonian of $BiFeO_3$ in a magnetic field $\mathbf{B} = B\mathbf{m}$ is [25]

$$\begin{aligned}\mathcal{H} = &- J_1 \sum_{\langle i,j \rangle} \mathbf{S}_i \cdot \mathbf{S}_j - J_2 \sum_{\langle i,j \rangle'} \mathbf{S}_i \cdot \mathbf{S}_j \\ &+ D_1 \sum_{\langle i,j \rangle} (\mathbf{z}' \times \mathbf{e}_{i,j}/a) \cdot (\mathbf{S}_i \times \mathbf{S}_j) + D_2 \sum_{\langle i,j \rangle} (-1)^{n_i} \mathbf{z}' \cdot (\mathbf{S}_i \times \mathbf{S}_j) \\ &- K \sum_i (\mathbf{z}' \cdot \mathbf{S}_i)^2 - 2\mu_B B \sum_i \mathbf{m} \cdot \mathbf{S}_i,\end{aligned} \tag{8.1}$$

where $\mathbf{e}_{i,j} = a\mathbf{x}$, $a\mathbf{y}$, or $a\mathbf{z}$ connects $\mathbf{R}_i$ with its nearest-neighbor $\mathbf{R}_j = \mathbf{R}_i + \mathbf{e}_{i,j}$ on an adjacent layer, as shown in figure 8.1. Hexagonal layers normal to $\mathbf{z}'$ are separated by $c = a/\sqrt{3}$ and are labeled by the integer $n_i = \mathbf{R}_i \cdot \mathbf{z}'/c$. Consequently, the D_2 sum alternates sign from one hexagonal layer to the next. The local DM interactions D_1 $(\mathbf{z}' \times \mathbf{e}_{i,j}/a)$ and D_2 $\mathbf{z}'$ are, respectively, perpendicular and parallel to $\mathbf{z}'$.

Ignoring the cycloidal harmonics $C_{l>1}$ produced by D_2 and K but including the tilt τ produced by D_2, the spin state propagating along the hexagonal axis $\mathbf{x}'$ (see figure 8.1) in zero field can be approximated by

$$S_{x'}(\mathbf{R}) = S(-1)^{n+1} \cos\tau \sin(2\pi\delta r), \tag{8.2}$$

$$S_{y'}(\mathbf{R}) = S \sin\tau \sin(2\pi\delta r), \tag{8.3}$$

$$S_{z'}(\mathbf{R}) = S(-1)^{n+1} \cos(2\pi\delta r), \tag{8.4}$$

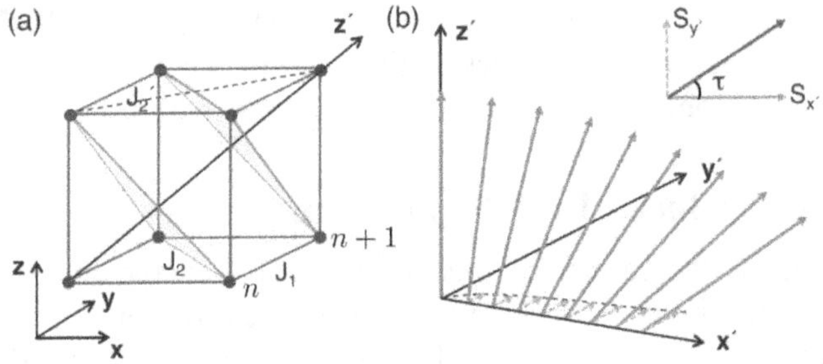

Figure 8.1. (a) The nearest-neighbor exchange J_1 between hexagonal layers n and $n + 1$ and the next-nearest-neighbor exchange J_2 within the hexagonal layers on the pseudo-cubic lattice for $BiFeO_3$ with $S = 5/2$ Fe^{3+} ions at the corners of the cube. Hexagonal layers are normal to $\mathbf{P}\|\mathbf{z}'$. Due to the weak rhombohedral distortion along $\mathbf{z}'$, the exchanges J_2 and J_2' may be slightly different. (b) The spin state of layer n in zero magnetic field is confined to the $x'z'$ plane except for a small canting. The variation of the canted component $S_{y'}$ along x' is shown by the dashed line. The projection of the canted spins onto the $x'y'$ plane is shown in the inset. Three magnetic domains with $\mathbf{x}_1' = [-1, 1, 0]$, $\mathbf{x}_2' = [1, 0, -1]$, or $\mathbf{x}_3' = [0, 1, -1]$ are allowed in a single FE domain with $\mathbf{P}\|\mathbf{z}'$. Reprinted with permission from [25].

where the integer $r = \mathbf{R} \cdot \mathbf{x}'\sqrt{2}/a$ ranges from 1 to $M_{\rm inc} = 1/\delta$ and the cycloidal period is $\lambda = a/\sqrt{2}\,\delta$. The tilted cycloid in a hexagonal layer is sketched in figure 8.1(b). In the language of section 4.3, $p = 1$, $M_0 = 2$, and $M = M_0 M_{\rm inc} = 2/\delta$. The higher harmonics of the cycloid are included just as for the distorted helix of Ga-doped $CuFeO_2$ in section 7.5.2.

The period λ of the cycloid increases with magnetic field [26] until the transition into a canted AF at B_c. In zero magnetic field, the number of spins along $\mathbf{x}'$ is $M_{\rm inc} \approx 222$ in each of two hexagonal layers indexed n and $n + 1$. The procedure for finding a self-consistent solution for the spin state is described in [25, 26]. In zero magnetic field, the parameters of the Hamiltonian are $J_1 = -\,5.32$ meV, $J_2 = -\,0.24$ meV, $D_1 = 0.18$ meV, $D_2 = 0.085$ meV, and $K = 0.0051$ meV. The second DM interaction tilts the cycloid by about $\tau \approx 0.45°$.

8.2.2 SWs in $BiFeO_3$

In the absence of DM interactions, the spin state of $BiFeO_3$ would be a G-type AF with wavevector $\mathbf{Q}_0 = (2\pi/a)(0.5, 0.5, 0.5)$ and $M_0 = 2$ sublattices. For the cycloidal spin state, the predicted dispersions of the SWs along $\mathbf{x}'$ at $\mathbf{Q} = \mathbf{Q}_0 + (2\pi/a)\sqrt{2}\, q\, \mathbf{x}'$ are shown in figure 8.2 for (a) $K = D_2 = 0$ or (b) $D_2 = 0.085$ meV and $K = 0.0051$ meV. Notice the similarity between the SW dispersions of this $M_{\rm inc} = 222$ cycloid and those of the $M_{\rm inc} = 20$ cycloid in figure 6.6(b). For the latter, SW branches starting from $H = 0.45$ ($q = \delta$) are gapped at $H = 0.5$ ($q = 0$) and 0.4 ($q = 2\delta$).

At this point, you might wonder whether INS can determine all the microscopic interactions of $BiFeO_3$. Performing INS measurements on the first-available single crystals, two groups [28, 29] obtained similar estimates for J_1 and J_2 from the SW spectra. Despite heroic efforts [30], however, INS measurements were unable to precisely determine the ‘small’ spin–orbit interaction terms D_1, D_2, and K responsible for the cycloid. As discussed in section 6.4, the details of the cycloidal spin state

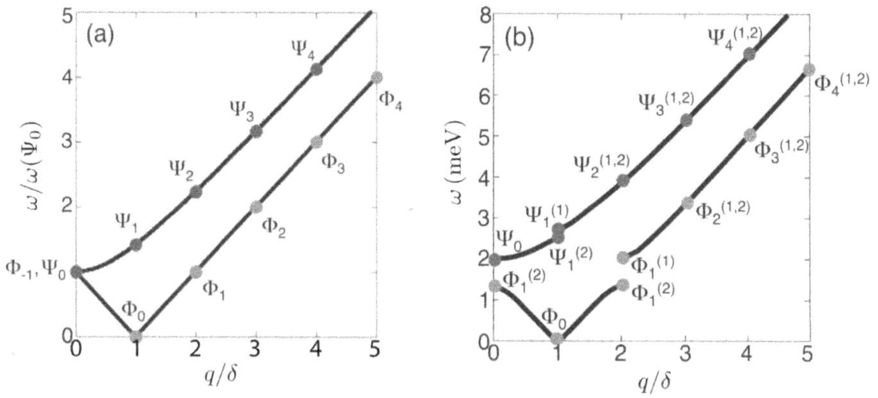

Figure 8.2. The mode spectra at multiples of the ordering wavevector δ in an extended zone scheme (a) without higher harmonics of the spin and (b) with higher harmonics for the predicted parameters of $BiFeO_3$. Modes are labeled as in-plane Φ_m and out-of-plane Ψ_m cycloidal modes. In the absence of higher harmonics, the energies in (a) are $\omega(\Phi_m) = |m|\omega(\Psi_0)$ and $\omega(\Psi_m) = \omega(\Psi_0)\sqrt{1 + m^2}$ [27]. Higher harmonics in (b) introduce mode splittings at wavevectors $q = n\delta$. Reprinted with permission from [25].

only affect the SW spectra at low energies near **Q**. So determining D_1, D_2, and K requires a frequency and wavevector resolution beyond what INS can currently offer. For example, INS lacks the momentum resolution to separate the SWs around $\mathbf{Q} = \mathbf{Q}_0 \pm (2\pi/a)\sqrt{2}\,\delta\mathbf{x}'$ separated by 0.006 rlu.

In contrast, with $q < 2 \times 10^{-10}$ rlu at 1 THz (recall that $\omega(0) = \omega(\mathbf{Q})$) and an energy resolution of about 0.01 meV, optical spectroscopy is ideally suited to this task. At multiples of δ, modes assigned to in-plane Φ_m and out-of-plane Ψ_m cycloidal modes [27] (see section 6.4) were observed by Raman [31] and THz spectroscopies [32] in zero field. In figure 8.3, the predicted and measured THz mode frequencies are plotted versus magnetic field for field orientations $\mathbf{m}$ = [0, 0, 1], [1, 1, 0], and [1, −1, 0]. Modes $\Phi_m^{(1,2)}$ and $\Psi_m^{(1,2)}$ are superpositions of the zero-field $\pm m$ modes introduced in section 6.4.

Because its frequency is so low, $\Phi_1^{(2)}$ was not detected when $\mathbf{m}$ = [1, 1, 0] and [1, −1, 0]. The predicted mode frequencies of the stable domain(s) are presented in the solid curves: domain 1 for $\mathbf{m}$ = [0, 0, 1] and [1, 1, 0] and domains 2 and 3 for $\mathbf{m}$ = [1, −1, 0]. For $\mathbf{m}$ = [0, 0, 1], the mode that dips below $\Phi_1^{(2)}$ arises from metastable domains 2 and 3, as indicated by the agreement with the predicted dashed curve. For zero field, the total number of spins in a unit cell is $M = M_0 M_{\text{inc}} = 444$. This number increases with field as δ decreases and $M_{\text{inc}} = 1/\delta$ increases.

Domains with $\mathbf{q}_{\text{dom}} = \mathbf{Q} - \mathbf{Q}_0 \perp \mathbf{B}$ are stable in a magnetic field and retain their orientation when the field is removed. Changes of domain structure were first observed in the THz spectra and interpreted as a change in the population of the three magnetic domains [26, 33]. Recent small-angle neutron-scattering measurements have shown that the domain wavevector $\mathbf{q}_{\text{dom}}$ rotates in the hexagonal plane perpendicular to the applied field at about 10 T [34, 35]. For $\mathbf{m}$ = [1, −1, 0], one domain is found with $\mathbf{q}_{\text{dom}} \| [-1, -1, 2]$ instead of two domains with $\mathbf{q}_{\text{dom}}$ along $\mathbf{x}_2'$

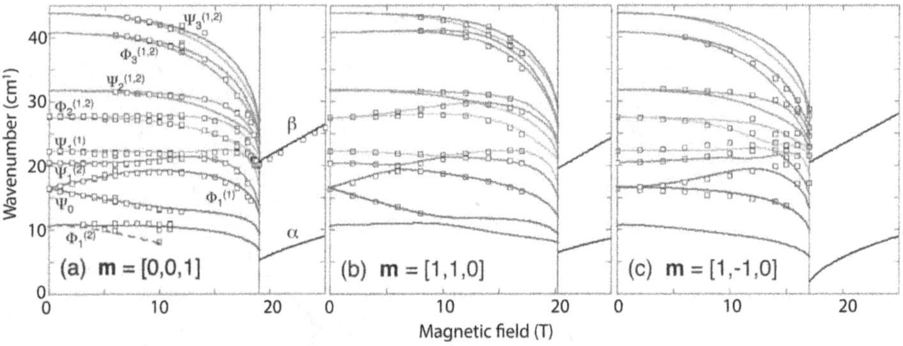

Figure 8.3. Theoretical mode energies (solid curves) and experimental data (squares) versus field for field orientations **m**= (a) [0, 0, 1], (b) [1, 1, 0], and (c) [1, −1, 0]. Solid vertical lines mark the transition to the canted AFM state with two AFM modes α and β. The dashed curve in (a) indicates the predicted $\Phi_1^{(2)}$ frequency for metastable domains 2 and 3. Reprinted with permission from [25].

and $\mathbf{x}_3'$. The rotation of $\mathbf{q}_{\rm dom}$ was not taken into account when calculating the mode frequencies and indeed small discrepancies were observed in figure 8.3(c) for $\mathbf{m} = [1, -1, 0]$.

The basics of optical absorption for a spin cycloid with $D_1 \neq 0$ and $K \neq 0$ were described in section 6.4. A cycloid generated by $D_1 > 0$ with $K = 0$ is purely harmonic. In that case, the in-plane and out-of-plane cycloidal modes do not mix. For a harmonic cycloid, $\Psi_1^{(1)}$ is active when $\mathbf{B}^\omega \| \mathbf{x}'$ and $\Psi_1^{(2)}$ is active when $\mathbf{B}^\omega \| \mathbf{z}'$. Based on the SC polarization, $\Psi_1^{(1)}$ becomes electric-dipole active when $\mathbf{E}^\omega \| \mathbf{y}'$ [36, 37], as shown in exercise 4.15. So $\Psi_1^{(1)}$ is an electromagnon.

Due to the higher harmonics generated by K and D_2 or by an applied magnetic field [26, 38], optically inactive modes are activated as they mix with the optically active ones. For example, Φ_2 becomes optically active for $\mathbf{B}^\omega \| \mathbf{y}'$ when $K > 0$. The second DM interaction $D_2 > 0$ activates Ψ_0 and $\Phi_1^{(1)}$ when $\mathbf{B}^\omega \| \mathbf{x}'$ and $\mathbf{y}'$, respectively. All modes in the frequency window shown in figure 8.3 are activated by a magnetic field. Nevertheless, Φ_0 has too low a frequency to be observed by THz spectroscopy.

The experimental work described above was performed on a crystal with a single FE domain grown by the flux method [39]. This crystal has a (0,0,1) face suitable for optical spectroscopy with $\mathbf{k} \| [0, 0, 1]$. However, a detailed study of the selection rules was not possible without a single FE domain crystal with faces perpendicular to the principal hexagonal directions. This may change in the near future since large single crystals can now be grown by a laser-diode heating floating-zone method [40].

8.3 An AF with strong easy-plane anisotropy

So far, every material we have discussed can be at least qualitatively understood using SW theory. But the multiferroic $Ba_2CoGe_2O_7$ exhibits more modes than SW theory can produce! SW theory fails for this material because it does not account for fluctuations of the spin amplitude in the presence of strong easy-plane anisotropy.

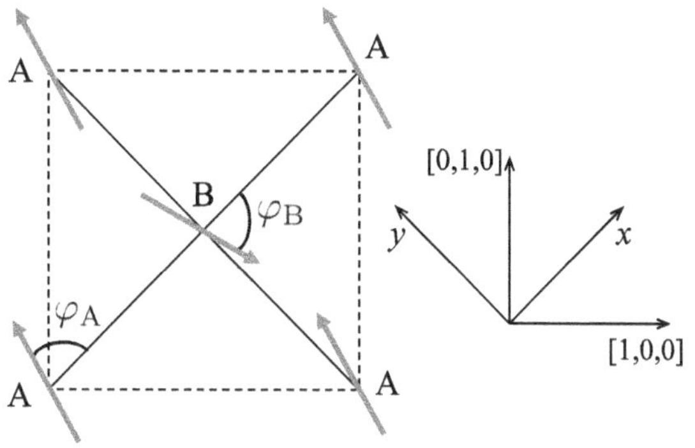

Figure 8.4. Spin model of $Ba_2CoGe_2O_7$ on a square lattice with two AF coupled spins on sites A and B. Spin canting is produced by an applied magnetic field along [1, 0, 0]. The canting angles are measured from the x-axis.

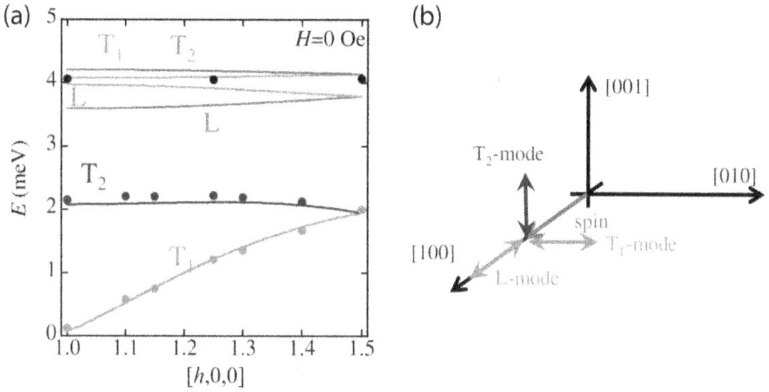

Figure 8.5. (a) Magnon dispersion relation for $\mathbf{Q} = [h, 0, 0]$ in zero magnetic field. (b) Schematics of spin fluctuations for each mode in zero magnetic field. A transverse fluctuation in the (0, 0, 1) plane is called a T_1 mode (THz modes b_-, d_- at $h = 1$ of Penc *et al* [43]). A transverse fluctuation with a component along [0, 0, 1] is called a T_2 (b_+, d_+) mode, and a longitudinal fluctuation along [1,0,0] is called an L ($c_\pm$) mode. Spins are along [1, 0, 0]. Reprinted with permission from [41].

8.3.1 Model for $Ba_2CoGe_2O_7$

As shown in figure 8.4, $Ba_2CoGe_2O_7$ can be considered to be an $S = 3/2$ square-lattice AF with easy-plane anisotropy. INS observed two SW branches [20]: an acoustic branch that vanishes at $\Gamma = (0.5, 0.5)$ and an optical branch with a frequency of 2.2 meV (0.53 THz) at Γ. The SW dynamics of $Ba_2CoGe_2O_7$ was first modeled by Zheludev *et al* [20] using an effective $S = 1/2$ spin Hamiltonian including anisotropic nearest-neighbor exchange couplings $J_x = J_y = J$ and J_z. This simple model suggested that the system is strongly anisotropic with $J \approx -0.50$ meV and $J_z \approx -0.19$ meV. However, THz spectroscopy observed an additional mode at 1 THz [8], which was later confirmed by INS [41] and is plotted in figure 8.5. Within conventional SW theory, the number of modes must equal the number of spins in the unit cell. Since $M = 2$ for $Ba_2CoGe_2O_7$, this third mode cannot be explained by conventional SW theory. Consequently, one or more assumptions of SW theory must be violated for this compound.

In the presence of strong easy-plane anisotropy $K < 0$, the spins lie predominantly in the xy plane and fluctuations out of the plane are suppressed. Since the spin amplitude is reduced, the spins behave as if $S_{\mathrm{eff}} < 3/2$. The resulting oscillations of the spin amplitude are not considered by conventional SW theory. To address this deficiency, two alternative approaches were proposed. The THz absorption spectra were first modeled by the exact diagonalization of a finite-size cluster [42] and then by multiboson SW theory [43]. Remarkably, the extra modes in the excitation spectra were predicted by multiboson SW theory [44] two years before they were experimentally observed [41] using INS!

Mutliboson SW theory [43] for $Ba_2CoGe_2O_7$ is based on the Hamiltonian of an $S = 3/2$ square lattice AF with strong easy-plane anisotropy $|K| > |J|$:

$$\begin{aligned}\mathcal{H} = &- J \sum_{<i,j>} (S_{ix}S_{jx} + S_{iy}S_{jy}) - J_z \sum_{<i,j>} S_{iz}S_{jz} \\ &- K \sum_i S_{iz}^2 + \mu_B \sum_i \{g_{xx}(B_x S_{ix} + B_y S_{iy}) + g_{zz} B_z S_{iz}\},\end{aligned} \tag{8.5}$$

where $\mathbf{B}$ is the magnetic field and the coordinate system is shown in figure 8.4. While the AF exchange coupling between nearest-neighbor spins is anisotropic, the single-ion anisotropy $K < 0$ favors the spins to lie in the xy plane. Experimental parameters are listed in table 8.1. The g-tensor has nonzero components $g_{xx} = g_{yy} = 2.3$ and $g_{zz} = 2.1$. Other interactions such as the DM interaction [42] and the anti-ferroelectric interaction between Co sites [45] (or both [41]) have only marginal effects on the SW spectra.

Before applying multiboson SW theory, the ground state is determined by minimizing the energy [43, 45, 46] as a function of the canting angles $\varphi_A = -\varphi_B$ and the parameter η. When $\eta \neq 1$, the spin amplitude is reduced from 3/2 due to the easy-plane anisotropy. The factorized variational wavefunction is given by

$$|\Psi\rangle = \prod_{i \in A} \prod_{j \in B} |A\rangle_i |B\rangle_j, \tag{8.6}$$

where i and j count the A and B sites (see figure 8.4). When a magnetic field is applied along the x-axis [46],

Table 8.1. Hamiltonian parameters of $Ba_2CoGe_2O_7$ from the exact diagonalization of a finite-size cluster and from multiboson SW theory by fitting the THz absorption [42, 43] and the INS spectra [41]. Parameters are in meV.

J	J_z	K	Ref
−0.175	−0.175	−1.4	[41]
−0.20	−0.16	−1.15	[42]
−0.208	−0.253	−1.034	[40]

$$|A\rangle = \frac{e^{-i\varphi_A S_A^z}}{\sqrt{2}\sqrt{3\eta^2+1}}\left\{|3/2\rangle + |-3/2\rangle + \sqrt{3}\eta(|1/2\rangle + |-1/2\rangle)\right\}, \tag{8.7}$$

where **z** is the spin quantization axis and the phase factor creates a unitary transformation into the local reference frame of spin A, which is rotated by φ_A about **z**. A similar expression is obtained for the B sites with the prefactor $\exp(-i\varphi_B S_B^z)$. In this new basis, the spin on an A site is

$$\langle A|\mathbf{S}_i|A\rangle = \frac{3\eta(\eta+1)}{3\eta^2+1}(\cos\varphi_A, \sin\varphi_A, 0) \equiv S_{\text{eff}}(\cos\varphi_A, \sin\varphi_A, 0), \tag{8.8}$$

with a similar expression for the spin on a B site. The ground-state variational parameters ϕ_A and η are solved from the relations

$$K = \frac{3(3\eta+1)(\eta^2-1)}{3\eta^2+1}|J|, \tag{8.9}$$

$$g_{xx}\,\mu_B B_x = \frac{24\eta(\eta+1)}{(3\eta^2+1)}|J|\cos\varphi_A. \tag{8.10}$$

While η is determined by K and J, $\varphi_A = -\varphi_B$ is controlled by the in-plane magnetic field. Since there is no in-plane anisotropy, the spins freely follow the magnetic field in the xy plane. A more general description of the ground states in $\{K, J_z, J\}$ parameter space was provided by Romhányi *et al* [46].

If $K = 0$, then $\eta = 1$ and $S_{\text{eff}} = 3/2$. For $\eta \neq 1$, $S_{\text{eff}} < 3/2$. Using the parameters for $Ba_2CoGe_2O_7$ [42, 43], $1.3 \leqslant S_{\text{eff}} \leqslant 1.35$. In the limit $|K/J| \to \infty$, $\eta \to \infty$,

$$|A\rangle \to e^{-i\varphi_A S_A^z}\frac{1}{\sqrt{2}}(|1/2\rangle + |-1/2\rangle), \tag{8.11}$$

and $S_{\text{eff}} \to 1$.

8.3.2 SWs in $Ba_2CoGe_2O_7$

In conventional SW theory, a single boson is associated with each spin to describe the transverse fluctuations about the classical ground state, regardless of the spin S. To account for additional spin degrees of freedom, the multiboson SW approach introduces more bosons at each site. Technical details about this theory are provided in [44].

Generally, multiboson SW theory introduces bosons $\alpha_m^\dagger$ that create the $S^z = m$ states so that $|m\rangle = \alpha_m^\dagger|0\rangle$, where $|0\rangle$ is the vacuum. For the planar two-site AF magnet, bosons $\alpha_m^\dagger$ (four at each site for $S = 3/2$) are transformed by an $SU(4)$ rotation into new bosons ($a^\dagger, b^\dagger, c^\dagger, d^\dagger$). This transformation ensures that the variational ground state of equation (8.7) is $|A\rangle = a_A^\dagger|0\rangle$ for the A sites and $|B\rangle = a_B^\dagger|0\rangle$ for the B sites. Since the number of bosons on each site is conserved,

$$\sum_m \alpha_m^\dagger \alpha_m = 1, \tag{8.12}$$

i.e. three independent bosons per site are associated with each spin $S = 3/2$.

Six modes are then created by the six bosons in the AF unit cell of $Ba_2CoGe_2O_7$. Four of these are transverse modes and two are longitudinal spin-stretching modes. The six SW modes labeled $b_\pm$, $c_\pm$, and $d_\pm$ and their frequencies are plotted in figure 8.6. THz absorption spectra are shown in figure 8.7.

Mode b_- is a Goldstone mode where the A and B spins rotate together in the easy-plane. When anisotropy is introduced ($K < 0$ or $J_z \neq J$), the out-of-plane mode b_+ is gapped. Although modes $b_\pm$ can be derived from standard SW theory, their frequencies depend differently on anisotropy in multiboson SW theory. In a standard SW calculation,

$$\hbar\omega(b_+) - \hbar\omega(b_-) = \sqrt{16|JK|S} = \sqrt{36|JK|}. \tag{8.13}$$

But in multiboson SW theory with $J = J_z$,

$$\hbar\omega(b_+) - \hbar\omega(b_-) = \sqrt{24|JK|}, \tag{8.14}$$

which is smaller by a factor of $\sqrt{2/3}$ [47].

Modes $c_\pm$ and $d_\pm$ cannot be derived from SW theory and are described in table 8.2. The $c_\pm$ spin-stretching modes have $\Delta \mathbf{S} \| \mathbf{y}$. For the c_+ mode, the spin is elongated

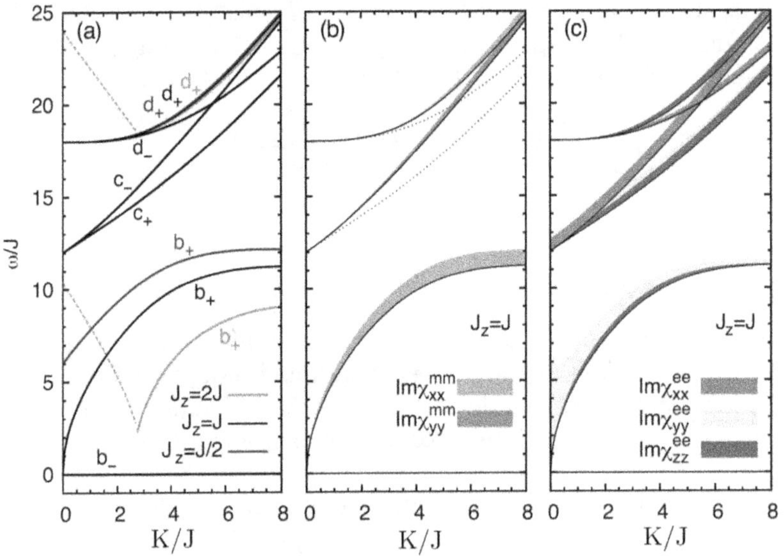

Figure 8.6. (a) Mode energies for three values of J_z/J in zero field. The $\omega = 0$ Goldstone mode is denoted by b_-. Only the b_+ and d_+ modes depend on J_z/J. Dashed lines indicate modes in the easy-axis AF state for $K/J < 2.7$ and $J_z = 2J$. (b) and (c) The imaginary part of the magnetic χ^{mm} and electric χ^{ee} dynamic susceptibilities for $J_z = J$. The shading above the lines represent the strength of the magnetic and electric response. Spins A and B are assumed to point in the $\pm\mathbf{y}$ directions. Only the b_- Goldstone mode has Im $\chi^{\mathrm{mm}} > 0$ when $\mathbf{h} = \mathbf{z}$. Reprinted with permission from [43].

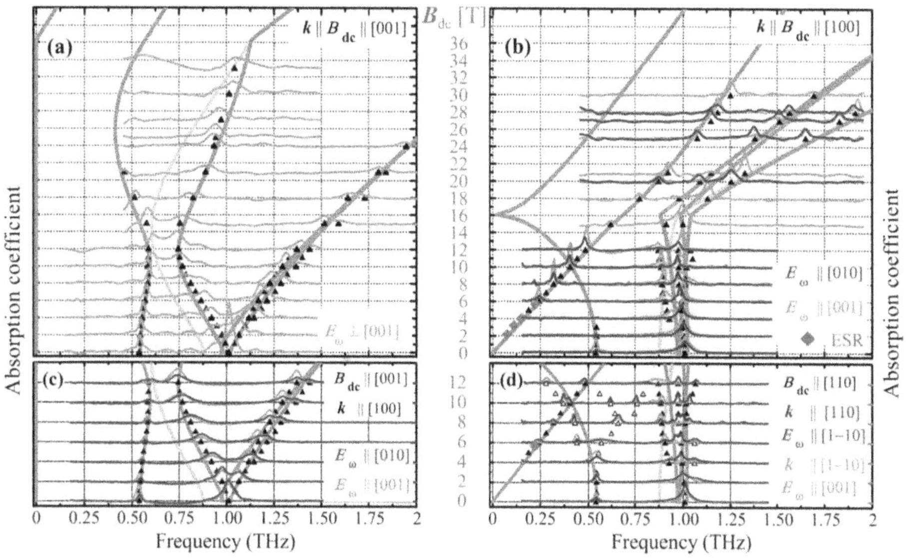

Figure 8.7. Magnetic-field dependence of the absorption spectra in $Ba_2CoGe_2O_7$ below 2 THz. The separation between horizontal grid lines corresponds to an absorption coefficient of 20 cm^{-1} in panels (a) and (c), and 30 cm^{-1} in (b) and (d). Black triangles (THz) and green diamonds (ESR) represent the SW resonances. Gray lines show the field dependence of the modes obtained in the multiboson SW approach. (d) For $B_{dc} \perp [001]$, modes not explained by theory are observed (open triangles) for some polarizations, two of which are shown. Measurements above 12 T were performed at the High Field Magnet Laboratory, Nijmegen, below 12 T with TeslaFIR [section 3.2.4], and ESR spectroscopy with multifrequency ESR spectrometer [48]. Reprinted with permission from [43].

Table 8.2. Amplitudes of SW modes $\Delta\mathbf{S}_i$ at sites $i = A, B$ and corresponding polarization amplitudes $\Delta\mathbf{P}_i$ in zero magnetic field calculated with $J = J_z$ and $K = 8J$. The ground-state spin configuration has $\mathbf{S}_A = S[0, 1, 0]$, $\mathbf{S}_B = -\mathbf{S}_A$, and $\varphi_A = -\varphi_B = \pi/2$. The polarization is then $\mathbf{P}_A = \mathbf{P}_B = P[0, 0, 1]$. The real and imaginary components of $\Delta\mathbf{M}$ ($\Delta\mathbf{P}$) oscillate as $\cos(\omega t)$ and $\sin(\omega t)$, respectively. Visualized modes are presented in the supplemental material to [43].

	ΔS^0	$\Delta\mathbf{S}_A/\Delta S^0$	$\Delta\mathbf{S}_B/\Delta S^0$	ΔP^0	$\Delta\mathbf{P}_A/\Delta P^0$	$\Delta\mathbf{P}_B/\Delta P^0$
b_-	155	(1, 0, 0)	(−1, 0, 0)	114	(0, 0, 1)	(0, 0, −1)
b_+	0.53	(1, 0, −0.92*i*)	(1, 0, 0.92*i*)	0.53	(−*i*, *i*, 0.46)	(*i*, *i*, 0.46)
c_-	0.41	(0, 1, 0)	(0, 1, 0)	0.89	(*i*, *i*, −0.9)	(*i*, −*i*, 0.9)
c_+	0.44	(0, 1, 0)	(0, −1, 0)	0.84	(*i*, *i*, −0.98)	(−*i*, *i*, −0.98)
d_-	0.31	(1, 0, 0)	(−1, 0, 0)	0.71	(−0.9*i*, 0.9*i*, 1)	(−0.9*i*, −0.9*i*, −1)
d_+	0.39	(1, 0, −0.42*i*)	(1, 0, 0.42*i*)	0.85	(−*i*, *i*, 0.9)	(*i*, *i*, 0.9)

or compressed in phase on sites A and B; for the c_- mode, the spin distortions on sites A and B are out-of-phase. Modes $d_\pm$ have the same pattern as $b_\pm$ but with different amplitudes. As seen from their dependence on J_z in figure 8.6 and from their amplitudes in table 8.2, modes b_+ and d_+ are the only out-of-plane modes.

If $K = 0$ with $S_{\text{eff}} = 3/2$, then modes $b_{\pm}$ are ordinary SWs and modes $c_{\pm}$ and $d_{\pm}$ become local spin excitations. The spectra of isolated spins are also recovered in the limit $K \to \infty$ with $\hbar\omega/K \to 0$ for two modes and $\hbar\omega \to 2K$ for the other four modes [42, 43].

When $K = 0$ and $J_z = J$ (figure 8.6(b)), modes $c_{\pm}$ and $d_{\pm}$ are not magnetic-dipole active and will not be visible in the INS spectra. With increasing K/J, however, $c_{\pm}$ and $d_{\pm}$ gain spectral weight and may become visible using INS [41, 44]. If they are electric-dipole active, modes $c_{\pm}$ may appear in optical measurements even when $K = 0$ and $J = J_z$. Modes $d_{\pm}$ may also be optically observed if $K/J > 0$ or $J_z \neq J$.

8.3.3 Selection rules and optical strengths

Non-reciprocal directional dichroism in the THz spectral range was first discovered in $Ba_2CoGe_2O_7$ [8, 9]. Directional optical anisotropy appears when the SW mode (i.e. the electromagnon) is both magnetic- and electric-dipole active. As emphasized in section 4.4, several spin interactions can produce the electric polarization operator $\mathbf{P}$. The dc magnetoelectric effect in $Ba_2CoGe_2O_7$ was studied experimentally [49] and described theoretically by Murakawa *et al* [50] based on metal-ligand *p–d* hybridization [51, 52], as in $CuFeO_2$. The same operator form for $\mathbf{P}$ was used to describe the ac magnetoelectric effect and non-reciprocal directional dichroism in $Ba_2CoGe_2O_7$ [8, 9, 42]. By taking into account additional constraints imposed by lattice symmetry and ignoring the small rotation of the oxygen tetrahedra, the polarization of $Ba_2CoGe_2O_7$ at site i is [42, 46]

$$\begin{aligned} P_{ix} &= -\,k(S_{ix}S_{iz} + S_{iz}S_{ix}),\\ P_{iy} &= k\left(S_{iy}S_{iz} + S_{iz}S_{iy}\right),\\ P_{iz} &= k\left(S_{iy}{}^2 - S_{ix}{}^2\right), \end{aligned} \tag{8.15}$$

where k has dimensions of μC cm.

The absorption of THz radiation by SWs is calculated using the dynamic susceptibilities χ^{mm} and χ^{ee} of equations (4.61) and (4.62). The imaginary parts of these susceptibilities are plotted in figures 8.6(b) and (c). Modes b_+ and d_+ are excited when $\mathbf{B}^{\omega}\|\mathbf{x}$ and the spin-stretching mode c_- is excited when $\mathbf{B}^{\omega}\|\mathbf{y}$. The zero-frequency Goldstone mode b_- is active when $\mathbf{B}^{\omega}\|\mathbf{z}$.

The spin dynamics is not affected by the magnetoelectric coupling because the spin Hamiltonian (8.5) does not depend on $\mathbf{P}$. So the role of the electric polarization is to couple the THz electric field $\mathbf{E}^{\omega}$ to the SW modes. As seen in table 8.2 and figure 8.6(c), all modes except b_- are electric-dipole active so that $\Delta\mathbf{P}_A + \Delta\mathbf{P}_B \neq 0$. Only mode b_- does not exhibit polarization oscillations in the xy plane.

The dynamic linear magnetoelectric susceptibilities χ^{em} and χ^{me} of equations (4.63) and (4.64) are determined from the difference of the absorption spectra measured for opposite propagation directions of THz radiation, as in equation (4.60). The non-reciprocal directional dichroism effect for $Ba_2CoGe_2O_7$ was described in [8, 9, 42] and was also found in two compounds with similar structures, $Ca_2CoSi_2O_7$ and $Sr_2CoSi_2O_7$ [11].

8.4 Prospects for the future

What does this work on $BiFeO_3$ and $Ba_2CoGe_2O_7$ imply about the advantages and shortcomings of conventional SW theory? The agreement between the predicted and measured mode frequencies of $BiFeO_3$ in a magnetic field indicates that SW theory is remarkably successful at describing the dynamics of that system. Nevertheless, the optical machinery developed in section 4.4 was only partially successful at describing the non-reciprocal directional dichroism of those SW modes in a magnetic field [53]. Discrepancies between theory and experiment might be related to the assumed form for the spin-induced polarization or, perhaps, to some assumption of SW theory.

As the extra observed SW mode in $Ba_2CoGe_2O_7$ makes abundantly clear, the failures of SW theory are typically quite spectacular. Due to strong easy-plane anisotropy, extra modes split from the ordinary SW modes. The spin-stretching $c_\pm$ modes appear as local spin excitations even when $K = 0$ and $J = J_z$. If inversion symmetry is broken, those modes become electric-dipole active (see figure 8.6(c)). Although the extra $c_\pm$ and $d_\pm$ modes can be detected with INS when $|K/J| > 0$ or $J \neq J_z$, their spectral weight only becomes significant when $|K/J|$ is sufficiently large. Work on $Ba_2CoGe_2O_7$ indicates that the extra modes appear at high frequencies, above the conventional SW modes.

Why not just apply multiboson theory as the default? The drawback of multiboson theory is that it dramatically increases the cost of the matrix diagonalization. With $2S$ bosons per site and M sites in the unit cell, the total number of degrees of freedom is now $2SM$ rather than M. Consider, for example, the cycloid of $BiFeO_3$ with $S = 5/2$ and $M = 444$ sites in zero field. The size of the $\underline{L}$ matrix would increase from 888×888 in conventional SW theory to 4440×4440 in multiboson theory. So the matrix becomes 25 times bigger!

Whether to search for spin-stretching modes using a multiboson theory before applying conventional SW theory should depend on the nature of the material and on experimental observations. A safe approach would be to apply conventional SW theory first. If extra modes appear at high frequencies, then it may be necessary to consider other options.

Despite its birth in the mid-1960s [54], the field of THz spectroscopy remains in its infancy or perhaps its early adolescence. The exciting discoveries of electromagnons and non-reciprocal directional dichroism underscore its importance in the exploration of multiferroic materials. As seen in the last section, this technique can also pinpoint the shortcomings in conventional SW theory. Hence, it may prove to be particularly useful in the study of low-spin or low-D materials where spin fluctuations are enhanced.

References

[1] Pimenov A, Mukhin A A, Ivanov V Y, Travkin V D, Balbashov A M and Loidl A 2006 Possible evidence for electromagnons in multiferroic manganites *Nat. Phys.* **2** 97–100

[2] Sushkov A B, Valdés Aguilar R, Park S, Cheong S-W and Drew H D 2007 Electromagnons in multiferroic YMn_2O_5 and $TbMn_2O_5$ *Phys. Rev. Lett.* **98** 027202

[3] Kida N, Takahashi Y, Lee J S, Shimano R, Yamasaki Y, Kaneko Y, Miyahara S, Furukawa N, Arima T and Tokura Y 2009 Terahertz time-domain spectroscopy of electromagnons in multiferroic perovskite manganites *J. Opt. Soc. Am.* B **26** A35–51

[4] Tokura Y and Kida N 2011 Dynamical magnetoelectric effects in multiferroic oxides *Philos. Trans. R. Soc. Lond.* A **369** 3679–94

[5] Rovillain P, de Sousa R, Gallais Y, Sacuto A, Méasson M A, Colson D, Forget A, Bibes M M, Barthélémy A and Cazayous M 2010 Electric-field control of spin waves at room temperature in multiferroic $BiFeO_3$ *Nat. Mater.* **9** 975

[6] Shuvaev A, Dziom V, Pimenov A, Schiebl M, Mukhin A A, Komarek A C, Finger T, Braden M and Pimenov A 2013 Electric field control of terahertz polarization in a multiferroic manganite with electromagnons *Phys. Rev. Lett.* **111** 227201

[7] Kuzmenko A M 2018 Switching of magnons by electric and magnetic fields in multiferroic borates *Phys. Rev. Lett.* **120** 027203

[8] Kézsmárki I, Kida N, Murakawa H, Bordács S, Onose Y and Tokura Y 2011 Enhanced directional dichroism of terahertz light in resonance with magnetic excitations of the multiferroic $Ba_2CoGe_2O_7$ oxide compound *Phys. Rev. Lett.* **106** 057403

[9] Bordács S 2012 Chirality of matter shows up via spin excitations *Nat. Phys.* **8** 734–8

[10] Takahashi Y, Shimano R, Kaneko Y, Murakawa H and Tokura Y 2012 Magnetoelectric resonance with electromagnons in a perovskite helimagnet *Nat. Phys.* **8** 121–5

[11] Kézsmárki I 2014 One-way transparency of four-coloured spin-wave excitations in multiferroic materials *Nat. Commun.* **5** 3203

[12] Kézsmárki I, Nagel U, Bordács S, Fishman R S, Lee J H, Yi H T, Cheong S-W and Rõõm T 2015 Optical diode effect at spin-wave excitations of the room-temperature multiferroic $BiFeO_3$ *Phys. Rev. Lett.* **115** 127203

[13] Park J-G, Le M D, Jeong J and Lee S 2014 Structure and spin dynamics of multiferroic $BiFeO_3$ *J. Phys.: Condens. Matter* **26** 433202

[14] Kuzmenko A M, Dziom V, Shuvaev A, Pimenov A, Schiebl M, Mukhin A A, Ivanov V Y U, Gudim I A, Bezmaternykh L N and Pimenov A 2015 Large directional optical anisotropy in multiferroic ferroborate *Phys. Rev.* B **92** 184409

[15] Bordács S, Kocsis V, Tokunaga Y, Nagel U, Rõõm T, Takahashi Y, Taguchi Y and Tokura Y 2015 Unidirectional terahertz light absorption in the pyroelectric ferrimagnet $CaBaCo_4O_7$ *Phys. Rev.* B **92** 214441

[16] Takahashi Y, Kibayashi S, Kaneko Y and Tokura Y 2016 Versatile optical magnetoelectric effects by electromagnons in $MnWO_4$ with canted spin-helix plane *Phys. Rev.* B **93** 180404

[17] Nii Y, Sasaki R, Iguchi Y and Onose Y 2017 Microwave magnetochiral effect in the non-centrosymmetric magnet CuB_2O_4 *J. Phys. Soc. Jpn.* **86** 024707

[18] Sosnowska I, Peterlin-Neumaier T and Steichele E 1982 Spiral magnetic ordering in bismuth ferrite *J. Phys. C: Solid State Phys.* **15** 483

[19] Sato T, Masuda T and Uchinokura K 2003 Magnetic property of $Ba_2CoGe_2O_7$ *Phys. B: Condens. Matter* **329-333** 880–1

[20] Zheludev A, Sato T, Masuda T, Uchinokura K, Shirane G and Roessli B 2003 Spin waves and the origin of commensurate magnetism in $Ba_2CoGe_2O_7$ *Phys. Rev.* B **68** 024428

[21] Kadomtseva A M, Zvezdin A K, Popov Y P, Pyatakov A P and Vorobev G P 2004 Space-time parity violation and magnetoelectric interactions in antiferromagnets *JETP Lett.* **79** 571–81

[22] Ederer C and Spaldin N A 2004 Magnetoelectrics: a new route to magnetic ferroelectrics *Nat. Mater.* **3** 849–51
[23] Pyatakov A P and Zvezdin A K 2009 Flexomagnetoelectric interaction in multiferroics *Eur. Phys. J.* B **71** 419–27
[24] Ohoyama K, Lee S, Yoshii S, Narumi Y, Morioka T, Nojiri H, Jeon G S, Cheong S-W and Park J-G 2011 High field neutron diffraction studies on metamagnetic transition of multiferroic $BiFeO_3$ *J. Phys. Soc. Jpn.* **80** 125001
[25] Fishman R S, Lee J H, Bordács S, Kézsmárki I, Nagel U and Rõõm T 2015 Spin-induced polarizations and nonreciprocal directional dichroism of the room-temperature multiferroic $BiFeO_3$ *Phys. Rev.* B **92** 094422
[26] Fishman R S 2013 Field dependence of the spin state and spectroscopic modes of multiferroic $BiFeO_3$ *Phys. Rev.* B **87** 224419
[27] de Sousa R and Moore J E 2008 Optical coupling to spin waves in the cycloidal multiferroic $BiFeO_3$ *Phys. Rev.* B **77** 012406
[28] Jeong J 2012 Spin wave measurements over the full Brillouin zone of multiferroic $BiFeO_3$ *Phys. Rev. Lett.* **108** 077202
[29] Matsuda M, Fishman R S, Hong T, Lee C H, Ushiyama T, Yanagisawa Y, Tomioka Y and Ito T 2012 Magnetic dispersion and anisotropy in multiferroic $BiFeO_3$ *Phys. Rev. Lett.* **109** 067205
[30] Jeong J, Le M D, Bourges P, Petit S, Furukawa S, Kim S-A, Lee S, Cheong S-W and Park J-G 2014 Temperature-dependent interplay of Dzyaloshinskii–Moriya interaction and single-ion anisotropy in multiferroic $BiFeO_3$ *Phys. Rev. Lett.* **113** 107202
[31] Cazayous M, Gallais Y, Sacuto A, de Sousa R, Lebeugle D and Colson D 2008 Possible observation of cycloidal electromagnons in $BiFeO_3$ *Phys. Rev. Lett.* **101** 037601
[32] Talbayev D, Trugman S A, Lee S, Yi H T, Cheong S-W and Taylor A J 2011 Long-wavelength magnetic and magnetoelectric excitations in the ferroelectric antiferromagnet $BiFeO_3$ *Phys. Rev. B* **83** 094403
[33] Nagel U, Fishman R S, Katuwal T, Engelkamp H, Talbayev D, Yi H T, Cheong S-W and Rõõm T 2013 Terahertz spectroscopy of spin waves in multiferroic $BiFeO_3$ in high magnetic fields *Phys. Rev. Lett.* **110** 257201
[34] Bordács S, Farkas D G, White J S, Cubitt R, DeBeer-Schmitt L, Ito T and Kézsmárki I 2018 Magnetic field control of cycloidal domains and electric polarization in multiferroic $BiFeO_3$ *Phys. Rev. Lett.* **120** 147203
[35] Fishman R S 2018 Pinning, rotation, and metastability of $BiFeO_3$ cycloidal domains in a magnetic field *Phys. Rev. B* **97** 014405
[36] Katsura H, Balatsky A V and Nagaosa N 2007 Dynamical magnetoelectric coupling in helical magnets *Phys. Rev. Lett.* **98** 027203
[37] Miyahara S and Furukawa N 2012 Nonreciprocal directional dichroism and toroidalmagnons in helical magnets *J. Phys. Soc. Jpn.* **81** 023712
[38] Fishman R S, Haraldsen J T, Furukawa N and Miyahara S 2013 Spin state and spectroscopic modes of multiferroic $BiFeO_3$ *Phys. Rev.* B **87** 134416
[39] Choi T, Lee S, Choi Y J, Kiryukhin V and Cheong S-W 2009 Switchable ferroelectric diode and photovoltaic effect in $BiFeO_3$ *Science* **324** 63
[40] Ito T, Ushiyama T, Yanagisawa Y, Kumai R and Tomioka Y 2011 Growth of highly insulating bulk single crystals of multiferroic $BiFeO_3$ and their inherent internal strains in the domain-switching process *Cryst. Growth Des.* **11** 5139–43

[41] Soda M, Matsumoto M, Månsson M, Ohira-Kawamura S, Nakajima K, Shiina R and Masuda T 2014 Spin-nematic interaction in the multiferroic compound $Ba_2CoSi_2O_7$ *Phys. Rev. Lett.* **112** 127205

[42] Miyahara S and Furukawa N 2011 Theory of magnetoelectric resonance in two-dimensional $S = 3/2$ antiferromagnet $Ba_2CoGe_2O_7$ via spin-dependent metal-ligand hybridization mechanism *J. Phys. Soc. Jpn.* **80** 073708

[43] Penc K *et al* 2012 Spin-stretching modes in anisotropic magnets: spin-wave excitations in the multiferroic $Ba_2CoGe_2O_7$ *Phys. Rev. Lett.* **108** 257203

[44] Romhányi J and Penc K 2012 Multiboson spin-wave theory for $Ba_2CoGe_2O_7$: a spin-3/2 easy-plane Néel antiferromagnet with strong single-ion anisotropy *Phys. Rev.* B **86** 174428

[45] Romhányi J, Lajkó M and Penc K 2011 Zero- and finite-temperature mean field study of magnetic field induced electric polarization in $Ba_2CoGe_2O_7$: effect of the antiferroelectric coupling *Phys. Rev.* B **84** 224419

[46] Romhányi J, Pollmann F and Penc K 2011 Supersolid phase and magnetization plateaus observed in the anisotropic spin-$\frac{3}{2}$ Heisenberg model on bipartite lattices *Phys. Rev.* B **84** 184427

[47] Akaki M, Yoshizawa D, Okutani A, Kida T, Romhányi J, Penc K and Hagiwara M 2017 Direct observation of spin-quadrupolar excitations in $Sr_2CoGe_2O_7$ by high-field electron spin resonance *Phys. Rev.* B **96** 214406

[48] Nagy K L, Quintavalle D, Fehér T and Jánossy A 2011 Multipurpose high-frequency ESR spectrometer for condensed matter research *Appl. Magn. Reson.* **40** 47–63

[49] Yi H T, Choi Y J, Lee S and Cheong S-W 2008 Multiferroicity in the square-lattice antiferromagnet of $Ba_2CoGe_2O_7$ *Appl. Phys. Lett.* **92** 212904

[50] Murakawa H, Onose Y, Miyahara S, Furukawa N and Tokura Y 2010 Ferroelectricity induced by spin-dependent metal-ligand hybridization in $Ba_2CoGe_2O_7$ *Phys. Rev. Lett.* **105** 137202

[51] Jia C, Onoda S, Nagaosa N and Han J H 2006 Bond electronic polarization induced by spin *Phys. Rev.* B **74** 224444

[52] Arima T-H 2007 Ferroelectricity induced by proper-screw type magnetic order *J. Phys. Soc. Jpn.* **76** 07370

[53] Fishman R S, Lee J H, Bordács S, Kézsmárki I, Nagel U and Rõõm T 2015 Spin-induced polarizations and nonreciprocal directional dichroism of the room-temperature multiferroic $BiFeO_3$ *Phys. Rev.* B **92** 094422

[54] Fleury P A and Loudon R 1968 Scattering of light by one- and two-magnon excitations *Phys. Rev.* **166** 514–30

Randy S Fishman, Jaime A Fernandez-Baca and Toomas Rõõm

Chapter 9

Conclusion

The man that invented the steam drill,
He figured he was mighty high and fine,
But John Henry sunk the steel down fourteen feet,
While the steam drill only made nine.
—Poem of John Henry, author unknown

In any contest of man versus machine with a closed set of rules, the machine will eventually win. This has already been demonstrated in games like Chess and Go, not to mention Jeopardy! Also recall that John Henry's victory over the steam drill was short lived. While the mathematics we use to describe the Universe is a closed system of rules, however, the Universe itself is certainly not. Consequently, the laws of physics attempt to solve a problem of ever growing complexity.

Computer codes that analyze dynamical spectra are becoming increasingly sophisticated. Currently, those codes minimize the difference between the measured and observed SW frequencies $\omega(\mathbf{q})$ (see appendix A for some guidelines). In the near future, analysis codes will almost certainly directly compare the measured and predicted $S(\mathbf{q},\omega)$. The transition from fitting frequencies to fitting intensities will mark a huge step forward in the analysis of INS and THz data. The ability to directly compare predicted and measured intensities will allow scientists to more easily study materials with incommensurate spin states where many excitations overlap to form a continuum, such as in Ga-doped $CuFeO_2$ [1] discussed in section 7.5.

However, analysis of INS intensities must overcome several hurdles. First, artifacts and noise must be removed from the INS data. Based on their different

doi:10.1088/978-1-64327-114-9ch9

wavevector periodicities, phonon modes must be excluded. Time-of-flight measurements at spallation sources produce an enormous amount of data throughout $\{\mathbf{q},\omega\}$ space. So the analyst must carefully choose which data slices to fit. Forgetting about sum rules for the moment, the predicted $S(\mathbf{q},\omega)$ is required to reproduce the relative but not the absolute spectral intensities. Hence, the overall scale of the spectrum is irrelevant. In digitizing $S(\mathbf{q},\omega)$, the pixel size must be small enough that the intensity is uniform over the whole pixel but large enough to avoid overtaxing the numerics. Fortunately, $S(\mathbf{q},\omega)$ can be computed in parallel for each wavevector $\mathbf{q}$.

There is little doubt that machine learning will play a huge role in the future analysis of INS data. Machine learning will be used to choose between different models with different sets of microscopic interactions. For two sets of models that describe the data equally well, preference should be given to the model with the smallest number of terms or the smallest number of interactions (e.g. exchanges, DM interactions, anisotropies, etc). Machine learning can also be used to estimate the probability that a given model accurately describes the data. Thus, future theoretical results based on INS data may include the relative probabilities of different microscopic models.

Another remaining challenge is to combine codes that construct the correct spin state with those that analyze the dynamical spectra. This requires two nested optimization loops: one to minimize the classical energy over spin configurations for a fixed set of interaction parameters and another to minimize χ^2 to obtain the interaction parameters that best describe the dynamical spectra. Publicly available codes such as SpinW[1] already provide simplistic energy minimization routines but (at last for now) without the higher harmonics considered in chapter 6.

But eventually, that gap will also be closed. What then is the future of SW analysis? At the risk of seeming arrogant, we believe that analysis of dynamical spectra will always be best handled by a human operator. Optimally, that human wrote his or her own set of codes to solve for the spin state and microscopic interactions. Even if a publicly available platform is employed, the human operator must at the very least understand the assumptions and limitations of that software.

As a cautionary tale, consider the lessons of density-functional theory. First-principles codes to evaluate the band structure and chemical bonding of a material are commercially available from several sources. The best practitioners in this field remain aware of the assumptions built into those codes and only use them where appropriate. Interpretation of first-principles results is just as important as running codes for a given material. Unfortunately, the ready availability of first-principles codes has also been responsible for lots of useless results.

Similar mistakes are bound to occur with SW analysis codes. Some scientists will employ those codes without understanding either their ingredients or their limitations. Sloppy science is the inevitable outcome. We are emboldened to think that the

[1] www.psi.ch/spinw/

majority of scientists will employ those analysis tools as, well, tools. They are meant to supplement critical thinking, not to replace it.

That is where this book comes in. We hope that it helps scientists to develop their own software platforms and to develop the intuition to know when those codes are producing reasonable results. We also hope that it helps experimentalists appreciate the utility of INS and THz techniques to study the spin states and microscopic interactions of interesting materials.

Several new spallation sources are currently in the planning or construction phases. Scheduled to launch in 2025, the European Spallation Source (ESS)[2] features many important upgrades over the SNS in Oak Ridge. The ESS should open the field of INS to a new community of European-based scientists.

One of the central challenges of time-of-flight instruments is the oversupply of data, popularly called 'information overload.' As in many other information fields, this abundance is both a blessing and a curse. A blessing because time-of-flight measurements provide all the information about a material we might possibly want, even in regions of $\{\mathbf{q},\omega\}$ space that were not considered important before the measurement. A curse because no one has the time or resources to sift through this mountain of data for useful information. Indeed, just as in other fields of information science, our ability to store information has greatly outpaced our ability to analyze it.

To risk a bad pun, the future of THz spectroscopy is bright. It deftly complements INS by providing greater resolution and allows scientists to probe small energies that affect the low-frequency SW spectrum. We believe that measurement of absorption line intensities and complex optical constants in addition to mode frequencies will become increasingly common. After all, an electromagnetic wave is an oscillation of both magnetic and electric fields. While magnetic-dipole coupling has the same form for every material, the spin-induced electric polarization may be different in different materials. Because the electric field does not couple directly to the oscillating magnetic moment of the SW but rather indirectly to its oscillating electric polarization, THz spectroscopy provides valuable information about the material-specific coupling of spin to charge. As new powerful THz radiation sources are developed, pump-probe and non-linear THz spectroscopy may provide even more detailed information about magnetic systems.

Like any human endeavor, this book certainly contains errors. Hopefully, those errors do not significantly undermine its intended purpose. Especially in the discussion of cycloids and helices in chapter 6, we have considered some problems that are not completely solved [2]. Perhaps, the remaining issues will be put to rest by one of our readers.

With these final words, we wish the reader good luck in his or her future experiments and analysis in a world of ever more complex and interesting magnetic materials.

[2] https://europeanspallationsource.se

References

[1] Haraldsen J T, Ye F, Fishman R S, Fernandez-Baca J A, Yamaguchi Y, Kimura K and Kimura T 2010 Multiferroic phase of doped delafossite $CuFeO_2$ identified using inelastic neutron scattering *Phys. Rev.* B **82** 020404

[2] Fishman R S, Rõõm T and deSousa R arXiv:1809.09680

Appendix A

Fitting SW frequencies

At least for the present, fitting routines compare the measured and predicted SW frequencies $\omega_n(\mathbf{q})$. Suppose that INS observes $O(\mathbf{q}) < M$ different modes at wavevector $\mathbf{q}$ below a frequency cutoff of ω_c with an experimental resolution $\Delta_{\rm exp}(\mathbf{q})$. Some of the M predicted modes may have small or negligible intensities $S^{(n)}(\mathbf{q})$. Other degenerate or nearly degenerate modes may not be distinguishable by INS. So it makes no sense to just blindly fit the lowest $O(\mathbf{q})$ predicted modes against the $O(\mathbf{q})$ observed modes at wavevector $\mathbf{q}$.

We have developed a rather simple fitting procedure. First, evaluate the M mode frequencies and their intensities. Second, order the predicted modes from lowest to highest in frequency. If $\omega_{n+1}(\mathbf{q}) - \omega_n(\mathbf{q}) < \Delta_{\rm exp}(\mathbf{q})$ then modes n and $n+1$ should be combined into a single mode with frequency

$$\omega_{n'}(\mathbf{q}) = \frac{S^{(n)}(\mathbf{q})\,\omega_n(\mathbf{q}) + S^{(n+1)}(\mathbf{q})\omega_{n+1}(\mathbf{q})}{S^{(n)}(\mathbf{q}) + S^{(n+1)}(\mathbf{q})}, \tag{A.1}$$

which uses a weighted average based on the individual mode intensities. The intensity of the consolidated mode is

$$S^{(n')}(\mathbf{q}) = S^{(n)}(\mathbf{q}) + S^{(n+1)}(\mathbf{q}). \tag{A.2}$$

This procedure can be extended to three or more modes if all fall within $\Delta_{\rm exp}(\mathbf{q})$ of each other.

Third, find the $P(\mathbf{q})$ different predicted modes with frequencies below the experimental cutoff of ω_c (and perhaps above some lower cutoff). Order those modes by their intensities $S^{(n)}(\mathbf{q})$, with the most intense modes coming first. Fourth, separate the $O(\mathbf{q})$ predicted modes with the highest intensities. Finally, fit the $O(\mathbf{q})$ predicted modes versus the $O(\mathbf{q})$ observed modes, both reordered from lowest to highest in frequency.

The advantage of this procedure is that the experimentalist does not have to specify O branches throughout the magnetic BZ. Rather, he or she can select only those modes with the highest intensities at each $\mathbf{q}$. If a mode is barely visible in the

doi:10.1088/978-1-64327-114-9ch10

inelastic spectrum, then leave it out. The analyst should remove duplicate modes before evaluating the $O(\mathbf{q})$ most intense modes below ω_c because the combined mode n' may make the list even when the separate modes n and $n+1$ would not. While a direct comparison between the measured and predicted SW intensities has not yet been implemented, this technique indirectly uses the predicted SW intensities to rank the SW frequencies.

A flow chart for this procedure is provided in figure A.1. What should you do if the number $P(\mathbf{q})$ of predicted modes below the cutoff ω_c is *lower* than the number $O(\mathbf{q})$ of measured modes? As seen in section 7.5, an extra set of modes can appear due to sample twinning. Modes may split when different parts of the sample are not aligned. If all experimental sources of error have been eliminated, then it may be time to consider the lessons learned in section 8.3: if one or more assumptions of SW theory has failed, then a more sophisticated approximation such as multiboson theory might be better suited to the problem.

Once the $O(\mathbf{q})$ mode frequencies have been evaluated at each wavevector $\mathbf{q}$, the goal is to minimize

$$\chi^2 = \frac{1}{N_{\mathrm{dat}} - N_{\mathrm{par}} + 1} \sum_{\mathbf{q}} \sum_{n=1}^{O(\mathbf{q})} \left(\frac{\omega_n^{\mathrm{th}}(\mathbf{q}) - \omega_n^{\mathrm{exp}}(\mathbf{q})}{\Delta_{\mathrm{exp}}(\mathbf{q})} \right)^2, \tag{A.3}$$

as a function of N_{par} fitting parameters (such as exchange, anisotropy, etc) with N_{dat} total data points in the two sums. See the flow chart in figure 1.3 for a bird's-eye view of the χ^2 minimization loop.

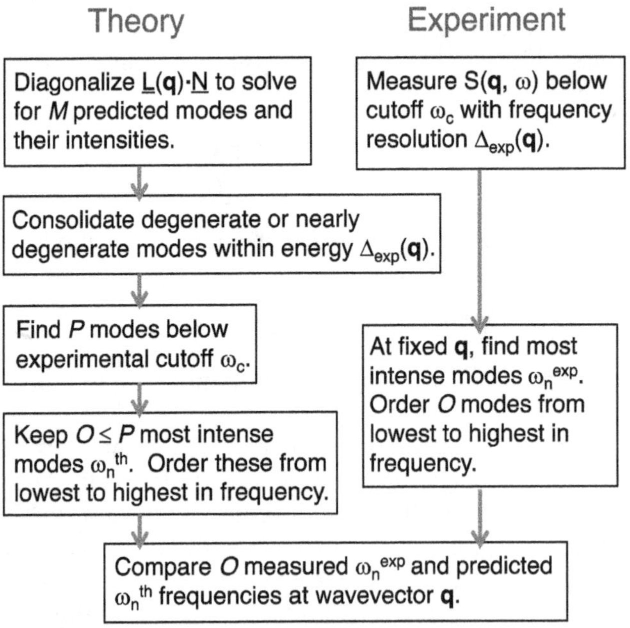
Theory
Experiment
Diagonalize L(q)·N to solve for M predicted modes and their intensities.
Measure S(q, ω) below cutoff ω_c with frequency resolution Δ_exp(q).
Consolidate degenerate or nearly degenerate modes within energy Δ_exp(q).
Find P modes below experimental cutoff ω_c.
At fixed q, find most intense modes ω_n^exp. Order O modes from lowest to highest in frequency.
Keep O ≤ P most intense modes ω_n^th. Order these from lowest to highest in frequency.
Compare O measured ω_n^exp and predicted ω_n^th frequencies at wavevector q.

Figure A.1. A flow chart for the recommended procedure to compare the measured and predicted SW frequencies at each wavevector **q**.

This procedure can be refined by allowing the overlap condition $\Delta_{\rm dup}(\mathbf{q})$ to be smaller than the experimental resolution $\Delta_{\rm exp}(\mathbf{q})$ if that improves χ^2. Also, the experimental resolution $\Delta^{(n)}_{\rm exp}(\mathbf{q})$ that enters χ^2 can be different for each mode $\omega_n(\mathbf{q})$. If indicated by the experiment, the cutoff ω_c can depend on wavevector $\mathbf{q}$.

These same considerations hold for the analysis of the $\mathbf{q} = 0$ THz spectroscopy data. One difficulty of fitting THz spectra is that the number of data points $N_{\rm dat}$ is much smaller than for INS because only $\mathbf{q} = 0$ frequencies are measured. However, THz spectroscopy compensates for that deficit with an experimental energy resolution $\Delta_{\rm exp}(\mathbf{q})$ about ten times smaller than for INS. This shortcoming can be also be addressed by measuring and fitting the THz frequencies over a range of magnetic fields.

Appendix B

Selected exercise solutions for chapter 2

Exercise 2.1. *Use equations* (2.6) *and* (2.8) *to show that for CL spins aligned along* $\pm\mathbf{z}$,

$$S_{xx}(\mathbf{q},\omega) = S_{yy}(\mathbf{q},\omega) \propto [n_\mathrm{B}(\omega(\mathbf{q})) + 1]\delta(\omega - \omega(\mathbf{q})) + n_\mathrm{B}(\omega(\mathbf{q}))\delta(\omega + \omega(\mathbf{q})). \quad (\mathrm{B}.1)$$

Use $n_\mathrm{B}(-\omega) = -1 - n_\mathrm{B}(\omega)$. Assume that there is no easy-plane anisotropy and that the system is inversion symmetric.

For spins aligned along $\mathbf{z}$, $S_{xx}(\mathbf{q},\omega) = S_{yy}(\mathbf{q},\omega)$ in the absence of easy-plane anisotropy. So equations (2.6) and (2.8) imply

$$\begin{aligned} S_{xx}(\mathbf{q},\omega) &= [n_\mathrm{B}(\omega(\mathbf{q})) + 1]S_{xx}(\mathbf{q})\,\delta(\omega - \omega(\mathbf{q})) \\ &\quad - [n_\mathrm{B}(-\omega(\mathbf{q})) + 1]S_{xx}(-\mathbf{q})\,\delta(\omega + \omega(-\mathbf{q})) \\ &= S_{xx}(\mathbf{q})\{[n_\mathrm{B}(\omega(\mathbf{q})) + 1]\delta(\omega - \omega(\mathbf{q})) + n_\mathrm{B}(\omega(\mathbf{q}))\delta(\omega + \omega(\mathbf{q}))\}, \end{aligned} \quad (\mathrm{B}.2)$$

where $S_{xx}(\mathbf{q}) = S_{xx}(-\mathbf{q})$ and $\omega(-\mathbf{q}) = \omega(\mathbf{q})$ due to the inversion symmetry.

doi:10.1088/978-1-64327-114-9ch11

Appendix C

Selected exercise solutions for chapter 3

Exercise 3.1. *Derive equation* (3.11) *for* $\Delta\alpha = \alpha(B, T) - \alpha(B_r, T_r)$. *Does this expression have the right dimensions?*

The definition of α was given in equation (3.1): $\Delta\alpha = -[\ln \mathcal{T} - \ln \mathcal{T}_r]/d$ where $\mathcal{T}$ and $\mathcal{T}_r$ are transmissions at $\{B, T\}$ and at reference field and temperature $\{B_r, T_r\}$. Using the definition of the measured transmission in equation (3.9), we have

$$\Delta\alpha = -\frac{1}{d}\left[\ln \frac{I(B, T)}{I_0(1 - \mathcal{R})^2} - \ln \frac{I(B_r, T_r)}{I_0(1 - \mathcal{R})^2}\right]. \tag{C.1}$$

Here, we assumed the intensity on the detector without the sample I_0 is independent of magnetic field and sample compartment temperature. Since $I(B, T)$ and $I(B_r, T_r)$ are measured at different times there should be no drift of I_0 in time. Then equation (3.11) follows from equation (C.1). The inverse of $\Delta\alpha$ has the same dimension as the thickness of the sample d.

Exercise 3.2. *On a single crystal sample, devise an experiment to show that a particular mode is an electromagnon with polarization* $\mathbf{E}^\omega$ *along* [010].

In the plane-parallel wave $\mathbf{k}$, $\mathbf{E}^\omega$, and $\mathbf{B}^\omega$ are orthogonal to each other. $\mathbf{E}^\omega$ is along [010] when the light propagation direction is $\mathbf{k} \parallel [100]$ or $\mathbf{k} \parallel [001]$. Measurements with both $\mathbf{k}$ directions are needed to exclude the possibility of magnetic-dipole transitions with selection rule $\mathbf{B}^\omega \parallel [001]$ or $\mathbf{B}^\omega \parallel [100]$.

Exercise 3.5. *Assume that the absorption of THz radiation below 100 cm*$^{-1}$ *is dominated by a magnetic-dipole active SW resonance at* $\bar{\omega}_0 = 50$ *cm*$^{-1}$ *and that the backgrounds* $\epsilon^b = 20$ *and* $\mu^b = 1$ *are both real. If the absorption due to this SW mode*

doi:10.1088/978-1-64327-114-9ch12

is $\alpha_{\mathrm{SW}}(\bar{\omega}_0) = 50\ \mathrm{cm}^{-1}$, *show that* $\mathrm{Im}\ \Delta\mu \ll \mu^b$ *so that the expansion of equation* (3.8) *is justified.*

For a magnetic-dipole active SW, equation (3.8) reads

$$\mathcal{N} \approx \sqrt{\epsilon^b \mu^b} + \sqrt{\frac{\epsilon^b}{\mu^b}} \frac{\Delta\mu(\bar{\omega})}{2}. \tag{C.2}$$

The size of $\Delta\mu(\bar{\omega}_0)$ is derived from equation (3.3):

$$\alpha_{\mathrm{SW}}(\bar{\omega}_0) = 4\pi\bar{\omega}\ \mathrm{Im}\ \Delta\mathcal{N}(\bar{\omega}_0), \tag{C.3}$$

where $\Delta\mathcal{N}(\bar{\omega}_0)$ is the second term of (C.2). If ϵ^b and μ^b are real we obtain

$$\mathrm{Im}\ \Delta\mu(\bar{\omega}_0) = \frac{\alpha_{\mathrm{SW}}(\bar{\omega}_0)}{2\pi\bar{\omega}} \sqrt{\frac{\mu^b}{\epsilon^b}}, \tag{C.4}$$

with a numerical value $\mathrm{Im}\ \Delta\mu(\bar{\omega}_0) = (2\sqrt{20}\,\pi)^{-1} \ll 1$.

IOP Concise Physics

Spin-Wave Theory and its Applications to Neutron Scattering and THz Spectroscopy

Randy S Fishman, Jaime A Fernandez-Baca and Toomas Rõõm

Appendix D

Selected exercise solutions for chapter 4

Exercise 4.1. *Verify the commutation relations of equation* (4.11). *Use the definitions of the spin operators in the local reference frame:*

$$\bar{S}_{iz} = S - a_i^\dagger a_i, \tag{D.1}$$

$$\bar{S}_{i+} = \bar{S}_{ix} + i\bar{S}_{iy} = \sqrt{2S}\sqrt{1 - \frac{a_i^\dagger a_i}{2S}}\, a_i, \tag{D.2}$$

$$\bar{S}_{i-} = \bar{S}_{ix} - i\bar{S}_{iy} = \sqrt{2S}\, a_i^\dagger \sqrt{1 - \frac{a_i^\dagger a_i}{2S}}. \tag{D.3}$$

Clearly, spin operators on different sites commute. Now

$$\begin{aligned}[\bar{S}_{i+}, \bar{S}_{i-}] &= 2S\left\{\sqrt{1 - \frac{a_i^\dagger a_i}{2S}}\, a_i a_i^\dagger \sqrt{1 - \frac{a_i^\dagger a_i}{2S}} - a_i^\dagger\left(1 - \frac{a_i^\dagger a_i}{2S}\right)a_i\right\} \\ &= 2S\left\{\left(1 - \frac{a_i^\dagger a_i}{2S}\right)(1 + a_i^\dagger a_i) - a_i^\dagger a_i + \frac{1}{2S}a_i^\dagger a_i^\dagger a_i a_i\right\}.\end{aligned} \tag{D.4}$$

Use

$$a_i^\dagger a_i^\dagger a_i a_i = a_i^\dagger a_i (a_i^\dagger a_i - 1) \tag{D.5}$$

to find

$$[\bar{S}_{i+}, \bar{S}_{i-}] = 2S\left\{1 - \frac{a_i^\dagger a_i}{S}\right\} = 2\bar{S}_{iz}. \tag{D.6}$$

doi:10.1088/978-1-64327-114-9ch13

Similarly,

$$[\bar{S}_{iz}, \bar{S}_{i+}] = -\sqrt{2S}\left\{a_i^\dagger a_i\sqrt{1-\frac{a_i^\dagger a_i}{2S}}\,a_i - \sqrt{1-\frac{a_i^\dagger a_i}{2S}}\,a_i a_i^\dagger a_i\right\}$$
$$= -\sqrt{2S}\sqrt{1-\frac{a_i^\dagger a_i}{2S}}\,\{a_i^\dagger a_i - (1+a_i^\dagger a_i)\}a_i = \bar{S}_{i+} \tag{D.7}$$

and

$$[\bar{S}_{iz}, \bar{S}_{i-}] = -\sqrt{2S}\left\{a_i^\dagger a_i a_i^\dagger\sqrt{1-\frac{a_i^\dagger a_i}{2S}} - a_i^\dagger\sqrt{1-\frac{a_i^\dagger a_i}{2S}}\,a_i^\dagger a_i\right\}$$
$$= -\sqrt{2S}\,a_i^\dagger\{(1+a_i^\dagger a_i) - a_i^\dagger a_i\}\sqrt{1-\frac{a_i^\dagger a_i}{2S}} = -\bar{S}_{i-}. \tag{D.8}$$

Putting this all together yields

$$[\bar{S}_{i\alpha}, \bar{S}_{j\beta}] = i\delta_{ij}\epsilon_{\alpha\beta\gamma}\bar{S}_{i\gamma}, \tag{D.9}$$

as required.

***Exercise 4.4.** *Why do the $a_i^\dagger a_i$ and $a_j^\dagger a_j$ terms in equation* (4.38) *not contribute to equation* (4.39) *when $\omega > 0$?*

To order S, the diagonal terms contribute

$$S\int dt\; e^{-i\omega t}\langle a_i^\dagger a_i e^{i\mathcal{H}t/\hbar}e^{-i\mathcal{H}t/\hbar}\rangle, \tag{D.10}$$

$$S\int dt\; e^{-i\omega t}\langle e^{i\mathcal{H}t/\hbar}a_i^\dagger a_i e^{-i\mathcal{H}t/\hbar}\rangle \tag{D.11}$$

to $S_{\alpha\beta}(\mathbf{q},\omega)$. The time dependence trivially vanishes from $\langle a_i^\dagger a_i e^{i\mathcal{H}t/\hbar}e^{-i\mathcal{H}t/\hbar}\rangle$. For the second,

$$\langle e^{i\mathcal{H}t/\hbar}a_i^\dagger a_i e^{-i\mathcal{H}t/\hbar}\rangle = \frac{1}{Z}\sum_n e^{-\beta E_n}\langle n|e^{i\mathcal{H}t/\hbar}a_i^\dagger a_i e^{-i\mathcal{H}t/\hbar}|n\rangle$$
$$= \frac{1}{Z}\sum_n e^{-\beta E_n}e^{iE_nt/\hbar}\langle n|a_i^\dagger a_i|n\rangle e^{-iE_nt/\hbar}, \tag{D.12}$$

where $|n\rangle$ are eigenstates of $\mathcal{H}$ with eigenvalue E_n and $Z = \sum_n \exp(-\beta E_n)$. So this term is also independent of time. Hence, the diagonal terms only contribute $\delta(\omega)$ to $S_{\alpha\beta}(\mathbf{q},\omega)$.

***Exercise 4.5.** *Use SW theory to prove equation (4.45) for the spin–spin correlation function at nonzero temperatures. Include the transition between a state with $m \geqslant 1$ SWs of frequency ω_n to one with $m \pm 1$ SWs of that frequency.*

Start with

$$\begin{aligned}
S_{\alpha\beta}(\mathbf{q},\omega) &= \frac{S}{4\pi M}\sum_{r,s=1}^{M}\int dt\, e^{-i\omega t}\Big\langle\Big\{V_{r\alpha}^{-}\, a_{\mathbf{q}}^{(r)} + V_{r\alpha}^{-}\, a_{-\mathbf{q}}^{(r)\dagger}\Big\}\\
&\times e^{i\mathcal{H}t}\Big\{V_{s\beta}^{-}\, a_{-\mathbf{q}}^{(s)} + V_{s\beta}^{+}\, a_{\mathbf{q}}^{(s)\dagger}\Big\}e^{-i\mathcal{H}t}\Big\rangle\\
&= \frac{S}{4\pi M}\sum_{r,s=1}^{M}\int dt\, e^{-i\omega t}\sum_{o,p}\langle o|\sum_{n=1}^{M}\Big[V_{r\alpha}^{-}\Big(X_{r,\,n}^{-1}(\mathbf{q})\alpha_{\mathbf{q}}^{(n)} + X_{r,\,n+M}^{-1}(\mathbf{q})\alpha_{-\mathbf{q}}^{(n)\,\dagger}\Big)\\
&+ V_{r\alpha}^{+}\Big(X_{r+M,\,n}^{-1}(\mathbf{q})\alpha_{\mathbf{q}}^{(n)} + X_{r+M,\,n+M}^{-1}(\mathbf{q})\alpha_{-\mathbf{q}}^{(n)\,\dagger}\Big)\Big]|p\rangle e^{i\omega_p t}\\
&\times\langle p|\sum_{l=1}^{M}\Big[V_{s\beta}^{-}\Big(X_{s,\,l}^{-1}(-\mathbf{q})\alpha_{-\mathbf{q}}^{(l)} + X_{s,\,l+M}^{-1}(-\mathbf{q})\alpha_{\mathbf{q}}^{(l)\,\dagger}\Big)\\
&+ V_{s\beta}^{+}\Big(X_{s+M,\,l}^{-1}(-\mathbf{q})\alpha_{-\mathbf{q}}^{(l)} + X_{s+M,\,l+M}^{-1}(-\mathbf{q})\alpha_{\mathbf{q}}^{(l)\,\dagger}\Big)\Big]|o\rangle e^{-i\omega_o t}\frac{e^{-\beta\omega_o}}{Z_o}.
\end{aligned}\qquad\text{(D.13)}$$

So the overall expectation value is taken over states o (with partition function Z_o) while we have inserted a complete set of states p.

Now there are two sets of states that must be included in the summations over states o and p. They are

$$\text{Set 1: } |o\rangle = |m;\, n,\, \mathbf{q}\rangle,\ |p\rangle = |m+1;\, n,\, \mathbf{q}\rangle, \qquad\text{(D.14)}$$

$$\text{Set 2: } |o\rangle = |m+1;\, n,\, -\mathbf{q}\rangle,\ |p\rangle = |m;\, n,\, -\mathbf{q}\rangle. \qquad\text{(D.15)}$$

In both cases, we restrict $l = n$ for SWs with frequency $\omega_n(\pm\mathbf{q})$. The index m then gives the *number* of SWs in an eigenstate. Set 1 starts with a state containing m SWs. Another is created to give an intermediate state with $m + 1$ SWs and then one is destroyed to return to a state with m SWs. Set 2 starts with a state containing $m + 1$ SWs. One is destroyed to produce an intermediate state sith m SWs and then one is created to return to a final state with $m + 1$ SWs. Thus, $m \geqslant 0$ for both sets.

We then find

$$\begin{aligned}
S_{\alpha\beta}(\mathbf{q},\omega) = \frac{S}{2M}\sum_{r,s=1}^{M}\sum_{n=1}^{M}\Big\{&P_1(\omega_n)\, W_{r\alpha}^{(n)}(\mathbf{q})\, W_{s\beta}^{(n)}(\mathbf{q})^*\,\delta(\omega-\omega_n(\mathbf{q}))\\
&+ P_2(\omega_n)\, W_{r\alpha}^{(n)}(-\mathbf{q})^*\, W_{s\beta}^{(n)}(-\mathbf{q})\,\delta(\omega+\omega_n(-\mathbf{q}))\Big\},
\end{aligned}\qquad\text{(D.16)}$$

where the probabilities for each set of terms is

$$\begin{aligned}
P_1(\omega) &= \frac{1}{Z}\sum_{m=0}^{\infty}e^{-\hbar\beta m\omega}\langle m|a|m+1\rangle\langle m+1|a^{\dagger}|m\rangle\\
&= \frac{1}{Z}\sum_{m=0}^{\infty}(m+1)e^{-\hbar\beta m\omega} = 1 + n_{\mathrm{B}}(\omega).
\end{aligned}\qquad\text{(D.17)}$$

$$P_2(\omega) = \frac{1}{Z}\sum_{m=0}^{\infty} e^{-\hbar\beta(m+1)\omega}\langle m+1|a^\dagger|m\rangle\langle m|a|m+1\rangle$$
$$= \frac{1}{Z}\sum_{m=0}^{\infty}(m+1)e^{-\hbar\beta(m+1)\omega} = n_B(\omega), \tag{D.18}$$

$$Z = \sum_{m=0}^{\infty} e^{-\hbar\beta m\omega} = \frac{1}{1-e^{-\hbar\beta\omega}}. \tag{D.19}$$

For both probabilities, $|m\rangle$ is an eigenstate containing m SWs with frequency ω. These relations use the definitions for the creation and destruction operators:

$$a|m\rangle = \sqrt{m}\,|m-1\rangle, \tag{D.20}$$

$$a^\dagger|m\rangle = \sqrt{m+1}\,|m+1\rangle, \tag{D.21}$$

which imply the commutation relation $[a, a^\dagger] = 1$. At $T = 0$, $P_1 = 1$ and $P_2 = 0$ so only set 1 contributes.

It follows that

$$S_{\alpha\beta}(\mathbf{q},\omega) = \frac{S}{2M}\sum_{n=1}^{M}\sum_{r,s=1}^{M}\Big\{(n_B(\omega_n)+1)W_{r\alpha}^{(n)}(\mathbf{q})\,W_{s\beta}^{(n)}(\mathbf{q})^*\,\delta(\omega-\omega_n(\mathbf{q}))$$
$$+\,n_B(\omega_n)W_{r\alpha}^{(n)}(-\mathbf{q})^*\,W_{s\beta}^{(n)}(-\mathbf{q})\,\delta(\omega+\omega_n(-\mathbf{q}))\Big\}. \tag{D.22}$$

Finally, use $n_B(-\omega) = -n_B(\omega) - 1$ to obtain equation (4.45).

Exercise 4.8. *Show that optical absorption $\alpha^\pm(\omega)$ given by equation (4.65) has units of cm^{-1} so that $\alpha^\pm(\omega)/\omega$ is dimensionless.*

Yes, electromagnetic units can be a real hassle. Recall that $c = 1/\sqrt{\mu_0\epsilon_0}$. Since the dimensions of $P/V\sim$ nC cm^{-2} have been factored into X, we find

$$A_n \sim \mu_B\sqrt{\frac{\mu_0}{\epsilon_0}}\frac{1}{\hbar c}\frac{\mathrm{C}}{\mathrm{cm}^2}. \tag{D.23}$$

But μ_B has dimensions of eV T^{-1} and

$$1\,\mathrm{T} = 10^5\frac{\mu_0}{4\pi}\frac{\mathrm{C}}{\mathrm{cm\,s}} \tag{D.24}$$

so

$$A_n \sim \frac{\mathrm{eV\,C}}{\mathrm{C\,cm}^2}\frac{\mathrm{cm\,s}}{\hbar c\sqrt{\mu_0\epsilon_0}} \sim \frac{\mathrm{eV\,s}}{\hbar\,\mathrm{cm}} \sim \frac{1}{\mathrm{cm}}. \tag{D.25}$$

Similarly,

$$B_n \sim \frac{C^2}{\text{cm}}\frac{1}{\hbar\epsilon_0 c} \sim \frac{\text{eV}}{\hbar c} \sim \frac{1}{\text{cm}}, \tag{D.26}$$

which uses the fact that C^2/ϵ_0 cm has dimensions of eV. Hence, $\alpha(\omega)$ has dimensions of cm^{-1}.

Exercise 4.12. *Derive equation (4.79) relating the spin and optical susceptibilities. Hint: Once you relate the imaginary parts, the Kramers–Kronig relation can be used to relate the real parts.*

Equation (4.78) implies that for any temperature,

$$\text{Im}\,\chi^{\text{mm}}(\omega) = \frac{4\pi\mu_0\mu_B^2}{\hbar V}\sum_{\alpha,\beta} h_\alpha h_\beta \,\text{Im}\,\chi_{\alpha\beta}^{\text{ss}}(\mathbf{q}=0,\omega). \tag{D.27}$$

Since both $\chi^{\text{mm}}(\omega)$ and $\sum_{\alpha,\beta} h_\alpha h_\beta \chi_{\alpha\beta}^{\text{ss}}(\mathbf{q}-0,\omega)$ are analytic functions of frequency in the upper-half complex plane and their imaginary parts are equal, their real parts must also be equal due to the Kramers–Kronig relation

$$\text{Re}\,\chi(\omega) = \frac{1}{\pi}\mathcal{P}\int_{-\infty}^{\infty} d\omega' \frac{\text{Im}\,\chi(\omega')}{\omega' - \omega}, \tag{D.28}$$

where $\mathcal{P}$ is the principal part of the integral. Consequently,

$$\chi^{\text{mm}}(\omega) = \frac{4\pi\mu_0\mu_B^2}{\hbar V}\sum_{\alpha,\beta} h_\alpha h_\beta \,\chi_{\alpha\beta}^{\text{ss}}(\mathbf{q}=0,\omega), \tag{D.29}$$

which is equation (4.79).

***Exercise 4.13.** *Prove equations (4.82)–(4.84) for the matrix elements of* **P**.

To get the required results, we transform

$$\begin{aligned} V_{r\beta}^- a_{\mathbf{q}=0}^{(r)} + V_{r\beta}^+ a_{\mathbf{q}=0}^{(r)\dagger} = \sum_n V_{r\beta}^-\Big(X_{rn}^{-1}(0)\alpha_{\mathbf{q}=0}^{(n)} + X_{r,\,n+M}^{-1}(0)\alpha_{\mathbf{q}=0}^{(n)\dagger}\Big) \\ + V_{r\beta}^+\Big(X_{r+M,\,n}^{-1}(0)\alpha_{\mathbf{q}=0}^{(n)} + X_{r+M,\,n+M}^{-1}(0)\alpha_{\mathbf{q}=0}^{(n)\dagger}\Big) \end{aligned} \tag{D.30}$$

so that

$$\begin{aligned} \langle n, \mathbf{q}=0|V_{r\beta}^- a_{\mathbf{q}=0}^{(r)} + V_{r\beta}^+ a_{\mathbf{q}=0}^{(r)\dagger}|0\rangle &= V_{r\beta}^- X_{r,\,n+M}^{-1}(0) + V_{r\beta}^+ X_{r+M,\,n+M}^{-1}(0) \\ &= W_{r\beta}^{(n+M)}(\mathbf{q}=0) - W_{r\beta}^{(n)}(\mathbf{q}=0)^*, \end{aligned} \tag{D.31}$$

where the last equality follows from exercise 4.6. The required results then follow easily.

Exercise 4.14. *Show that each matrix element of* $\underline{X}$ *and* $\underline{X}^{-1}$ *scales like* $1/\sqrt{M}$. *What does that imply about the scaling of* $S_{\alpha\alpha}^{(n)}(\mathbf{q})$, $W_{r\alpha}^{(n)}(\mathbf{q})$, $\Delta S_{r\alpha}^{(n)}(\mathbf{q}, t)$, *and* $\langle n, \mathbf{q}|\, M_\alpha|0\rangle$ *with* N_u *and M? How about the magnetic optical absorption strength* $\alpha_{\mathrm{m}}(\omega)$*?*

Consider the normalization condition

$$\underline{X}(\mathbf{q}) \cdot \underline{N} \cdot \underline{X}(\mathbf{q})^\dagger = \underline{N} \tag{D.32}$$

or

$$\sum_{l=1}^{M} X_{n,\,l}(\mathbf{q}) N_{ll} X_{l,\,m}(\mathbf{q})^* = N_{nm}. \tag{D.33}$$

It follows that $X_{n,\,l}(\mathbf{q})$ must scale like $1/\sqrt{M}$. Similarly for $X_{l,\,n}^{-1}(\mathbf{q})$. Therefore, $W_{r\alpha}^{(n)}(\mathbf{q})$ also scales like $1/\sqrt{M}$.

Using equation (4.42), we find that $S_{\alpha\beta}^{(n)}(\mathbf{q})$ scales like $(1/M)M^2(1/\sqrt{M})^2 \sim 1$. Similarly for $\Delta S_{r\alpha}^{(n)}(\mathbf{q}, t)$ from equation (4.92). Then equation (4.72) implies that

$$\langle n, \mathbf{q}|M_\alpha|0\rangle \sim \frac{\sqrt{N_u}M}{\sqrt{M}} = \sqrt{N}, \tag{D.34}$$

which uses $N_u = N/M$. So the matrix element of an extensive quantity like $M_\alpha \sim N$ between the ground state $|0\rangle$ and an excited state $|n, \mathbf{q}\rangle$ scales like $\sqrt{N}$.

***Exercise 4.15.** *Derive equations* (4.87–4.88) *using* $V_{r\alpha}^{\pm}$ *for the simple cycloid.*

For a simple cycloid in the xz plane, the components of $\Delta S_{r\alpha} = S_{r+1,\,\alpha} - S_{r-1,\,\alpha}$ are

$$\Delta S_{rx} = 2S\cos\left(\frac{2\pi r}{M}\right)\sin\left(\frac{2\pi}{M}\right), \tag{D.35}$$

$$\Delta S_{ry} = 0, \tag{D.36}$$

$$\Delta S_{rz} = -2S\sin\left(\frac{2\pi r}{M}\right)\sin\left(\frac{2\pi}{M}\right). \tag{D.37}$$

So equations (4.82)–(4.84) imply that

$$\langle n, 0|P_x|0\rangle = 0, \tag{D.38}$$

$$\langle n, 0|P_y|0\rangle = \lambda a\sqrt{2N_u}S^{3/2}\sin\left(\frac{2\pi}{M}\right)\sum_{r=1}^{M}\cos\left(\frac{2\pi r}{M}\right)W_{ry}^{(n)}(0)^*, \tag{D.39}$$

$$\langle n,\, 0|P_z|0\rangle = \lambda a\sqrt{2N_u}\,S^{3/2}\sin\left(\frac{2\pi}{M}\right)\sum_{r=1}^{M}\left\{\cos\left(\frac{2\pi r}{M}\right)W_{rz}^{(n)}(0)^* + \sin\left(\frac{2\pi r}{M}\right)W_{rx}^{(n)}(0)^*\right\}. \tag{D.40}$$

Now use the cycloidal angle $\theta_r = 2\pi r/M$ to find $V_{rx}^{\pm} = \cos\theta_r$, $V_{ry}^{\pm} = \pm i$, and $V_{rz}^{\pm} = -\sin\theta_r$. Therefore,

$$W_{rx}^{(n)} = \cos\theta_r\left(X_{r+M,\,n}^{-1}(0) + X_{r,\,n}^{-1}(0)\right), \tag{D.41}$$

$$W_{ry}^{(n)} = i\left(X_{r+M,\,n}^{-1}(0) - X_{r,\,n}^{-1}(0)\right), \tag{D.42}$$

$$W_{rz}^{(n)} = -\sin\theta_r\left(X_{r+M,\,n}^{-1}(0) + X_{r,\,n}^{-1}(0)\right). \tag{D.43}$$

Consequently,

$$\langle n,\, 0|P_x|0\rangle = \langle n,\, 0|P_z|0\rangle = 0, \tag{D.44}$$

$$\langle n,\, 0|P_y|0\rangle = -\,i\lambda a\sqrt{2N_u}\,S^{3/2}\sin\left(\frac{2\pi}{M}\right)\sum_{r=1}^{M}\cos\left(\frac{2\pi r}{M}\right) \times\left(X_{r+M,\,n}^{-1}(0)^* - X_{r,\,n}^{-1}(0)^*\right), \tag{D.45}$$

as required. So the spin-current polarization matrix elements are perpendicular to the cycloidal plane and to **Q**.

Exercise 4.17. *Prove equation (4.97):* $\underline{P}'(\mathbf{q}) = \underline{P}(-\mathbf{q})^\star$.
Notice that

$$\begin{aligned}&\sum_{\mathbf{q}}{}'\left\{a_{\mathbf{q}}^{(r)\dagger}L_{r,\,s}(\mathbf{q})a_{\mathbf{q}}^{(s)} + a_{-\mathbf{q}}^{(r)}L_{r+M,\,s+M}(\mathbf{q})a_{-\mathbf{q}}^{(s)\,\dagger}\right\}\\ &= \sum_{\mathbf{q}}{}' a_{\mathbf{q}}^{(r)\dagger}a_{\mathbf{q}}^{(s)}\left(L_{r,\,s}(\mathbf{q}) + L_{s+M,\,r+M}(-\mathbf{q})\right)\\ &= \sum_{\mathbf{q}}{}' a_{\mathbf{q}}^{(r)\dagger}a_{\mathbf{q}}^{(s)}\left(L_{r,\,s}(\mathbf{q}) + L_{r+M,\,s+M}(-\mathbf{q})^*\right),\end{aligned} \tag{D.46}$$

where the last equality uses the Hermiticity of $\underline{L}(\mathbf{q})$. Based on the well-known 'spread it around' law, the two quantities in parenthesis must be equal. Therefore,

$$L_{r+M,\,s+M}(\mathbf{q}) = L_{r,\,s}(-\mathbf{q})^* \tag{D.47}$$

and $\underline{P}'(\mathbf{q}) = \underline{P}(-\mathbf{q})^\star$.

Exercise 4.19. *What is the relation between $\chi^{\mathrm{me}}(\omega)$ and $\chi^{\mathrm{em}}(\omega)$?*

Since both the electric polarization **P** and the magnetization **M** are observables, they are given by Hermitian operators. So we can write

$$\chi^{\mathrm{em}}(\omega) = \sum_n \frac{A_n}{\omega_n - \omega - i\varepsilon}, \tag{D.48}$$

$$\chi^{\mathrm{me}}(\omega) = \sum_n \frac{A_n^*}{\omega_n - \omega - i\varepsilon}, \tag{D.49}$$

where

$$A_n = \sqrt{\frac{\mu_0}{\epsilon_0}} \frac{1}{\hbar V} \langle 0|\mathbf{e} \cdot \mathbf{P}|n, \mathbf{q} = 0\rangle\langle n, \mathbf{q} = 0|\mathbf{h} \cdot \mathbf{M}|0\rangle \tag{D.50}$$

and ε is a positive infinitesimal quantity. Therefore,

$$\chi^{\mathrm{em}}(\omega) = \mathcal{P}\sum_n \frac{A_n}{\omega_n - \omega} + i\pi \sum_n \delta(\omega - \omega_n)A_n, \tag{D.51}$$

$$\chi^{\mathrm{me}}(\omega) = \mathcal{P}\sum_n \frac{A_n^*}{\omega_n - \omega} + i\pi \sum_n \delta(\omega - \omega_n)A_n^*, \tag{D.52}$$

where $\mathcal{P}$ is the principal part of the summation. Because A_n is complex, the two terms on the right-hand side are *not* the real and imaginary parts of $\chi^{\mathrm{em}}(\omega)$ and $\chi^{\mathrm{me}}(\omega)^*$. If the first and second terms in equation (D.51) are written as $\chi^{(1)}(\omega)$ and $\chi^{(2)}(\omega)$, then

$$\chi^{\mathrm{em}}(\omega) = \chi^{(1)}(\omega) + i\chi^{(2)}(\omega), \tag{D.53}$$

$$\chi^{\mathrm{me}}(\omega) = \chi^{(1)}(\omega)^* + i\chi^{(2)}(\omega)^*. \tag{D.54}$$

That is why the optical absorption involves the combination

$$\mathrm{Im}\,\{\chi^{\mathrm{me}}(\omega) + \chi^{\mathrm{em}}(\omega)\} = 2\,\mathrm{Re}\,\chi^{(2)}(\omega), \tag{D.55}$$

which is real.

Exercise 4.20. *If polarization per unit volume has dimensions of $\mu C\ cm^{-2}$, what are the dimensions of the coupling constant λ?*

Since

$$\mathbf{P} = \lambda \sum_i \mathbf{e}_{i,\, i+1} \times (\mathbf{S}_i \times \mathbf{S}_{i+1}) \tag{D.56}$$

and $\mathbf{e}_{i,\, j} = \mathbf{R}_j - \mathbf{R}_i$ has dimensions of cm, λ must have dimensions of μC cm^{-3}.

Exercise 4.21. *By explicitly evaluating the exchange contributions to $\mathcal{H}$, prove equations (4.119)–(4.123). Show that those expressions are consistent with equations (4.97)–(4.98).*

As obtained previously,

$$\begin{aligned}\mathcal{H} \;=\; & -\frac{1}{2}\sum_{i,j} J_{ij}\mathbf{S}_i \cdot \mathbf{S}_j \to \frac{1}{2}\sum_{r,s}\sum_{u,\mathbf{q}} z_{rs}^{(u)} J_{rs}^{(u)} \Bigg\{ \left(a_{\mathbf{q}}^{(r)\dagger} a_{\mathbf{q}}^{(r)} S_s + a_{\mathbf{q}}^{(s)\dagger} a_{\mathbf{q}}^{(s)} S_r\right) F_{zz}(r,\, s) \\ & - \frac{\sqrt{S_r S_s}}{2}\, \Gamma_{rs}^{(u)}(\mathbf{q}) \left[a_{\mathbf{q}}^{(r)} a_{\mathbf{q}}^{(s)\dagger} G_1(r,\, s)^* + a_{\mathbf{q}}^{(r)} a_{-\mathbf{q}}^{(s)} G_2(r,\, s) \right] \\ & - \frac{\sqrt{S_r S_s}}{2}\, \Gamma_{rs}^{(u)}(\mathbf{q})^* \left[a_{\mathbf{q}}^{(r)\dagger} a_{\mathbf{q}}^{(s)} G_1(r,\, s) + a_{\mathbf{q}}^{(r)\dagger} a_{-\mathbf{q}}^{(s)\dagger} G_2(r,\, s)^* \right] \Bigg\}.\end{aligned} \tag{D.57}$$

For exchange interactions $J_{rr}^{(u)}$ within the same sublattice $r = s$, $G_1(r,\, r) = 2$, $G_2(r,\, r) = 0$, and $F_{zz}(r,\, r) = 1$. Therefore,

$$\mathcal{H} \to \sum_r S_r \sum_{u,\mathbf{q}} z_{rr}^{(u)} J_{rr}^{(u)} a_{\mathbf{q}}^{(r)\dagger} a_{\mathbf{q}}^{(r)} \left\{ 1 - \Gamma_{rr}^{(u)}(\mathbf{q}) \right\}, \tag{D.58}$$

because $\Gamma_{rr}(\mathbf{q}) = \Gamma_{rr}(-\mathbf{q})^*$. So

$$L_{rr}(\mathbf{q}) \to L_{rr}(\mathbf{q}) + \frac{S_r}{2} \sum_u z_{rr}^{(u)} J_{rr}^{(u)} \left\{ 1 - \Gamma_{rr}^{(u)}(\mathbf{q}) \right\}, \tag{D.59}$$

$$L_{r+M,\, r+M}(\mathbf{q}) \to L_{r+M,\, r+M}(\mathbf{q}) + \frac{S_r}{2} \sum_u z_{rr}^{(u)} J_{rr}^{(u)} \left\{ 1 - \Gamma_{rr}^{(u)}(\mathbf{q}) \right\}. \tag{D.60}$$

If $r \neq s$ then we can read off from equation (D.57) that

$$L_{rr}(\mathbf{q}) \to L_{rr}(\mathbf{q}) + \frac{S_s}{4} \sum_u z_{rs}^{(u)} J_{rs}^{(u)} F_{zz}(r,\, s), \tag{D.61}$$

$$L_{r+M,\, r+M}(\mathbf{q}) \to L_{r+M,\, r+M}(\mathbf{q}) + \frac{S_s}{4} \sum_u z_{rs}^{(u)} J_{rs}^{(u)} F_{zz}(r,\, s), \tag{D.62}$$

$$L_{rs}(\mathbf{q}) \to L_{rs}(\mathbf{q}) - \frac{\sqrt{S_r S_s}}{4} \sum_u z_{rs}^{(u)} J_{rs}^{(u)}\, \Gamma_{rs}^{(u)}(\mathbf{q})^\star\, G_1(r,\, s), \tag{D.63}$$

$$L_{r+M,\,s+M}(\mathbf{q}) \rightarrow L_{r+M,\,s+M}(\mathbf{q}) - \frac{\sqrt{S_r S_s}}{4} \sum_u z_{rs}^{(u)} J_{rs}^{(u)}\, \Gamma_{rs}^{(u)}(\mathbf{q})^\star\, G_1(r,\,s)^\star, \tag{D.64}$$

$$L_{r,\,s+M}(\mathbf{q}) \rightarrow L_{r,\,s+M}(\mathbf{q}) - \frac{\sqrt{S_r S_s}}{4} \sum_u z_{rs}^{(u)} J_{rs}^{(u)}\, \Gamma_{rs}^{(u)}(\mathbf{q})^\star\, G_2(r,\,s)^\star, \tag{D.65}$$

$$L_{r+M,\,s}(\mathbf{q}) \rightarrow L_{r+M,\,s}(\mathbf{q}) - \frac{\sqrt{S_r S_s}}{4} \sum_u z_{rs}^{(u)} J_{rs}^{(u)}\, \Gamma_{rs}^{(u)}(\mathbf{q})^\star\, G_2(r,\,s). \tag{D.66}$$

Exercise 4.22. *Derive the classical field $\mathbf{A}_i$ in equation (4.105) for a DM interaction that couples neighboring spins in 1D. Take the spins $\mathbf{S}_i = S(\sin\theta_i,\, 0,\, \cos\theta_i)$ to lie in the xz plane.*

The energy involving site i is

$$\begin{aligned} - DS^2\{\sin(\theta_{i+1} - \theta_i) + \sin(\theta_i - \theta_{i-1})\} &= - DS^2\{(\sin\theta_{i+1} - \sin\theta_{i-1})\cos\theta_i \\ &\quad + (\cos\theta_{i-1} - \cos\theta_{i+1})\sin\theta_i\} \\ &= - \mathbf{A}_i \cdot \mathbf{S}_i = -SA_x \sin\theta_i - SA_z \cos\theta_i \end{aligned} \tag{D.67}$$

with

$$A_x = -DS(\cos\theta_{i+1} - \cos\theta_{i-1}), \tag{D.68}$$

$$A_z = DS(\sin\theta_{i+1} - \sin\theta_{i-1}). \tag{D.69}$$

Of course, $A_y = 0$. Bear in mind that exchange and other interactions also contribute to $\mathbf{A}$.

***Exercise 4.23.** *Use the unitary matrix $\underline{U}_i$ given by equation (4.2) to derive equations (4.127)–(4.129).*

Transforming into the local reference frame of each spin,

$$\begin{aligned} \mathbf{S}_r \cdot \mathbf{S}_s &= \sum_{\alpha,\beta,\gamma} (U_r^{-1})_{\alpha\beta} \bar{S}_{r\beta}\, (U_s^{-1})_{\alpha\gamma} \bar{S}_{s\gamma} = \sum_{\alpha,\beta,\gamma} \bar{S}_{r\beta} (U_r)_{\beta\alpha}\, (U_s^{-1})_{\alpha\gamma} \bar{S}_{s\gamma} \\ &= \bar{S}_r \cdot \underline{F}(r,\,s) \cdot \bar{S}_s, \end{aligned} \tag{D.70}$$

where

$$\underline{F}(r,\,s) = \underline{U}_r \cdot \underline{U}_s^{-1} \tag{D.71}$$

$$F_{\alpha\beta}(r, s) = \sum_{\gamma}(U_r)_{\alpha\gamma}(U_s^{-1})_{\gamma\beta} = \sum_{\gamma}(U_r)_{\alpha\gamma}(U_s)_{\beta\gamma}. \tag{D.72}$$

Using the unitary matrix from equation (4.2), these matrix elements are

$$F_{xx}(r, s) = \cos\theta_r \cos\theta_s \cos(\phi_r - \phi_s) + \sin\theta_r \sin\theta_s, \tag{D.73}$$

$$F_{yy}(r, s) = \cos(\phi_r - \phi_s), \tag{D.74}$$

$$F_{zz}(r, s) = \sin\theta_r \sin\theta_s \cos(\phi_r - \phi_s) + \cos\theta_r \cos\theta_s, \tag{D.75}$$

$$F_{xy}(r, s) = F_{yx}(s, r) = \cos\theta_r \sin(\phi_r - \phi_s), \tag{D.76}$$

$$F_{xz}(r, s) = F_{zx}(s, r) = \cos\theta_r \sin\theta_s(\cos(\phi_r - \phi_s) - 1), \tag{D.77}$$

$$F_{yz}(r, s) = F_{zy}(s, r) = -\sin\theta_s \sin(\phi_r - \phi_s). \tag{D.78}$$

The second-order terms in the creation and annihilation operators for the dot product are

$$\begin{aligned}
\mathbf{S}_r \cdot \mathbf{S}_s \to & -\left(S_r a_s^{\dagger} a_s + S_s a_r^{\dagger} a_r\right)F_{zz}(r, s) + \frac{\sqrt{S_r S_s}}{2}\left(a_r + a_r^{\dagger}\right)\left(a_s + a_s^{\dagger}\right)F_{xx}(r, s) \\
& - \frac{\sqrt{S_r S_s}}{2}\left(a_r - a_r^{\dagger}\right)\left(a_s - a_s^{\dagger}\right)F_{yy}(r, s) \\
& - i\frac{\sqrt{S_r S_s}}{2}\left(a_r + a_r^{\dagger}\right)\left(a_s - a_s^{\dagger}\right)F_{xy}(r, s) \\
& - i\frac{\sqrt{S_r S_s}}{2}\left(a_r - a_r^{\dagger}\right)\left(a_s + a_s^{\dagger}\right)F_{yx}(r, s) = -\left(S_r a_s^{\dagger} a_s + S_s a_r^{\dagger} a_r\right)F_{zz}(r, s) \\
& + \frac{\sqrt{S_r S_s}}{2} a_r a_s\left\{F_{xx}(r, s) - F_{yy}(r, s) - i\left(F_{xy}(r, s) + F_{yx}(r, s)\right)\right\} \\
& + \frac{\sqrt{S_r S_s}}{2} a_r^{\dagger} a_s^{\dagger}\left\{F_{xx}(r, s) - F_{yy}(r, s) + i\left(F_{xy}(r, s) + F_{yx}(r, s)\right)\right\} \\
& + \frac{\sqrt{S_r S_s}}{2} a_r^{\dagger} a_s\left\{F_{xx}(r, s) + F_{yy}(r, s) - i\left(F_{xy}(r, s) - F_{yx}(r, s)\right)\right\} \\
& + \frac{\sqrt{S_r S_s}}{2} a_r a_s^{\dagger}\left\{F_{xx}(r, s) + F_{yy}(r, s) + i\left(F_{xy}(r, s) - F_{yx}(r, s)\right)\right\}.
\end{aligned} \tag{D.79}$$

If $J_{rs}^{(u)}$ couples spins on different sublattices $r \neq s$, then

$$
\begin{aligned}
-\sum_{i,j} J_{ij}\mathbf{S}_i\cdot\mathbf{S}_j \to \sum_{r,s}\sum_{u,\mathbf{q}} z_{rs}^{(u)} J_{rs}^{(u)} \Big\{ & \left(a_{\mathbf{q}}^{(r)\dagger}a_{\mathbf{q}}^{(r)}S_s + a_{\mathbf{q}}^{(s)\dagger}a_{\mathbf{q}}^{(s)}S_r\right)F_{zz}(r,s) \\
& -\Gamma_{rs}^{(u)}(\mathbf{q})^*\frac{\sqrt{S_rS_s}}{2}a_{\mathbf{q}}^{(r)\dagger}a_{\mathbf{q}}^{(s)}\Big[F_{xx}(r,s)+F_{yy}(r,s) \\
& \quad - i\big(F_{xy}(r,s)-F_{yx}(r,s)\big)\Big] \\
& -\Gamma_{rs}^{(u)}(\mathbf{q})\frac{\sqrt{S_rS_s}}{2}a_{\mathbf{q}}^{(r)}a_{\mathbf{q}}^{(s)\dagger}\Big[F_{xx}(r,s)+F_{yy}(r,s) \\
& \quad + i\big(F_{xy}(r,s)-F_{yx}(r,s)\big)\Big] \\
& -\Gamma_{rs}^{(u)}(\mathbf{q})\frac{\sqrt{S_rS_s}}{2}a_{\mathbf{q}}^{(r)}a_{-\mathbf{q}}^{(s)}\Big[F_{xx}(r,s)-F_{yy}(r,s) \\
& \quad - i\big(F_{xy}(r,s)+F_{yx}(r,s)\big)\Big] \\
& -\Gamma_{rs}^{(u)}(\mathbf{q})^*\frac{\sqrt{S_rS_s}}{2}a_{\mathbf{q}}^{(r)\dagger}a_{-\mathbf{q}}^{(s)\dagger}\Big[F_{xx}(r,s)-F_{yy}(r,s) \\
& \quad + i\big(F_{xy}(r,s)+F_{yx}(r,s)\big)\Big]\Big\}.
\end{aligned} \tag{D.80}
$$

Since

$$
\begin{aligned}
-\sum_{i,j} J_{ij}\mathbf{S}_i\cdot\mathbf{S}_j \to \sum_{r,s}\sum_{u,\mathbf{q}} z_{rs}^{(u)} J_{rs}^{(u)} \Big\{ & \left(a_{\mathbf{q}}^{(r)\dagger}a_{\mathbf{q}}^{(r)}S_s + a_{\mathbf{q}}^{(s)\dagger}a_{\mathbf{q}}^{(s)}S_r\right)F_{zz}(r,s) \\
& -\frac{\sqrt{S_rS_s}}{2}\Gamma_{rs}^{(u)}(\mathbf{q})\Big[a_{\mathbf{q}}^{(r)}a_{\mathbf{q}}^{(s)\dagger}G_1(r,s)^* + a_{\mathbf{q}}^{(r)}a_{-\mathbf{q}}^{(s)}G_2(r,s)\Big] \\
& -\frac{\sqrt{S_rS_s}}{2}\Gamma_{rs}^{(u)}(\mathbf{q})^*\Big[a_{\mathbf{q}}^{(r)\dagger}a_{\mathbf{q}}^{(s)}G_1(r,s) + a_{\mathbf{q}}^{(r)\dagger}a_{-\mathbf{q}}^{(s)\dagger}G_2(r,s)^*\Big]\Big\},
\end{aligned} \tag{D.81}
$$

we obtain

$$
\begin{aligned}
G_1(r,s) &= F_{xx}(r,s)+F_{yy}(r,s) - i\{F_{xy}(r,s)-F_{yx}(r,s)\} \\
&= (\cos\theta_r\cos\theta_s + 1)\cos(\phi_r-\phi_s) + \sin\theta_r\sin\theta_s \\
&\quad - i\sin(\phi_r-\phi_s)\{\cos\theta_r+\cos\theta_s\},
\end{aligned} \tag{D.82}
$$

$$
\begin{aligned}
G_2(r,s) &= F_{xx}(r,s)-F_{yy}(r,s) - i\{F_{xy}(r,s)+F_{yx}(r,s)\} \\
&= (\cos\theta_r\cos\theta_s - 1)\cos(\phi_r-\phi_s) + \sin\theta_r\sin\theta_s \\
&\quad - i\sin(\phi_r-\phi_s)\{\cos\theta_r-\cos\theta_s\},
\end{aligned} \tag{D.83}
$$

which completes the exercise!

***Exercise 4.24** *Use the general unitary matrix to generalize the forms for $G_1(r, s)$ and $G_2(r, s)$.*

We used Maple to get the answer. You should too!

Appendix E

Selected exercise solutions for chapter 5

Exercise 5.2. *Evaluate the SW stiffness D for a FM on a square lattice from the expression* $\omega(\mathbf{q}) = \Delta + Dq^2$ *near* $\mathbf{q} = 0$ *for* $K \neq 0$.

Near $\mathbf{q} = 0$,

$$\Gamma_{\mathbf{q}} = \frac{1}{2}\left\{\cos(q_x a) + \cos(q_y a)\right\} \approx 1 - \frac{1}{4}(qa)^2. \tag{E.1}$$

So

$$\omega(\mathbf{q}) = 2S\left\{2J(1 - \Gamma_{\mathbf{q}}) + K\right\} \approx SJ\{(qa)^2 + 2K/J\}. \tag{E.2}$$

Hence, $\Delta = 2SK$ is the SW gap and $D = SJa^2$ is the SW stiffness.

Exercise 5.3. *Use the measured SW stiffness for face-centered cubic Ni in figure 1.1 of Chapter 1 to estimate its nearest-neighbor exchange coupling J. Using the full expression for the SW frequency (not just the quadratic expansion), evaluate the predicted SW frequencies at wavevector 0.5* Å^{-1} *in directions* [1, 0, 0], [1, 1, 0], *and* [1, 1, 1]. *How do these compare with the predicted value from the quadratic expansion?*

Since a face-centered cubic lattice has $z = 12$ nearest neighbors, the SW frequency is given by

$$\hbar\omega(\mathbf{q}) = 12JS(1 - \Gamma_{\mathbf{q}}), \tag{E.3}$$

doi:10.1088/978-1-64327-114-9ch14

where we assume that the anisotropy is negligible. For a face-centered cubic lattice

$$\begin{aligned}\Gamma_{\mathbf{q}} = \frac{1}{6}\Big\{&\cos\big((q_x + q_y)a/\sqrt{2}\big) + \cos\big((q_x - q_y)a/\sqrt{2}\big) \\ &+ \cos\big((q_x + q_z)a/\sqrt{2}\big) + \cos\big((q_y + q_z)a/\sqrt{2}\big) \\ &+ \cos\big((q_x - q_z)a/\sqrt{2}\big) + \cos\big((q_y - q_z)a/\sqrt{2}\big)\Big\},\end{aligned} \tag{E.4}$$

where a is the cubic lattice constant. For small q,

$$\Gamma_{\mathbf{q}} \approx 1 - \frac{q^2a^2}{6}, \tag{E.5}$$

$$\hbar\omega(\mathbf{q}) \approx Dq^2, \tag{E.6}$$

with SW stiffness $D = 2JSa^2$. The lattice constant of Ni is about $a \approx 3.52$ Å and $S = 1$. So if $D \approx 433$ meV Å^2, then $J \approx 17.5$ meV.

For the different directions,

$$\begin{aligned}[1, 0, 0]:\quad &\mathbf{q} = 0.5\,(1, 0, 0)\text{Å}^{-1}, \\ &\Gamma_{\mathbf{q}} = \frac{1}{3}\Big\{2\cos(q_x a/\sqrt{2}) + 1\Big\} \approx 0.547, \\ &\hbar\omega(\mathbf{q}) \approx 95.13 \text{ meV},\end{aligned} \tag{E.7}$$

$$\begin{aligned}[1, 1, 0]:\quad &\mathbf{q} = 0.353\,(1, 1, 0)\text{Å}^{-1}, \\ &\Gamma_{\mathbf{q}} = \frac{1}{6}\Big\{\cos(\sqrt{2}q_x a) + 4\cos(q_x a/\sqrt{2}) + 1\Big\} \approx 0.560, \\ &\hbar\omega(\mathbf{q}) \approx 92.38 \text{ meV},\end{aligned} \tag{E.8}$$

$$\begin{aligned}[1, 1, 1]:\quad &\mathbf{q} = 0.289\,(1, 1, 1)\text{Å}^{-1}, \\ &\Gamma_{\mathbf{q}} = \frac{1}{2}\Big\{\cos(\sqrt{2}q_x a) + 1\Big\} \approx 0.567, \\ &\hbar\omega(\mathbf{q}) \approx 91.00 \text{ meV}.\end{aligned} \tag{E.9}$$

Based on the quadratic expansion, $\omega = Dq^2 = 108$ meV, which is higher than any of the above.

Exercise 5.4. *For a FM with spins aligned along* **z**, *would easy-axis anisotropy K and a magnetic field* **B** *compete with one another if the magnetic field was along* **–z** *instead of along* **z**?

If the magnetic field was along $-\mathbf{z}$, then the spins would also flip. So once again the anisotropy and the magnitude of the field would add up: $K' = K + \mu_B|B|/S$. They would not compete.

Exercise 5.6. *Generalize the results for a FM on a square lattice by adding the next-nearest-neighbor coupling J'.*

A square lattice has $z = 4$ nearest neighbors and $z' = 4$ next-nearest neighbors but there is still only one sublattice for a FM. Using equation (4.119), we find

$$\hbar\omega(\mathbf{q}) = 4JS(1 - \Gamma_\mathbf{q}) + 4J'S(1 - \Gamma'_\mathbf{q}), \tag{E.10}$$

where

$$\Gamma_\mathbf{q} = \frac{1}{2}\left\{\cos(q_x a) + \cos(q_y a)\right\}, \tag{E.11}$$

$$\Gamma'_\mathbf{q} = \frac{1}{2}\left\{\cos\left((q_x + q_y)\sqrt{2}a\right) + \cos\left((q_x - q_y)\sqrt{2}a\right)\right\}. \tag{E.12}$$

Notice that $\Gamma_{\mathbf{q}=0} = \Gamma'_{\mathbf{q}=0} = 1$.

Exercise 5.7. *Show that the 'mother-of-all sum rules' stated in equation (4.53) is obeyed by the FM on a square lattice.*

For a FM on a square lattice, equation (5.18) gave

$$S^{(1)}_{xx}(\mathbf{q}) = S^{(1)}_{yy}(\mathbf{q}) = \frac{S}{2}, \tag{E.13}$$

and $S^{(1)}_{zz}(\mathbf{q}) = 0$. From equation (4.41),

$$S_{\alpha\beta}(\mathbf{q},\omega) = \sum_n S^{(n)}_{\alpha\beta}(\mathbf{q})\, \delta(\omega - \omega_n(\mathbf{q})). \tag{E.14}$$

So

$$\sum_\alpha \int d\omega S_{\alpha\alpha}(\mathbf{q},\omega) = S \tag{E.15}$$

and

$$\frac{1}{N}\sum_{\alpha,\mathbf{q}} \int d\omega S_{\alpha\alpha}(\mathbf{q},\omega) = S. \tag{E.16}$$

But our dynamical expression for $S_{\alpha\beta}(\mathbf{q},\omega)$ does not include the static contribution $S^2\delta(\omega)\delta_{\alpha z}\,\delta_{\beta z}$. Adding that contribution, we have

$$\frac{1}{N}\sum_{\alpha,\mathbf{q}} \int d\omega\; S_{\alpha\alpha}(\mathbf{q},\omega) = S(S+1), \tag{E.17}$$

which is the 'mother-of-all sum rules'.

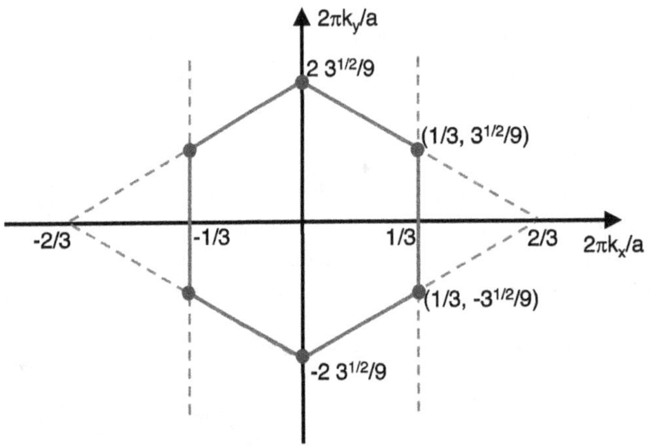

Figure E.1. The first BZ of the honeycomb or hexagonal lattices.

Exercise 5.8. *In the limit $J_2 \to 0$, the FM chain with alternating exchange develops a flat, zero-frequency mode. Physically, why does this SW branch appear?*

When $J_2 = 0$, the chain breaks up into a system of dimers, pairs of spins coupled by J_1. Each dimer can rotate freely with respect to the others. This degree of freedom produces the flat zero-frequency mode.

Exercise 5.12. *Show that the 'Dirac' points of the honeycomb lattice lie at the corners of the first structural BZ.*

The reciprocal lattice vectors of the honeycomb lattice are the same as for the hexagonal lattice in exercise 4.28. The boundaries of the first structural BZ are given by the Wigner–Seitz conditions $\mathbf{k} \cdot \mathbf{b}_n = \pm|\mathbf{b}_n|^2$ for $n = 1$ and 2. Now $|\mathbf{b}_1| = |\mathbf{b}_2| = 4\pi/3a$. So the conditions are

$$k_x \pm \sqrt{3}k_y = \pm\frac{4\pi}{3a}. \tag{E.18}$$

We get a similar condition by bisecting another reciprocal lattice vector $\mathbf{b}' = \mathbf{b}_1 + \mathbf{b}_2 = 4\pi\mathbf{x}/3a$ with $k_x = \pm 2\pi/3a$. Not surprisingly, the first BZ is the hexagon sketched in figure E.1.

In units of $2\pi/a$, the six 'Dirac' points are at $(\pm 1/3, \sqrt{3}/9)$, $(\pm 1/3, -\sqrt{3}/9)$, and $(0, \pm 2\sqrt{3}/9)$. At each point, $\Gamma_\mathbf{k} = 0$. These points lie at the six corners of the BZ, as shown in figure E.1.

Exercise 5.13. *Show that the ratio of intensities of modes 1 and 2 on the honeycomb lattice as either of the 'Dirac' points $(1/3, \pm\sqrt{3}/9)$ is approached from the x-axis equals 1/3.*

Recall from equations (5.42) and (5.43) that

$$S_{yy}^{(1)}(\mathbf{q}) = S_{zz}^{(1)}(\mathbf{q}) = \frac{S}{4}\left\{1 - \frac{\mathrm{Re}(\Gamma_\mathbf{q})}{|\Gamma_\mathbf{q}|}\right\}, \tag{E.19}$$

$$S_{yy}^{(2)}(\mathbf{q}) = S_{zz}^{(2)}(\mathbf{q}) = \frac{S}{4}\left\{1 + \frac{\mathrm{Re}(\Gamma_\mathbf{q})}{|\Gamma_\mathbf{q}|}\right\}, \tag{E.20}$$

for modes 1 and 2 with spins along $\mathbf{x}$ and $S_{xx}^{(n)}(\mathbf{q}) = 0$. Along $\mathbf{q} = (1/3, K)$,

$$\begin{aligned}\Gamma_\mathbf{q} &= \frac{1}{3}\{e^{i2\pi/3} + e^{-i2\pi/3}e^{-i\sqrt{3}K\pi} + e^{-i2\pi/3}e^{i\sqrt{3}K\pi}\}\\ &= \frac{e^{i2\pi/3}}{3}\{1 - 2\cos(\sqrt{3}K\pi)\},\end{aligned} \tag{E.21}$$

which vanishes at $K = \pm\sqrt{3}/9$. Then

$$\frac{\mathrm{Re}(\Gamma_\mathbf{q})}{|\Gamma_\mathbf{q}|} = -\frac{1}{2} \tag{E.22}$$

and

$$\frac{S_{yy}^{(1)}(\mathbf{q})}{S_{yy}^{(2)}(\mathbf{q})} = \frac{S_{zz}^{(1)}(\mathbf{q})}{S_{zz}^{(2)}(\mathbf{q})} = \frac{1}{3}, \tag{E.23}$$

as required.

Exercise 5.14. *Evaluate the SW frequencies for an AF on a square lattice with easy-plane anisotropy E but K = 0 and B = 0. Does the Goldstone mode survive?*

The Hamiltonian is

$$\mathcal{H} = -J\sum_{\langle i,j\rangle} \mathbf{S}_i \cdot \mathbf{S}_j - E\sum_i {S_{ix}}^2. \tag{E.24}$$

For $J < 0$ and $E < 0$, the spins are assumed to lie along $\pm\mathbf{z}$.

It follows that

$$\underline{\mathcal{L}} = \underline{L} \cdot \underline{N} = \frac{S}{2}\begin{pmatrix} -4J - E & 0 & E & -4J\Gamma_\mathbf{q} \\ 0 & -4J - E & -4J\Gamma_\mathbf{q} & E \\ -E & 4J\Gamma_\mathbf{q} & 4J + E & 0 \\ 4J\Gamma_\mathbf{q} & -E & 0 & 4J + E \end{pmatrix}. \tag{E.25}$$

The SW frequencies are then

$$\hbar\omega_{1,2}(\mathbf{q}) = 2S\left\{4J^2(1 - \Gamma_\mathbf{q}^2) + 2JE(1 \pm \Gamma_\mathbf{q})\right\}^{1/2}, \tag{E.26}$$

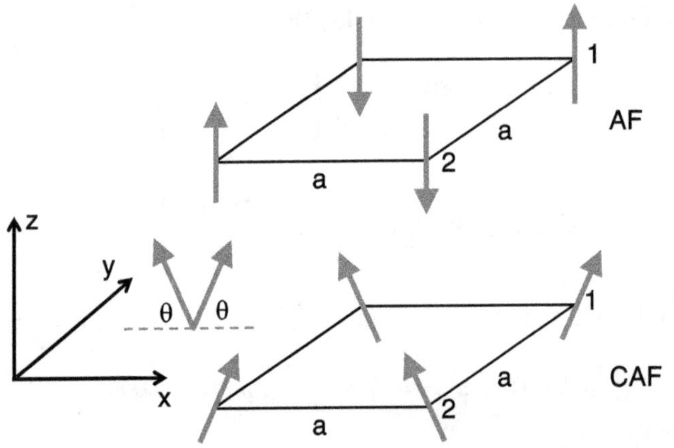

Figure E.2. AF and CAF states on a square lattice.

with $\omega_1(\mathbf{q})$ corresponding to the positive sign and $\omega_2(\mathbf{q})$ to the negative sign. so the two degenerate modes now split into two distinct modes. At $\mathbf{Q} = (0.5, 0.5)$, $\Gamma_\mathbf{q} = -1$ and $\omega_1(\mathbf{Q}) = 0$ so a Goldstone mode (associated with spin rotations in the yz plane) survives, as seen in figure 5.8(b).

Exercise 5.15. *Can you think of some other interaction term that would break the degeneracy of the two SW modes for an AF on a square lattice in zero field?*

Splitting the degenerate modes of an AF in zero field requires terms like $a_\mathbf{q}a_{-\mathbf{q}}$ or $a_\mathbf{q}b_{-\mathbf{q}}$ (and their complex conjugates). Easy-plane anisotropy produces the first kind of term. All terms in the second category would also tilt the spins away from $\mathbf{z}$. We cannot think of any that would work.

Exercise 5.17. *Evaluate the canted moment along* $\mathbf{z}$ *above* $B_{\rm SF}$ *for the AF on a square lattice in a magnetic field with easy-axis anisotropy. Derive* $B_{\rm SF}$ *from the balance between the anisotropy and Zeeman energies. Does the lower SW frequency* $\omega(\mathbf{Q})$ *of the AF precisely vanish as* $B_{\rm SF}$ *is approached from below?*

We will compare the energies of the AF and canted AF (CAF) states in figure E.2 for a magnetic field B and easy-axis anisotropy K along $\mathbf{z}$. The AF state does not gain any energy from the magnetic field so

$$\frac{E_{\rm AF}}{N} = 2JS^2 - KS^2. \tag{E.27}$$

For the CAF with canting angle θ,

$$\frac{E_{\rm CAF}}{N} = -B'S^2 \sin\theta - KS^2 \sin^2\theta + 2JS^2(1 - 2\sin^2\theta), \tag{E.28}$$

where $B' = 2\mu_B B/S$. While the CAF gains some Zeeman energy from the magnetic field, it loses anisotropy and exchange energy. Minimizing E_{CAF} with respect to $\sin\theta$ gives the condition

$$\sin\theta = \frac{1}{2}\frac{B'}{4|J| - K}. \tag{E.29}$$

At this angle,

$$\frac{E_{\mathrm{CAF}}}{N} = 2JS^2 - \frac{S^2}{4}\frac{B'^2}{4|J| - K}. \tag{E.30}$$

Hence,

$$\frac{\Delta E}{N} = \frac{E_{\mathrm{AF}} - E_{\mathrm{CAF}}}{N} = -KS^2 + \frac{1}{4}\frac{B'^2 S^2}{4|J| - K}. \tag{E.31}$$

This energy difference is positive ($E_{\mathrm{CAF}} < E_{\mathrm{AF}}$) when $B' > B_{\mathrm{SF}}{}' = 2\mu_B B_{\mathrm{SF}}/S$ where

$$B_{\mathrm{SF}}{}' = 2\{(4|J| - K)K\}^{1/2}, \tag{E.32}$$

which vanishes like $\sqrt{|J|K}$ as $K \to 0$.

Now recall that the lower SW frequency for an AF on a square lattice is

$$\hbar\omega(\mathbf{q}) = 2S\left\{(2|J| + K)^2 - (2|J|\Gamma_{\mathbf{q}})^2\right\}^{1/2} - B'S, \tag{E.33}$$

so that

$$\hbar\omega(\mathbf{Q}) = 2S\{(4|J| + K)K\}^{1/2} - B'S. \tag{E.34}$$

At $B' = B_{\mathrm{SF}}{}'$,

$$\hbar\omega(\mathbf{Q}) = 2SK^{1/2}\{(4|J| + K)^{1/2} - (4|J| - K)^{1/2}\} \approx S\frac{K^{3/2}}{|J|^{1/2}}. \tag{E.35}$$

Thus, the SW frequency is nonzero at the spin-flip field B_{SF}.

Exercise 5.19. *To obtain* $\underline{L}(\mathbf{q})$ *for the AF state, we took the spins* $r = 1$ *and* 2 *to have* $\cos\theta_r = \pm 1$, $\cos\phi_r = 1$, *and* $\sin\phi_r = 0$. *Of course, the value of* ϕ_r *should not matter when the spins are aligned along* $\pm\mathbf{z}$. *But what would happen to the E terms in equation (5.47) for* $\underline{L}(\mathbf{q})$ *if* ϕ_r *were arbitrary? Could this change have any effect on the inelastic spectra?*

According to equations (4.138), (4.139), and (4.140) with $\phi_r \neq 0$ but $\phi_r = 0$ or π,

$$L_{rr} \to L_{rr} - \frac{ES}{2}, \tag{E.36}$$

$$L_{r+M,\,r+M} \to L_{r+M,\,r+M} - \frac{ES}{2}, \tag{E.37}$$

$$L_{r,\,r+M} \rightarrow L_{r,\,r+M} - \frac{ES}{2}\,e^{\pm 2i\phi_r}, \tag{E.38}$$

$$L_{r+M,\,r} \rightarrow L_{r+M,\,r} - \frac{ES}{2}\,e^{\mp 2i\phi_r}, \tag{E.39}$$

where the upper sign is for the up spins ($r = 1$) and the lower sign is for the down spins ($r = 2$). Clearly, these additional phase factors can have no physical effects on the inelastic spectra. But be careful to also replace $G_2(1, 2) = G_2(2, 1) = -2$ by

$$G_2(1, 2) = G_2(2, 1) = -2e^{i(\phi_1-\phi_2)}, \tag{E.40}$$

so that equation (5.47) becomes

$$\underline{L} = \frac{S}{2}\begin{pmatrix} A_+ & 0 & -Ee^{2i\phi_1} & 4J\Gamma_{\mathbf{q}}e^{-i(\phi_1-\phi_2)} \\ 0 & A_- & 4J\Gamma_{\mathbf{q}}e^{-i(\phi_1-\phi_2)} & -Ee^{-2i\phi_2} \\ -Ee^{-2i\phi_1} & 4J\Gamma_{\mathbf{q}}e^{i(\phi_1-\phi_2)} & A_+ & 0 \\ 4J\Gamma_{\mathbf{q}}e^{i(\phi_1-\phi_2)} & -Ee^{2i\phi_2} & 0 & A_- \end{pmatrix}, \tag{E.41}$$

$$A_\pm = -4J + 2K - E \pm \frac{2\mu_B B}{S}. \tag{E.42}$$

Then the spectra will not depend on the choice of ϕ_r.

Exercise 5.20. *Using the expression for the AF SW frequencies on a cubic lattice, evaluate the width of the band in the powder spectrum (maximum minus the mimimum frequency). Do the same for a FM.*

For an AF on a cubic lattice,

$$\hbar\omega(\mathbf{q}) = 2S\left\{(3|J| + K)^2 - (3|J|\Gamma_{\mathbf{q}})^2\right\}^{1/2}. \tag{E.43}$$

So the minimum frequency at $\Gamma_{\mathbf{q}} = \pm 1$ is

$$\hbar\omega_{\min} = 2S\{(3|J| + K)^2 - 9J^2\}^{1/2} \approx 2\sqrt{6}\,S\sqrt{|J|K}. \tag{E.44}$$

The maximum frequency at $\Gamma_{\mathbf{q}} = 0$ is

$$\hbar\omega_{\max} = 6|J|S + 2SK \tag{E.45}$$

so the width

$$\hbar\Delta\omega = \hbar(\omega_{\max} - \omega_{\min}) \approx 2S\left(3|J| + K - \sqrt{6|J|K}\right). \tag{E.46}$$

For $S = 5/2$, $K = 0.2$ meV, and $J = -1$ meV, $\hbar\omega_{\min} \approx 5.48$ meV, $\hbar\omega_{\max} = 16$ meV, and $\hbar\Delta\omega = 10.52$ meV.

For a FM on a cubic lattice, the SW frequencies are

$$\hbar\omega(\mathbf{q}) = 6JS(1 - \Gamma_{\mathbf{q}}) + 2KS. \tag{E.47}$$

So $\hbar\omega_{\max} = 12JS + 2KS$ at $\Gamma_{\mathbf{q}} = -1$ and $\hbar\omega_{\min} = 2KS$ at $\Gamma_{\mathbf{q}} = 1$ with $\hbar\Delta\omega = 12JS$ independent of K. For $S = 5/2$, $K = 0.2$ meV, and $J = 1$ meV, $\hbar\omega_{\max} = 31$ meV, $\hbar\omega_{\min} = 1$ meV, and $\hbar\Delta\omega = 30$ meV.

***Exercise 5.21.** *Consider the G-type AF in figure 5.11(a) with different exchanges $J_{ab} < 0$ in the ab plane and $J_c < 0$ along the c axis. Also add the diagonal interaction J_d between next-nearest neighbors on adjacent planes separated by $\Delta z = \pm a$ and easy-axis anisotropy K along* **z**. *Show that the G-type AF becomes unstable with respect to the A- or C-type AFs when $J_d < -|J_c|/4$ or $J_d < -|J_{ab}|/2$, respectively, regardless of K. When $K = 0$, show that the SW frequency of the G-type AF becomes imaginary when either of those conditions is satisfied*

Compare the energies of the G-type, A-type, and C-type AFs in figure 5.11:

$$\frac{E_G}{N} = +2J_{ab}S^2 + J_cS^2 - 4J_dS^2 - KS^2, \tag{E.48}$$

$$\frac{E_A}{N} = -2J_{ab}S^2 + J_cS^2 + 4J_dS^2 - KS^2, \tag{E.49}$$

$$\frac{E_C}{N} = +2J_{ab}S^2 - J_cS^2 + 4J_dS^2 - KS^2. \tag{E.50}$$

So $E_C < E_G$ when $J_d < -|J_c|/4$ and $E_A < E_G$ when $J_d < -|J_{ab}|/2$. Notice that the easy-axis anisotropy does not affect these conditions.

Now evaluate the SW frequencies of the G-type AF. While $J_{ab} < 0$ and $J_c < 0$ couple spins on different sublattices, J_d couples spins on the same sublattice (up or down). So

$$L_{11} = L_{22} = L_{33} = L_{44} = -2J_{ab}S - J_cS + 4J_dS\{1 - \Gamma_d(\mathbf{q})\} + KS, \tag{E.51}$$

$$L_{14} = L_{23} = L_{32} = L_{41} = 2J_{ab}S\Gamma_{ab}(\mathbf{q}) + J_cS\Gamma_c(\mathbf{q}), \tag{E.52}$$

where

$$\Gamma_{ab}(\mathbf{q}) = \frac{1}{2}\left\{\cos(q_xa) + \cos(q_ya)\right\}, \tag{E.53}$$

$$\Gamma_c(\mathbf{q}) = \cos(q_za), \tag{E.54}$$

$$\begin{aligned}\Gamma_d(\mathbf{q}) = \frac{1}{4}\Big\{&\cos\big((q_x + q_z)a\big) + \cos\big((q_x - q_z)a\big)\\ &+ \cos\big((q_y + q_z)a\big) + \cos\big((q_y - q_z)a\big)\Big\}.\end{aligned} \tag{E.55}$$

The doubly degenerate SW frequency is given by

$$\hbar\omega(\mathbf{q}) = 2S\{(2|J_{ab}|(1+\Gamma_{ab}(\mathbf{q})) + |J_c|(1+\Gamma_c(\mathbf{q})) + 4J_d(1-\Gamma_d(\mathbf{q})) + K) \times (2|J_{ab}|(1-\Gamma_{ab}(\mathbf{q})) + |J_c|(1-\Gamma_c(\mathbf{q})) + 4J_d(1-\Gamma_d(\mathbf{q})) + K)\}^{1/2}, \tag{E.56}$$

At high-symmetry points, the SW frequency is given by:

$$\hbar\omega(0,0,0) = 2S\{(4|J_{ab}| + 2|J_c| + K)K\}^{1/2}, \tag{E.57}$$

$$\hbar\omega(0.5,0.5,0.5) = 2S\{K(4|J_{ab}| + 2|J_c| + K)\}^{1/2}, \tag{E.58}$$

$$\hbar\omega(0.5,0.5,0) = 2S\{(2|J_c| + 8J_d + K)(4|J_{ab}| + 8J_d + K)\}^{1/2}, \tag{E.59}$$

$$\hbar\omega(0,0,0.5) = 2S\{(4|J_{ab}| + 8J_d + K)(2|J_c| + 8J_d + K)\}^{1/2}. \tag{E.60}$$

Notice that $\omega(0,0,0.5) = \omega(0.5,0.5,0)$ become imaginary when $J_d + K/8 < -|J_{ab}|/2$ or $J_d + K/8 < -|J_c|/4$. When $K = 0$, these are the same conditions obtained above from the relative energies of the three AF phases,

You may wonder what happens when

$$-\frac{K}{8} - \frac{|J_{ab}|}{2} < J_d < -\frac{|J_{ab}|}{2} \tag{E.61}$$

and

$$-\frac{K}{8} - \frac{|J_c|}{4} < J_d < -\frac{|J_c|}{4}. \tag{E.62}$$

Then the G-type phase is *locally* stable but *globally* unstable. So the SWs of the G-type AF are well defined even though the G-type AF does not have the lowest possible energy!

Appendix F

Selected exercise solutions for chapter 6

Exercise 6.1. *Would the SW spectrum of an AF chain with alternating DM interactions change if D switched sign?*

If D changes sign, then $\theta_1 \rightarrow -\theta_1$ and $\theta_2 \rightarrow -\theta_2$. It is then easy to see that all matrix elements of $\underline{L}$ remain unchanged. Of course, flipping D corresponds to translating the chain along **z** by a. So the SW spectrum cannot change.

Exercise 6.2. *Add easy-plane anisotropy E to the AF chain with alternating DM interactions and re-evaluate the SW intensity.*

With the term

$$-E\sum_i S_{ix}{}^2, \tag{F.1}$$

the spins favor the yz plane when $E < 0$. This term does not affect the tilts of the spins within that plane.

Based on the shortcuts of equations (4.141)–(4.144), we use ($\eta = \pi/2$, $\delta = 0$, and $\phi_r = \pi/2$) $\mathbf{A} = (0, -1, 0)$ to modify $\underline{L}$ by

$$L_{rr} \rightarrow L_{rr} - \frac{ES}{2}, \tag{F.2}$$

$$L_{r+2,\,r+2} \rightarrow L_{r+2,\,r+2} - \frac{ES}{2}, \tag{F.3}$$

doi:10.1088/978-1-64327-114-9ch15

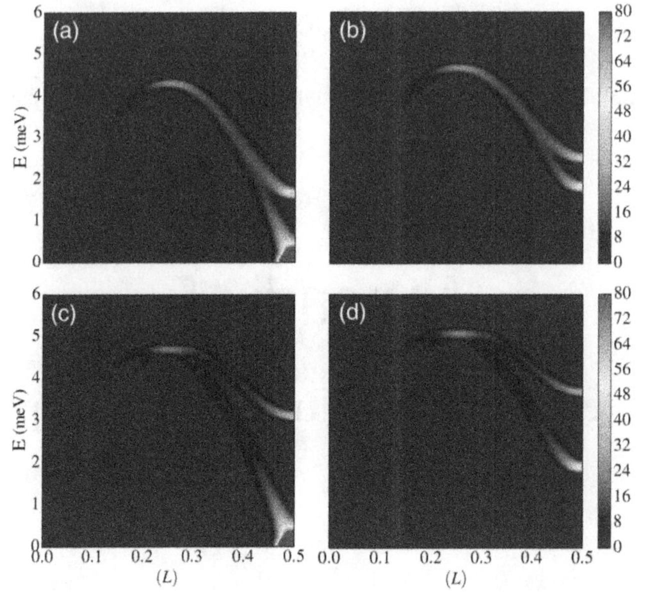

Figure F.1. The inelastic spectrum of an AF-coupled, canted chain with $S = 2$, $J = -1$ meV, $D = 0.4$ meV, and (a) $K = 0$ and $E = 0$, (b) $K = 0.1$ meV and $E = 0$, (c) $K = 0$ and $E = -0.2$ meV, or (d) $K = 0.1$ meV and $E = -0.2$ meV.

$$L_{r,\,r+2} \rightarrow L_{r,\,r+2} + \frac{ES}{2}, \tag{F.4}$$

$$L_{r+2,\,r} \rightarrow L_{r+2,\,r} + \frac{ES}{2}. \tag{F.5}$$

The inelastic spectra are plotted in figure F.1, where the top panels take $E = 0$ and the bottom panels take $E = -0.2$ meV. Notice that the dominant effects of easy-plane anisotropy are to lift the upper band, which contains spin fluctuations predominantly along **y** and to suppress the intensity of that mode. The lower mode, with fluctuations along **x** and **z**, is hardly affected by the easy-plane anisotropy.

Exercise 6.4. *Revisit the AF on a square lattice in the xy plane with field applied along* **z***, parallel to the easy-axis anisotropy K. Are the SW frequencies* $\omega(\mathbf{Q})$ *continuous at* B_{SF}*? What is the effect of easy-plane anisotropy E perpendicular to* **y***?*

We evaluate the spin dynamics using

$$L_{11} = L_{33} = -\frac{KS}{2}(1 - 3\cos^2\theta_1) + 2JSF_{zz} - \frac{ES}{2} + B'\cos\theta_1, \tag{F.6}$$

$$L_{22} = L_{44} = -\frac{KS}{2}(1 - 3\cos^2\theta_2) + 2JSF_{zz} - \frac{ES}{2} + B'S\cos\theta_2, \qquad \text{(F.7)}$$

$$L_{12} = L_{21} = L_{34} = L_{43} = -JSG_1\Gamma(\mathbf{q}), \qquad \text{(F.8)}$$

$$L_{14} = L_{41} = L_{23} = L_{32} = -JSG_2\Gamma(\mathbf{q}), \qquad \text{(F.9)}$$

$$L_{13} = L_{31} = -\frac{KS}{2}\sin^2\theta_1 + \frac{ES}{2}, \qquad \text{(F.10)}$$

$$L_{24} = L_{42} = -\frac{KS}{2}\sin^2\theta_2 + \frac{ES}{2}, \qquad \text{(F.11)}$$

where $K > 0$ is the easy-axis anisotropy along **z**, $E < 0$ is the easy-plane anisotropy perpendicular to **y**, and $B' = 2\mu_B B/S$ is the scaled magnetic field. Recall that

$$\Gamma(\mathbf{q}) = \frac{1}{2}\big(\cos(q_x a) + \cos(q_y a)\big). \qquad \text{(F.12)}$$

Since the spins lie in the xz plane, $\phi_r = 0$ and

$$F_{zz} = \cos(\theta_1 - \theta_2), \qquad \text{(F.13)}$$

$$G_1 = \cos\theta_1\cos\theta_2 + 1, \qquad \text{(F.14)}$$

$$G_2 = \cos\theta_1\cos\theta_2 - 1. \qquad \text{(F.15)}$$

In the AF phase, $\theta_1 = 0$ and $\theta_2 = \pi$. In the canted AF phase above the spin-flop field

$$B'_{SF} = 2\sqrt{K(4|J| - K)}, \qquad \text{(F.16)}$$

$\theta_1 = -\theta_2$ is given by

$$\cos\theta_1 = \frac{B'}{2(4|J| - K)}. \qquad \text{(F.17)}$$

The resulting SW frequencies at $\mathbf{Q} = (0.5, 0.5)$ are plotted in figure F.2 for $S = 5/2$, $J = -1$ meV, $K = 0.1$ meV, and either $E = 0$ or -0.05 meV.

In both cases, the lower SW frequency is discontinuous at B_{SF}. For $E = 0$, the system supports a Goldstone mode above B_{SF} due to the rotational symmetry about **z**. That Goldstone mode is lifted by easy-plane anisotropy $E < 0$. For $E = 0$, there is no mode splitting at $B = 0$ and the mode frequencies split linearly with field. For $E < 0$, the SW frequencies split at $B = 0$.

Since the lower SW mode nearly vanishes at the spin-flop transition, $B_{SF} \approx \Delta/2\mu_B$ where Δ is the SW gap in zero field.

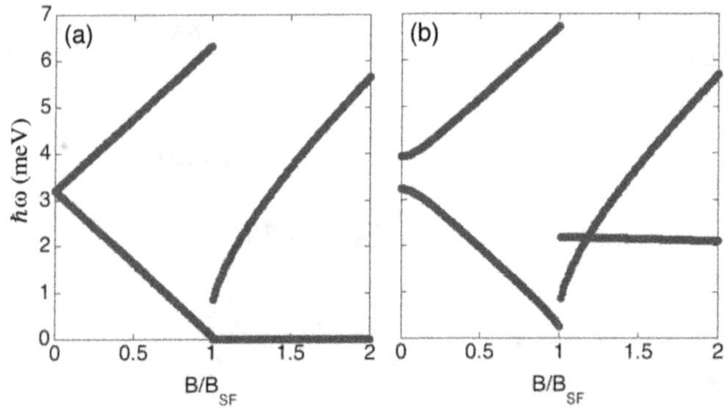

Figure F.2. The $\mathbf{q} = \mathbf{Q}$ SW frequencies of a square lattice $S = 5/2$ AF with magnetic field applied perpendicular to the plane along $\mathbf{z}$. Coupling constants are $J = -1$ meV and $K = 0.1$ meV with easy-plane anisotropy (a) $E = 0$ or (b) -0.05 meV perpendicular to the spin plane.

Exercise 6.5. *For a cycloid generated by CE in 1D with FM nearest-neighbor exchange $J_1 > 0$, is there a sudden change in the inelastic spectrum as $|J_2|$ falls below $J_1/4$?*

As $|J_2|$ decreases towards $J_1/4$, $Q \rightarrow 0$ so the period Ma of the cycloid diverges. When $|J_2| = J_1/4$, the cycloid transforms into a FM. The SW spectrum changes continuously during this transformation. You can already see the resemblance between figure 6.4(b) for a cycloid with $M = 20$ and figure 5.2(a) for a FM, both with $J_1 = 1$ meV.

Exercise 6.6. *How would easy-plane anisotropy affect the inelastic spectra for a cycloid produced by CE? Would a Goldstone mode still survive? Try it for $p = 1$ and $M = 5$ with $J_1 = 1$ meV.*

A Goldstone mode survives due to rotational symmetry in the spin plane. But as seen in figure F.3, the other Goldstone mode is lifted by the easy-plane anisotropy $E < 0$.

Exercise 6.9. *Consider the sequence of Φ_n and Ψ_n cycloidal modes in the limit where the period of the cycloid diverges or $Q \rightarrow Q_0 = \pi/a$. How many distinct modes survive? What happens to the coefficients $\xi_2^{(n)}$ and $\rho_2^{(n)}$ in this limit?*

In the long-wavelength limit, $\xi_2^{(n)} \rightarrow 0$ and $\rho_2^{(n)} \rightarrow 0$, so we recover the forms

$$\delta \mathbf{S}_r\left(\Phi_{\pm 1}, Q\right) = \sqrt{S}\left(-1\right)^r \mathbf{t}\left(ra\right)\xi_1^{(n)} e^{\pm 2\pi i n \delta r}, \tag{F.18}$$

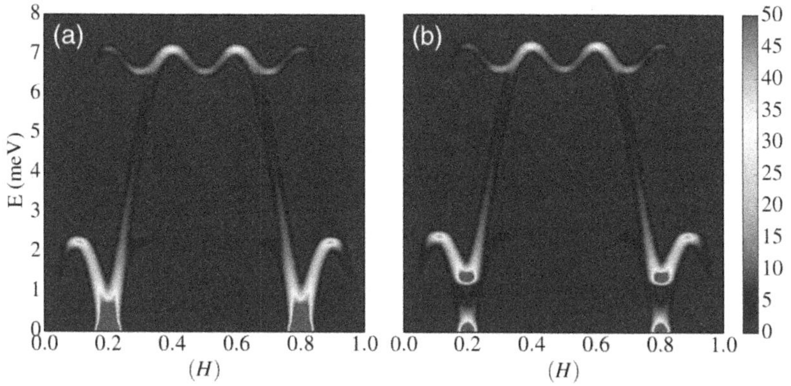

Figure F.3. The inelastic spectra of a cycloid produced by competing exchange with $J_1 = 1$ meV, $S = 5/2$, $p = 1$, $M = 5$, and either (a) $E = 0$ or (b) $E = -0.05$ meV.

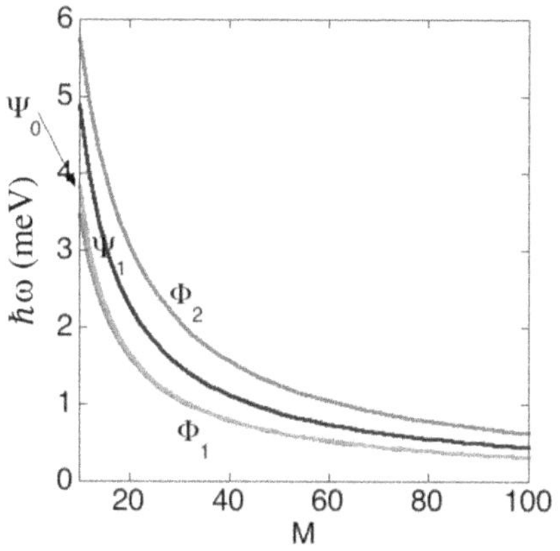

Figure F.4. The dependence of the first few mode frequencies on M (even M only with $p = 2$) for the same parameters as in figure 6.7: $J = -1$ meV and $S = 5/2$. Recall that $\omega(\Psi_0) = 0$ for all M.

$$\delta \mathbf{S}_r(\Psi_{\pm 1}, Q) = \sqrt{S}\,(-1)^r \mathbf{y} \rho_1^{(n)} \mathrm{e}^{\pm 2\pi i n \delta r}, \tag{F.19}$$

predicted by deSousa and Moore.

As seen in figure F.4, each mode frequency $\omega_n(Q)$ scales like $1/M$ as $Q \to Q_0 = \pi/a$. So all mode frequencies $\omega_n(Q)$ vanish in this limit. Based on the form for the spin disturbances, two distinct modes survive as $\delta \to 0$:

$$\delta \mathbf{S}_r(\Phi_{\pm 1}, Q_0) = \sqrt{S}\,(-1)^r \mathbf{t}(ra), \tag{F.20}$$

$$\delta \mathbf{S}_r(\Psi_{\pm 1}, Q_0) = \sqrt{S}\,(-1)^r \mathbf{y}. \tag{F.21}$$

For $\Phi_{\pm 1}$, the spins oscillate along the cycloidal tangent $\mathbf{t}(ra)$; for $\Psi_{\pm 1}$, the spins oscillate along $\mathbf{y}$. For any q, the total in-plane intensity $S_{xx}^{(1)}(q) + S_{zz}^{(1)}(q)$ of mode 1 approaches the out-of-plane intensity $S_{yy}^{(2)}(q)$ of mode 2 as $Q \to Q_0$ or $\delta \to 0$. Neither of the modes is optically active in the continuum limit.

Exercise 6.13. *Calculate the energy for the spin helix on a TLA with three exchange constants. Show that* $\mathbf{Q} = (0.2, 0.2)_h$ *minimizes this energy when* $J_1 = -0.1855$ *meV,* $J_2 = -0.10$ *meV, and* $J_3 = -0.15$ *meV.*

For a cycloid without harmonics, the energy is the same for every site. So if

$$\mathbf{S}(\mathbf{R}) = S(\cos(\mathbf{Q}\cdot\mathbf{R}), \sin(\mathbf{Q}\cdot\mathbf{R}), 0), \tag{F.22}$$

then the energy per site is

$$\begin{aligned}\frac{E}{N} =& -J_1S^2\Big\{\cos(Q_xa) + \cos\left(Q_xa/2 + \sqrt{3}\,Q_ya/2\right)\\&+\cos\left(Q_xa/2 - \sqrt{3}\,Q_ya/2\right)\Big\}\\&-J_2S^2\Big\{\cos(\sqrt{3}\,Q_ya) + \cos\left(3Q_xa/2 + \sqrt{3}\,Q_ya/2\right)\\&+\cos\left(3Q_xa/2 - \sqrt{3}\,Q_ya/2\right)\Big\}\\&-J_3S^2\Big\{\cos(2Q_xa) + \cos\left(Q_xa + \sqrt{3}\,Q_ya\right)\\&+\cos\left(Q_xa - \sqrt{3}\,Q_ya\right)\Big\}.\end{aligned} \tag{F.23}$$

For a fixed set of J_i, minimize the energy as a function of $\mathbf{Q}$. For $J_1 = -0.1855$ meV, $J_2 = -0.1$ meV, and $J_3 = -0.15$ meV, $\mathbf{Q} = (0.4, 0) = (0.2, 0.2)_h$ regardless of the spin S.

Exercise 6.14. *Evaluate the inelastic spectra for the 120° phase with ordering wavevector* $\mathbf{Q} = (0.33, 0.33)_h$ *when* $J_1 = -1$ *meV and all other exchanges are neglected but with easy-plane anisotropy E.*

The Hamiltonian for this problem is

$$\mathcal{H} = -J_1\sum_{\langle i,j\rangle}\mathbf{S}_i\cdot\mathbf{S}_j - E\sum_i S_{iz}{}^2, \tag{F.24}$$

where $J_1 < 0$ and $E < 0$. The 120° phase is sketched in figure F.5 along with the notation for the lattice sites 1, 2, and 3. Any other spin configuration obtained by a uniform rotation about $\mathbf{z}$ would also work. Since the spins lie in the xy plane, the vector $\mathbf{A}$ used to obtain the contributions of the anisotropy (see equations (4.138)–(4.140)) to $\underline{L}$ is $\mathbf{A} = (-1, 0, 0)$. The exchange interactions require the functions

$$\Gamma_{12}(\mathbf{q}) = \frac{1}{3}\left\{e^{iq_xa} + e^{i(-q_x+\sqrt{3}q_y)a/2} + e^{i(-q_x-\sqrt{3}q_y)a/2}\right\}, \tag{F.25}$$

$$\Gamma_{13}(\mathbf{q}) = \frac{1}{3}\left\{e^{-iq_xa} + e^{i(q_x+\sqrt{3}q_y)a/2} + e^{i(q_x-\sqrt{3}q_y)a/2}\right\}, \tag{F.26}$$

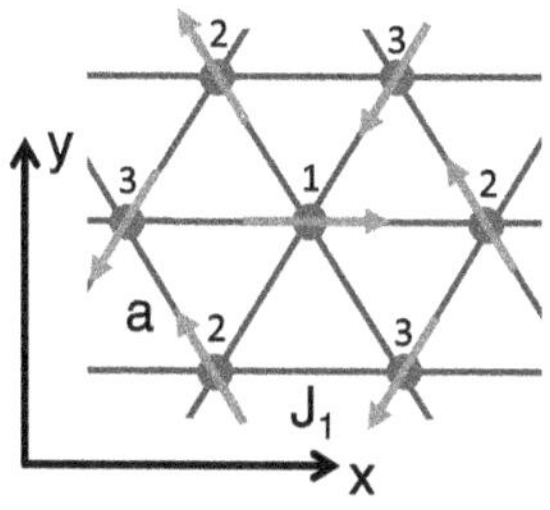

Figure F.5. The 120° phase on a triangular lattice.

$$\Gamma_{23}(\mathbf{q}) = \Gamma_{12}(\mathbf{q}), \tag{F.27}$$

and $\Gamma_{lm}(\mathbf{q}) = \Gamma_{ml}(\mathbf{q})^*$.

Using the shortcuts of equations (4.141)–(4.144), we find

$$L_{11} = L_{44} = \frac{3J_1S}{2}\{F_{zz}(1, 2) + F_{zz}(1, 3)\} - \frac{ES}{2}, \tag{F.28}$$

$$L_{22} = L_{55} = \frac{3J_1S}{2}\{F_{zz}(2, 1) + F_{zz}(2, 3)\} - \frac{ES}{2}, \tag{F.29}$$

$$L_{33} = L_{66} = \frac{3J_1S}{2}\{F_{zz}(3, 1) + F_{zz}(3, 2)\} - \frac{ES}{2}, \tag{F.30}$$

$$L_{14} = L_{25} = L_{36} = L_{41} = L_{52} = L_{63} = -\frac{ES}{2}. \tag{F.31}$$

For $r \neq s$ and $r, s = 1, 2$, or 3, we obtain matrix elements

$$L_{rs} = L_{r+3,\, s+3} = -\frac{3J_1S}{4}\,\Gamma_{rs}\, G_1(r, s), \tag{F.32}$$

$$L_{r,\, s+3} = L_{r+3,\, s} = -\frac{3J_1S}{4}\,\Gamma_{rs}\, G_2(r, s), \tag{F.33}$$

where

$$F_{zz}(r, s) = \cos(\phi_r - \phi_s) = -\frac{1}{2}, \tag{F.34}$$

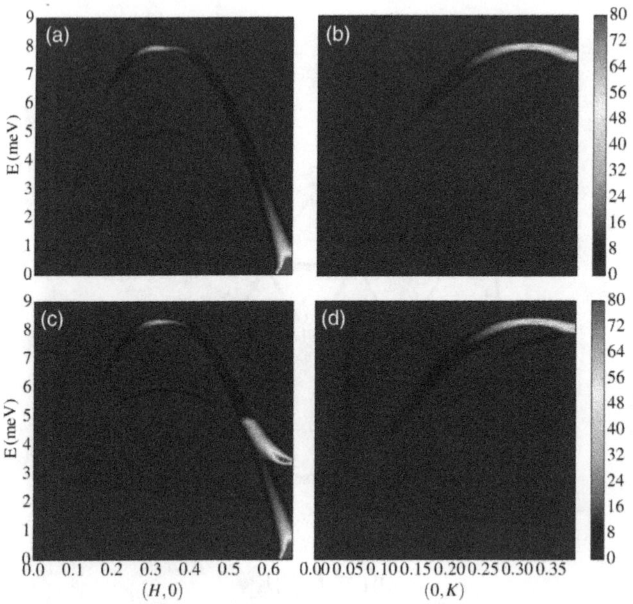

Figure F.6. Inelastic intensity of a 120° phase with $S = 5/2$, $J_1 = -1$ meV along (a) and (c) $(H, 0)$ and (b) and (d) $(0, K)$. The top panels (a) and (b) take $E = 0$ and the lower (c) and (d) take $E = -0.2$ meV.

$$G_1(r, s) = 1 + \cos(\phi_r - \phi_s) = \frac{1}{2}, \tag{F.35}$$

$$G_2(r, s) = 1 - \cos(\phi_r - \phi_s) = \frac{3}{2}. \tag{F.36}$$

which uses the functions defined equations (4.127)–(4.129).

In orthorhombic notation, $\mathbf{Q} = (2/3, 0)$ in units of $2\pi/a$ (see exercise 4.28). We numerically solve for the SW frequencies using $S = 5/2$, $J_1 = -1$ meV, and either $E = 0$ or $E = -0.2$ meV. The results along $(H, 0)$ and $(0, K)$ are plotted in figure F.6. Despite appearances, there are three non-degenerate modes for $\mathbf{q} \parallel (H, 0)$. One of the two highest modes contains spin fluctuations solely along $\mathbf{z}$. The other two modes contain fluctuations in the xy plane.

When $E = 0$, all three modes are Goldstone modes that vanish at $\mathbf{Q}$, as seen in figure F.6(a). When $E < 0$, only one Goldstone mode survives due the remaining rotational invariance around $\mathbf{z}$. The other two mode frequencies are nonzero at $\mathbf{Q}$, as seen in figure F.6(c). There is not so large a difference between the mode frequencies along $(0, K)$ in figures F.6(b) and (d), except near $K = 0$ where the intensity is quite weak. Recall that the edge of the first BZ (see exercise 5.12) is at $K = 2\sqrt{3}/9 \approx 0.385$.

www.ingramcontent.com/pod-product-compliance
Lightning Source LLC
Chambersburg PA
CBHW080538220326
41599CB00032B/6303